VOLCANISM IN AUSTRALASIA

A collection of papers in honour of
the late G.A.M. Taylor, G.C.

Edited by

R.W. JOHNSON

Bureau of Mineral Resources, Geology, and Geophysics (Canberra)

ELSEVIER SCIENTIFIC PUBLISHING COMPANY
Amsterdam — Oxford — New York
1976

ELSEVIER SCIENTIFIC PUBLISHING COMPANY
335 Jan van Galenstraat
P.O. Box 211, Amsterdam, The Netherlands

AMERICAN ELSEVIER PUBLISHING COMPANY, INC.
52 Vanderbilt Avenue
New York, New York 10017

With 5 colour plates, 180 figs and 29 tables

ISBN: 0-444-41462-2

Printed in The Netherlands

G.A.M. TAYLOR

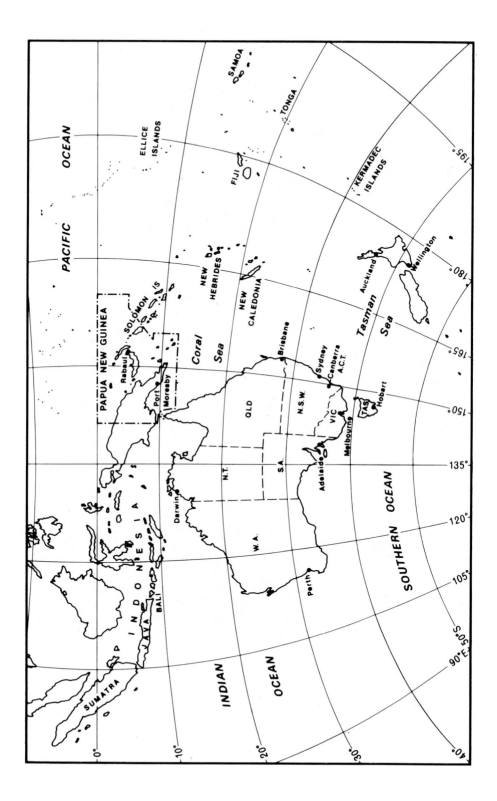

PREFACE

The range of subjects covered in this volume, and the relative proportions of papers covering different topics, are a close indication of G.A.M. Taylor's particular interests in volcanism. Because his overriding concern was for an understanding of the eruptive mechanisms of island-arc volcanoes in the Australasian region — and especially in Papua New Guinea — it is appropriate that most of the papers presented here deal with Australasian eruptions and with the Quaternary geology of Melanesian volcanoes. However, Mr Taylor's outlook was much broader, and during the last years of his life he was involved with other subjects, such as seismic refraction studies of island-arcs, the petrology of volcanic rocks, the practicalities of establishing volcanic surveillance networks, and geological field studies of ancient volcanic formations. The papers presented here reflect this broad overview of volcanism in Australasia.

All the papers (except one) deal with specific areas in the region, and the chosen order of presentation is therefore a geographic one, progressing clockwise from Australia to New Zealand by way of Indonesia, Papua New Guinea, the Solomon Islands, and Tonga. This route takes the reader from an intra-continental environment into the complex island-arc systems that essentially define the northeastern and eastern margins of the Indo-Australian plate.

Thanks are extended to the following referees who provided reviews of papers submitted for the volume: M.J. Abbott, R.J. Arcalus, P.E. Baker, D.H. Blake, C.D. Branch, R.N. Brothers, P.R.L. Browne, J.R. Cleary, R.J.S. Cooke, K.A. Crook, D. Denham, W.R. Dickinson, J.C. Dooley, D.J. Ellis, F.J. Fitch, J.G. Fitton, P.W. Francis, A.S. Furumoto, D.H. Green, T.H. Green, G.W. Grindley, P. Jakeš, R.H. Johnson, J.G. Jones, W.C. Lacey, I.B. Lambert, E. Löffler, G.A. Macdonald, R. Macdonald, D.E. Mackenzie, J.S. Milsom, W.R. Morgan, K.W. Muirhead, I.A. Nicholls, W.H. Oldham, B.S. Oversby, W.D. Palfreyman, H. Pichler, P.E. Pieters, M.J. Roobol, K.J. Seers, I.E.M. Smith, R.L Stanton, R.N. Thompson, R. Varne, G.P.L. Walker, J.F.G. Wilkinson, and eight others who prefer to remain anonymous. The editorial assistance of K.A. Townley and W.H. Oldham is also gratefully acknowledged. Camera-ready copy was set by the *Canberra Times.*

R.W. Johnson
October, 1975

Canberra

MEMORIAL — G.A.M. TAYLOR

by N.H. Fisher

This volume is dedicated to the memory of G.A.M. Taylor, whose premature death whilst engaged on field work on Manam volcano in August, 1972, was a great shock to his many friends and colleagues. Ever since graduation, Mr Taylor's whole professional career had been devoted to the study of volcanoes and volcanic centres, mainly of Papua New Guinea and Australia, but also of other countries, and he was well known internationally for his contributions to volcanology and particularly for his study of Mount Lamington after the disastrous eruption of 21 January, 1951.

George Anthony Morgan Taylor, always known as Tony Taylor, was born in Moree, northern New South Wales, on 30 October 1917. His father was an Englishman who had come to Australia in the early 1900's, and settled at Weemelah, about 75 km northwest of Moree and 25 km east of the Queensland border town of Mungindi. Tony Taylor received his early education at the local primary school, and then attended Maitland High School near Newcastle, but in 1936 he moved to Sydney and at the end of that year passed his Matriculation Examination from Sydney Boys' High School.

At school his favourite subject was chemistry and he maintained at home a 'laboratory' which his two younger sisters were occasionally allowed to visit as a rare treat to observe his 'experiments'. He was a member of the Young Naturalists and keenly interested in all aspects of nature study, including geology, though it must be admitted that the alluvial plains of the Moree-Mungindi district are not a very fertile field for the study of rocks. This interest was to develop later.

After matriculation he joined the Broken Hill Proprietary Company (BHP) as a staff trainee analytical chemist, and remained in the employment of this company until he enlisted in the Second A.I.F. in April, 1942. His position at BHP involved shift work, with on-the-job training, and attendance four nights a week at a chemistry diploma course at the Newcastle Branch of the Sydney Technical College, a course which he had not completed at the time of his enlistment. This schedule did not leave much time for other activities but he played Rugby Union Football for a couple of years as a winger with one of the teams of the Newcastle Wanderers Club. At school he had played football and had also represented the school in athletics — running and high jump.

During his period with BHP it was discovered that Tony Taylor had a degree of colour-blindness and as some of the analytical tests depended on accurate perception of fine colour differences this imposed some limitations on his future as an analytical chemist.

On 29 April 1942 he enlisted in the Australian Imperial Force and was posted to the 2/4 Ordnance Stores Company. He spent most of the next three years in North Queensland, where his first interest in volcanoes was aroused by the numerous volcanic forms and evidences of fairly recent activity that are so abundant in the Cairns Hinterland.

In May 1945 he embarked in Brisbane on the 'Duntroon' for Jacquinot Bay, New Britain, and moved on 9 September, after the Japanese surrender, into

Rabaul where he remained until the end of October 1946. Here he had the opportunity to observe at length the many manifestations of volcanism around Rabaul Harbour.

On 8 January 1947 Tony Taylor was discharged from the Army with the rank of Warrant Officer 1st Class and in March of the same year he began a Science Degree Course at the University of Sydney under the Commonwealth Government's Post-War Reconstruction Training Scheme. He completed the requirements for a Bachelor of Science degree in three years and joined the Bureau of Mineral Resources of the Australian Government's Department of National Development as a Geologist Grade 1 on 20 March, 1950.

The Bureau of Mineral Resources had been established in May/June 1946 and one of its responsibilities was to provide geological and volcanological services for the Administration of the Territory of Papua and New Guinea. Lack of available geologists in Australia and absence of geological training in universities during the war years had made it very difficult for the BMR to build up its staff in its early years, until the post-war graduates started to become available. Consequently the Volcanological Observatory in Rabaul, which had been started in 1940 and destroyed during the war, had not been re-established and this was Taylor's first task when he was posted to New Guinea in April 1950. The Observatory was rebuilt on the same site, over the same instrument cellars, which were intact and usable, the Benioff seismographs destined for the Observatory which had been on loan for several years to Queensland University, were installed, and systematic observations of the Rabaul volcanic centres resumed. Re-establishment of the Rabaul surveillance system occupied most of Taylor's attention, but the volcanologist's responsibilities included the provision of advice on the activity of all of Papua and New Guinea's volcanoes. One of his early field visits was to Bougainville Island to examine Balbi and Bagana volcanoes and to report on the possibility of Peléan-type eruptions from the latter, one of the Territory's most active volcanoes. But it was the catastrophic and disastrous eruption of Mount Lamington in Eastern Papua on 21 January, 1951, that catapulted Taylor, normally one of the most reserved and retiring of men, into public prominence and provided the opportunity to put into effect the studies of the effects of volcanic eruptions of various types that he had been assiduously pursuing.

Mount Lamington was not previously known even to be a volcanic site, and although preliminary eruptive activity began several days before the great Peléan outburst, inadequate communications and transport difficulties prevented any expert advice arriving on the scene before the main eruption. Taylor therefore did not reach Popondetta, the nearest point of access to Mount Lamington, until immediately after the disaster, when the bodies of 3000 recent inhabitants lay dead around the fertile northern slopes of Mount Lamington, surrounding the village of Higaturu. The volcano was relatively quiet but nobody knew what type of activity it was likely to produce in the immediate future, and the tasks of care and rehabilitation of the injured and displaced people, assessment of the damage and of the future of the area, as well as disposal of the dead, had to proceed immediately.

An observation post, equipped with seismographs, was set up at Sangara Plantation, 16 km from the volcano, and systematic observations commenced. Excursions were made into the crater area to examine the growing lava dome and the distribution of products of the main eruption, and to collect samples. Of particular interest was the correlation of the growth of the lava dome with the almost continuous seismic activity recorded on the seismographs at the Observation Post.

Much of Taylor's activity during the next two years was devoted to the surveillance of Mount Lamington, studying the later eruptions, advising the Administration on the location of evacuation camps, the movement of people in the devastated areas, the return of village groups to their areas and the resumption of agricultural activity. Some of the later eruptions of the nuée ardente type were particularly fearsome and awe-inspiring, hurtling down the valley of the Ambogo River in huge clouds and threatening to completely engulf and destroy the Observation Post. It was Taylor's steadfastness and devotion to his work during this troubled period that led to his being recommended for, and awarded in April 1952, the George Cross, the citation for which read as follows.

> CITATION: Mount Lamington, in Papua, began to erupt on the night of 18th January 1951. Three days later there was a violent eruption when a large part of the northern side of the mountain was blown away and steam and smoke poured from the gap for a considerable time afterwards. The area of extreme damage extended over a radius of about eight miles, while people near Higaturu, nine miles from the volcano, were killed by the blast or burned to death. This and subsequent eruptions caused the death of some 4,000* persons, and considerable damage. Dust and ash filled every stream and tank for some miles around, and there was urgent need of food, water and medical supplies. Rescue parties were hampered by suffocating pumice dust and sulphurous fumes, and hot ashes on the ground, and the advance post of relief workers at Popondetta was threatened with destruction by other eruptions during several days following. Further tremors and explosions occurred during February. As late as March 5th a major eruption occurred which threw as far as two miles pieces of volcanic dome, 15 ft. by 12 ft. by 10 ft. and caused a flow of pumice and rocks for a distance of nine miles, the whole being so hot as to set fire to every tree in its path.
>
> For a prolonged period Mr Taylor showed conspicuous courage in the face of great danger. He arrived at Mount Lamington on the day following the main eruption and from that day onwards, over a period of several months he visited the crater by aircraft almost daily, and on many other occasions on foot. On some occasions he stayed at the foot of the volcano throughout the night. During the whole of this period the volcano was never entirely quiet. Several eruptions took place without any warning or any indications from the seismographical data which he had collected. Without regard for his personal safety he entered the danger area again and again, each time at great risk, both in order to ensure the safety of the rescue and working parties and in order to obtain scientific information relating to this type of volcano, about which little was known. His work saved many lives, for as a result of his investigations in the danger zone he was able, when necessary, to warn rehabilitation parties and ensure that they were prevented from entering an area which he so fearlessly entered himself. (London Gazette: 22nd April 1952.)

* Although the exact number of dead was never determined, the usually quoted figure, and the one used by Taylor, is almost 3000. Ed.

During 1951, in addition to his Mount Lamington work and general oversight of the Rabaul surveillance system, Taylor was called upon again to investigate reported activity of volcanoes on Bougainville, and to examine other volcanoes in Papua to check against possible sympathetic reactivation in response to the Mount Lamington eruption. He also was made available by the Australian authorities, at the request of the Condominium Administration of the New Hebrides, to examine and report on Ambrym volcano which had been erupting for a year causing great devastation, and to advise on the safety of the inhabitants still remaining on the island.

During the next few years Taylor continued his general studies of the New Guinea volcanoes, making specific field investigations wherever and whenever signs of eruptive activity became apparent. The first to show such signs was the long dormant Langila volcano at the western end of New Britain, in 1952. Eruptions of Long Island and Bam took place the following year, Tuluman, a submarine volcano near the southwestern tip of Manus, and Langila in 1954, and Manam began an eruptive period of major proportions in 1956 after a dormancy of 11 years.

A particularly exciting incident took place in June 1955 when Taylor and one of the local people attempted in an improvised boat to examine the then active Motmot island crater in the southern part of the lake that occupies the caldera of Long Island, Lake Wisdom. An explosion occurring as the frail craft was approaching the crater placed the expedition at considerable risk and presented the intrepid investigators with unsolicited samples of the ejecta.

During the period 1954 to 1956 Taylor spent a considerable part of his time in Canberra, studying the Mount Lamington seismograms and other records, carrying out laboratory work and experiments of various kinds, and writing a comprehensive report on the Mount Lamington 1951 eruption. This report was published in 1958 as Bulletin No. 38 of the Bureau of Mineral Resources and ranks as probably the most complete and authoritative analysis of a major Peléan-type cataclysmic eruption and subsequent phenomena that has been written. In 1957 he was awarded the degree of Master of Science by Sydney University in recognitiion of his Mount Lamington work.

Returning early in 1957 to Rabaul, Taylor found the Manam eruption increasing in severity, culminating in a series of climactic eruptions during the first three months of 1958. Two instrument stations were established on the island, and the 3000 inhabitants were evacuated to the mainland in December 1957. They were returned to the island in August the following year.

Meanwhile the development of the Rabaul system of Observation Posts was proceeding, with improvements in the seismic instrumentation and the installation of tiltmeters. Observations posts were established at Sulphur Creek, and later in the caldera wall near Tavurvur, in a tunnel at Taviliu overlooking Vulcan, and at Rabalankaia. Activity at Manam continued with another series of major eruptions in March-May 1960, but evacuation of the island was avoided by defining danger areas and maintaining a 24-hour vigilance. Bam Island and Langila erupted in 1958. Advice was given to the administrations of the British Solomon Islands Protectorate and of the New Hebrides on volcano surveillance systems.

For some years Taylor had been studying the possible effects of luni-solar phenomena in triggering off eruptions and also the possible correlation between regional seismic acitivity and the incidence of eruptions at the volcanic centres of Papua and New Guinea. In 1959 he attended a meeting in Paris of the International Association of Volcanology (IAV) and presented a paper on the prediction concepts, including those mentioned above, used on Manam volcano. During this overseas trip he examined volcano research and surveillance centres at Vesuvius, in Japan, and in the Philippines.

In February 1961 Taylor was promoted to the position of Senior Resident Geologist, Port Moresby, and became responsible for supervision of all official geological and volcanological work in Papua and New Guinea. Here he was able to advance plans to upgrade the volcano surveillance instrumentation at Rabaul by establishing a telemetered seismic network connecting the Central Observatory with the Observation Posts situated, as mentioned above, at strategic locations around the caldera, and also to establish permanent volcanological observatories at Manam, Mount Lamington, and Esa-ala in eastern Papua. In 1961 he attended a meeting of the IAV in Japan.

In 1963 Taylor became the first scientist to make the difficult descent to the caldera floor of Karkar volcano to examine the craters and volcanic products within the caldera. This volcano had last erupted in 1895 and was to become active again in 1974.

At the end of 1963 Taylor transferred his headquarters to Canberra, but continued to devote most of his attention to Papua New Guinea volcanological problems and spent a considerable amount of time in Papua New Guinea on field work and on development of volcanological stations. During the succeeding years he played a major part in planning and carrying out crustal study projects in the New Britain region. In these projects shots were fired at sea at locations along two extended lines at right angles to each other and the resultant seismic waves measured at a number of field stations set up especially for the purpose. The 1967 exercise involved the use of two ships and twelve field parties, the 1969 one three ships and eighteen field parties. Co-operating institutions as well as the Papua New Guinea resident staff and the Bureau of Mineral Resources included the University of Hawaii, University of Queensland, Australian National University, and the Queensland Geological Survey. These projects made an important contribution to the knowledge of tectonics and crustal structure in the Melanesian region and to the determination of travel times of seismic waves and thereby the precise location of points of origin of local earthquakes.

In 1966 Taylor examined recent lava flows and volcanic centres in North Queensland and Victoria, and the following year prepared for the IAV a report on work in Palaeovolcanology and Plutonism in Australia and for the Upper Mantle Committee a progress report on volcanological work and its results. Much of his time was occupied with preparation of data sheets and maps concerning over 800 post-Miocene volcanoes in Australia and Melanesia in time for a meeting of the IAV at Zurich as part of the IAV's compilation of post-Miocene volcanoes of the World.

An interesting project during 1967 was the investigation of a very high temperature (at times over 600°C) thermal vent which suddenly appeared in the middle of a landslip area at 'the Crater', at Koranga, near Wau. The occurrence lasted only about three months, and the gases appeared more likely to have been generated by oxidising sulphides at depth rather than by genuine rejuvenating volcanicity.

During 1968 Taylor was the convenor of a symposium on palaeovolcanology in Australia and Papua New Guinea organised by the Geological Society of Australia; he contributed a paper on post-Miocene volcanoes in Papua New Guinea, and edited the volume that was subsequently published.

The crustal study projects and various field investigations in Papua New Guinea occupied most of Taylor's time during the next few years. He prepared a booklet for a volcanological excursion in Papua New Guinea, attended the ANZAAS Congress in Port Moresby in August 1970, and led the post-congress excursion. The same year he paid, by invitation, a visit to New Zealand to examine and discuss volcano surveillance techniques and to present a paper at the Annual Meeting of the New Zealand Geological Survey. In Papua New Guinea he took part, together with resident staff, in the investigation of a full-scale eruption of Ulawun on New Britain, advised the Administration on the return of village population to affected areas, and prepared a report on the eruption. He also continued his long-term study of Manam volcano, about which he had accumulated a tremendous amount of data during its many years of eruptive activity. His death on August 19, 1972, from a heart attack, whilst actually carrying out field work on this volcano, prevented the completion of what would no doubt have been another major contribution to volcanological literature.

Taylor was quiet and reserved by nature. He was not normally loquacious although he was always willing to discuss at length the subjects in which he was especially interested, particularly, of course, volcanoes, volcanism and volcanic prediction. He considered carefully what he wished to say and was tenacious in argument where his convictions were concerned. He was a careful writer with the special gift of presenting his material in an interesting, even absorbing way, and many of his papers make fascinating reading. He had the ability to convey to the reader a special sense of excitement and involvement in the subject being discussed. He had the patience that is often characteristic of quiet men and pursued his aims with persistence and diligence. He was a good craftsman, with a wide understanding of construction methods and instrument operation, a most useful accomplishment in a volcanologist engaged in the development of surveillance systems in remote locations, where he has to plan, supervise installation, adjust and improvise to meet the requirements of the situation. He never allowed considerations of comfort or personal safety to affect his approach to his volcanological studies, and in fact tended to pay insufficient attention to his material requirements whilst on field excursions.

Taylor had a deep and continuing interest in, almost an obsession with, volcanology, and the main thrust of his work was devoted towards establishing and improving methods of prediction of volcanic eruptions and putting them into practical effect. His work was his main hobby but he was a keen fisherman, played occasional golf and tennis, and was a successful gardener.

Taylor had a strong, somewhat dry, sense of humour. He was always well respected by his colleagues and regarded with deep affection by these fortunate enough to penetrate his quiet reserve and gain his friendship. He was married on April 4, 1956, to Lindsay Hudson; he was a devoted family man, and is mourned by his wife, two sons and a daughter.

CONTENTS

FLOOD BASALTS OF PROBABLE EARLY CAMBRIAN AGE IN NORTHERN AUSTRALIA

R. J. BULTITUDE

Bureau of Mineral Resources, P.O. Box 378, Canberra City, A.C.T. 2601

ABSTRACT

The Antrim Plateau Volcanics, a flood-basalt sequence of probable early Cambrian age, crop out over an area of about 35,000 sq km in the East Kimberley and Victoria River regions of northern Australia, and are known from drill-hole intersections to underlie most of the Daly River and northern Wiso Basins containing predominantly Lower Palaeozoic sediments. Small, isolated outcrops of basic volcanic rocks exposed around the northern and western margins of the Georgina Basin and the eastern margin of the Daly River Basin probably originally formed part of the same extensive lava field but have become separated from the main mass of volcanics by erosion and a cover of younger sedimentary rocks. Little-altered, massive rocks from the central parts of flows consist essentially of labradorite, clinopyroxene, opaque minerals, and a mesostasis of devitrified glass or a quartz-alkali feldspar residuum. Pseudomorphs of olivine are present in a few flows, but the great majority are olivine-free. The overall tholeiitic character of the suite is indicated by the common presence of modal pigeonite and quartz. Chemically the lavas form a homogeneous group; all have hypersthene in the norm and many also bear normative quartz. They range in composition from saturated olivine tholeiite to quartz tholeiite and tholeiitic andesite. Distinctive chemical characteristics of most of these lavas are their relatively high SiO_2 (between about 52 and 54 percent), low TiO_2 (between about 0.8 and 1.4 percent), and total Fe (between about 8 and 11 percent, as FeO) contents compared with other flood-basalt sequences. They are not high-alumina types, a characteristic shared with most flood basalts.

INTRODUCTION

The Antrim Plateau Volcanics are an extensive, but hitherto little-studied, sequence of flood basalts* in northern Australia (Figs 1, 2). They form one of the more extensive accumulations of flood basalts in the world (Table 1) and this paper represents a first attempt to provide a comprehensive account of their geology and petrology.

Extensive inland erosional plains covered by black and grey cracking clay soils, benches, mesas, buttes, and low rounded residual hills are the characteristic landforms developed on the essentially flat-lying (regional dips less than 1º) lavas and interbedded sediments in the eastern Victoria River region (Fig. 2). In contrast, the volcanics in the East Kimberley and western Victoria River regions have been more deeply dissected and the country is more rugged. Plateaux with boulder-strewn mesas and hills up to about 100 m high have formed where the lava flows are flat-lying or gently dipping (regional dips less than 5º; Fig. 3). Valley walls are steep and are generally broken by structural benches. Cuestas and hogbacks are well developed on the moderately dipping (20º-30º) lava flows around the western margin of the Hardman Basin (Fig. 4).

Antrim Plateau Volcanics in the East Kimberley region were first described by Hardman (1885) who was government geologist attached to the Kimberley Survey Expeditions. The first recorded observations on the geology of the Victoria River hinterland were made by the explorer A.C. Gregory (1858). H.I. Jensen (1915) described the volcanic rocks in the Victoria River region and referred to reported discoveries of native copper from several places in the volcanics. A.B. Edwards (Edwards & Clarke, 1940) analysed and described basic volcanic rocks collected by

---

*Following common practice the term 'basalt' is used in a general sense throughout the text to describe the lavas of the Antrim Plateau Volcanics and their correlatives.

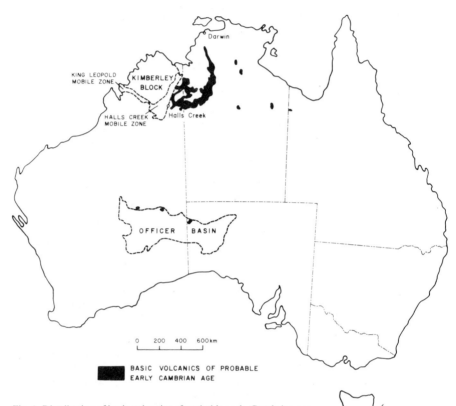

Fig. 1: Distribution of basic volcanics of probable early Cambrian age
in northern and southwestern Australia.

Clarke in 1927 from the East Kimberley region and concluded that the lavas
formed a single homogeneous petrographic province. In 1945, Matheson &
Teichert (1948) made a geological reconnaissance of the Cambrian rocks in the
East Kimberley region. Their report contains an excellent bibliography and
summary of the literature up to 1945. Pyroclastic intercalations in the volcanics
were reported by Traves (1955). More recently, Dunn & Brown (1969) correlated
probable early Cambrian basic volcanics in other parts of the Northern Territory
and northwestern Queensland with the Antrim Plateau Volcanics.

STRATIGRAPHY

The Antrim Plateau Volcanics consist predominantly of basaltic lava flows,
and minor flow breccia and agglomerate. Explosive volcanism appears to have
played only a very minor role in the igneous activity. Thin beds (generally less than
10 m thick) of quartz sandstone, siltstone, chert (commonly stromatolitic),
sedimentary breccia, and limestone are intercalated with the lava flows in many
places. Most of the interbeds are of small extent.

The formation is separated from the underlying Precambrian rocks by a
marked erosional and regional angular unconformity. Locally, the lavas rest
conformably on thin layers of sandstone or conglomerate mapped as Lower

Cambrian, because of their unconformable relations with underlying Proterozoic strata.

The volcanics are overlain by early Middle Cambrian sediments consisting predominantly of shallow-water marine limestone, dolomite, shale, siltstone, and sandstone with an aggregate thickness of less than 600 m. South of Wyndham the Antrim Plateau Volcanics and the overlying Blatchford Formation are reported to be separated by a low-angle unconformity (Kaulback & Veevers, 1969). Traves (1955) found basalt pebbles in basal shales overlying the volcanics in this area. Farther south the volcanics are overlain by the Negri Group, now mainly confined to the Hardman, Rosewood, and Argyle Basins. Regional angular unconformity between the Headleys Limestone (the basal formation of the Negri Group) and the volcanics has not been observed by the writer (however, see Hardman, 1885), but the lava field was eroded prior to the deposition of the Headleys Limestone. In several places around the margins of the Hardman and Rosewood Basins, Headleys Limestone is underlain by thin, poorly exposed, irregular lenses of predominantly sedimentary breccia, siltstone, and sandstone ranging from about 1 m to 7 m thick. The breccias consist of large, angular to sub-rounded clasts (2-60 cm across) of intensely altered vesicular basalt in a matrix of mainly soft dark red-brown calcareous siltstone and fine-grained red-brown friable sandstone.

Weathering profiles up to 6 m thick have been reported (unpublished BMR Record 1964/104) in volcanics underlying Headleys Limestone around the northern and western margins of the Hardman Basin implying that, although the two formations appear to be conformable, there had been a period of subaerial weathering and possible erosion of the lava pile before the Headleys Limestone was deposited.

Extent, thickness, and correlation

The formation has been extensively eroded since the Lower Cambrian, and much of the lava field covered by younger, predominantly Palaeozoic sediments (Fig. 2). Volcanics forming the eastern belt of Traves (1955) are exposed along the western margins of the Wiso and Daly River Basins. Volcanics of the western belt (Traves, 1955) crop out extensively in the East Kimberley and western Victoria River regions. Continuity of outcrop between the eastern and western belts cannot be established because of an extensive veneer of laterite and superficial deposits in the southern part of the area. However, there is sufficient borehole data to indicate that the volcanics are continuous below surface. The laterite is believed to have formed during periods of intense weathering in the Tertiary (Traves, 1955; Hays, 1967).

The formation is thickest in the East Kimberley region, where a maximum of 1,500 m is attained west of Halls Creek (Roberts et al., 1968, 1972). Around the western margin of the Hardman Basin, total thicknesses range from about 470 m in the south (Gemuts & Smith, 1968) to about 960 m in the north. South of Wyndham the formation is up to about 1,250 m thick. These thicknesses are much greater than those indicated by stratigraphic drilling for the eastern belt of volcanics (less than 250 m), and those measured south and north of Tanami (less than 30 m; Blake, in press; Hodgson, in press a, b).

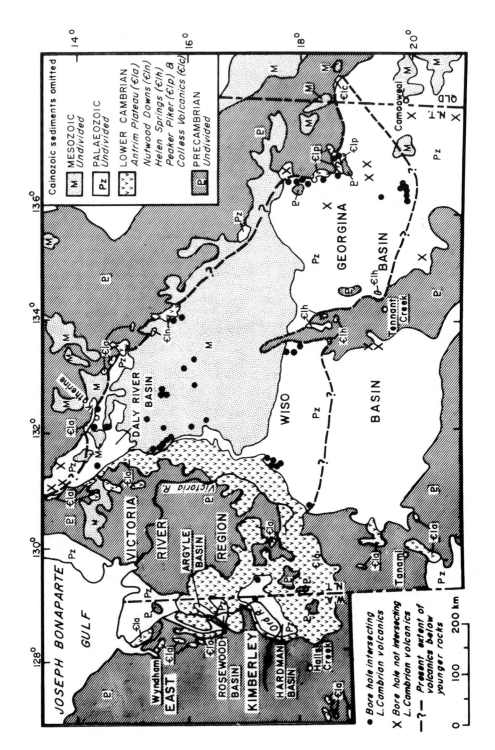

Small outcrops of basic lavas and interbedded sediments mapped as Peaker Piker Volcanics (Smith & Roberts, 1963), Colless Volcanics (Carter, Brooks & Walker, 1961), Helen Springs Volcanics (Randal & Brown, 1969) and Nutwood Downs Volcanics (Dunn, 1963a) are exposed around the northern and western margins of the Georgina Basin and the southeastern margin of the Daly River Basin (Fig. 2). These formations may have originally formed part of the same extensive lava field as the Antrim Plateau Volcanics and have been separated from the main mass of volcanics by erosion and a cover of younger sedimentary rocks. The lavas are similar petrographically and chemically to Antrim Plateau lavas. The formations are unconformably or disconformably overlain by early Middle Cambrian sediments and are separated from underlying late Proterozoic rocks by marked angular unconformities, or, in the case of the Nutwood Downs Volcanics, are conformable on sandstone mapped as Lower Cambrian.

The Nutwood Downs Volcanics have a maximum recorded thickness of 122 m (Dunn, 1963a); the Helen Springs Volcanics, 37 m (Randal & Brown, 1969); the Peaker Piker Volcanics, 37 m (Smith & Roberts, 1963); and the Colless Volcanics, 61 m (Carter, Brooks & Walker, 1961).

Borehole data indicate that lavas of the Antrim Plateau Volcanics underlie most of the Daly River and northern Wiso Basins (Fig. 2). Basic volcanics correlated with the Peaker Piker Volcanics have been intersected below early Middle Cambrian sediments in a stratigraphic hole, in mineral exploration holes, and in water-bores drilled in the northern part of the Georgina Basin (unpublished BMR Records 1970/114, 1972/74); but their subsurface extent is not known. Several wells and bores drilled in this part of the basin have bottomed in Precambrian basement rocks without intersecting basic volcanics. The eastern sequence of volcanics appears to have been relatively thin and it is possible that, in places, all traces of the volcanics were eroded before the deposition of the Middle Cambrian sediments and that the lavas are now preserved mainly as discontinuous sheets. Alternatively the lava flows may have originally been confined mainly to topographic depressions.

The Antrim Plateau Volcanics and their correlatives are exposed over an area of about 35,000 sq km (Fig. 2) and extend over about a further 115,000 sq km beneath the sediments of the Hardman, Rosewood, Argyle, northern Wiso, and Daly River Basins. Small erosional remnants west of Halls Creek (Roberts et al., 1968, 1972), north and south of Tanami (Blake, in press; Hodgson, in press a, b), around the eastern margin of the Daly River Basin (Dunn, 1963a, b; Randal, 1963, 1969), south and north of Wyndham, and in the central Victoria River region indicate that the area covered by lavas was formerly much more extensive. Lava flows also underlie parts of the northern Georgina Basin (Fig. 2). It is therefore likely that the lava field originally had an area of at least 300,000-400,000 sq km.

Small outcrops of tholeiitic basalt have also been described from around the margins of the Officer Basin (Peers, 1969; Lowry et al., 1972; Krieg et al., in press) several hundreds of kilometres south and south-southwest of the area investigated by the writer (Fig. 1). Basic volcanics have also been intersected in holes drilled in

Fig. 2 (opposite): Distribution of Antrim Plateau Volcanics and correlative formations in northern Australia.

R.J. BULTITUDE

Table 1. Comparison between Antrim Plateau Volcanics and other selected flood-basalt sequences.

Unit	Age	Maximum reported thickness (metres)	Present-day extent (sq km)	Reported lithologies of the igneous rocks	Selected references
Antrim Plateau Volcanics (northern Australia)	early Cambrian	1,500+	150,000+	tholeiitic basalt and tholeiitic andesite	this paper
Columbia River Group (USA)	Miocene-Pliocene	3,000+	130,000	tholeiitic basalt	Waters, 1961; Schmincke, 1967; Swanson, 1967; Snavely et al., 1973; Wright et al., 1973
Deccan Traps (India)	late Cretaceous — early Tertiary	3,000	510,000	tholeiitic basalt; minor acid and alkaline lavas and intrusives	Sukheswala & Poldervaart, 1958; Bose, 1972; Macdonald, 1972
Serra Geral Formation (South America)	early Cretaceous (mainly 120-126 m.y.)	3,000+	1,200,000	tholeiitic basalt and dolerite; minor acid and alkaline variants	Amaral et al., 1966; McDougall & Rüegg, 1966; Macdonald, 1972
Kaoko basalts (South-West Africa)	early Cretaceous	900	?	tholeiitic basalt and dolerite	Siedner & Miller, 1968
Karroo basalts (southeastern Africa)	mainly late Triassic-early Jurassic	9,000	140,000	tholeiitic basalt; minor rhyolite, nephelinite, limburgite, teschenite, shoshonite	Cox, 1971, 1972
Portage Lake Lava Series (southern Lake Superior region, USA)	late Proterozoic	5,000	?	tholeiitic basalt; minor andesite and rhyolite	Butler & Burbank, 1929; White, 1960; Jolly & Smith, 1972
North Shore Volcanic Group (northern Lake Superior region, USA)	late Proterozoic	7,500	?	olivine tholeiite to quartz tholeiite; subordinate basaltic andesite, andesite, trachybasalt, latite and rhyolite	Konda & Green, 1974
Trap rocks of the Siberian Platform (USSR)	late Carboniferous-early Triassic	2,000-3,000	1,500,000	basaltic lavas, tuffs, intrusives, minor alkaline variants	Nalivkin, 1973
Coppermine River Group (northern Canada)	late Proterozoic	4,300		tholeiitic basalt	Baragar, 1969
Volcanics of eastern Greenland	Palaeocene-Eocene	7,000	60,000	mainly quartz-normative tholeiites; minor alkalic and nephelinitic types; rare tuff and rhyolite	Brooks, 1973; Fawcett et al., 1973
Volcanics of western Greenland	Palaeocene-Eocene	10,000	?	picrite, quartz- and olivine-normative tholeiites; minor trachyte, rhyolite and alkaline variants	Brooks, 1973; Fawcett et al., 1973
Plateau basalts of Iceland	Tertiary	10,000	?	predominantly basaltic lava flows, minor rhyolite, andesite and intrusives	Einarsson, 1973

the Officer Basin, and seismic data suggest that they underlie most (at least 130,000 sq km; M.J. Jackson, pers. comm., 1975) of the basin (Lowry et al., 1972; Krieg et al., in press). The lavas are probably early Cambrian (Compston, 1974) and may also correlate with the Antrim Plateau Volcanics, which they resemble petrographically and chemically (unpublished data).

Age

The age of the Antrim Plateau Volcanics must be inferred from the stratigraphic evidence, because no diagnostic fossils have been found in the interbedded sediments. The most definitive age limits are set in the East Kimberley region where lavas unconformably overlie the Albert Edward Group (Dow & Gemuts, 1969) and are overlain by the fossiliferous Blatchford Formation and Negri Group. Measurements of rubidium and strontium isotopic abundances in shales from the Albert Edward Group have yielded minimum ages of deposition of 666 ±43 m.y. and 654 ±48 m.y. for the sediments (Bofinger, 1967). Furthermore, the Albert Edward Group overlies the Duerdin Group, which contains glacial rocks deposited during a glacial epoch that affected widespread areas in Australia during the late Adelaidean (Dow, 1965). The Blatchford Formation contains the oldest faunal assemblage of the Ordian Stage (the oldest stage of the Middle Cambrian) and is slightly older than the Negri Group, the lower part of which is also Ordian (Öpik, 1967).

The available data, therefore, restrict the age of the Antrim Plateau Volcanics to between late Adelaidean and early Middle Cambrian. Because of the nature of the contacts with the overlying and underlying rocks, the volcanics are regarded as being little older than the overlying sediments and, therefore, most probably early Cambrian. However, the possibility that volcanism commenced in some·areas in the late Adelaidean cannot be rejected on the available evidence. Walter (1972) suggested a late Precambrian, probably Vendian, age for silicified stromatolites from chert interbeds in Antrim Plateau Volcanics from the southern part of the area.

The Nutwood Downs Volcanics are conformably underlain by the Bukulara Sandstone, which rests on eroded Proterozoic sediments as a shallow-dipping capping (Dunn, 1963a). Sandstone 'dykes' in the basal lava flow (Dunn, 1963a) suggest it was extruded before the underlying Bukulara Sandstone was lithified. The Bukulara Sandstone contains *Scolithus*-like structures which are generally considered to be confined to the Phanerozoic (Plumb & Derrick, in press).

Ages ranging from 395 ±10 m.y. to 511 ±12 m.y. have been obtained on 12 lava samples (10 from the Antrim Plateau Volcanics and one from each of the Helen Springs and Nutwood Downs Volcanics) submitted to the Australian Mineral Development Laboratories for isotopic age determination by the K-Ar whole-rock method. The most likely explanation for these anomalous Upper Cambrian to Siluro-Devonian ages is that the samples have lost varying amounts of radiogenic argon since their formation (A. Webb, in unpublished Australian Mineral Development Laboratories report).

Structure

The lavas and interbedded sediments of the Antrim Plateau Volcanics are mainly flat-lying or gently dipping (less than 5⁰) and show little deformation, except

Fig. 3: Terraced topography developed on a succession of flat-lying lava flows of the Antrim Plateau Volcanics, southeastern margin of Hardman Basin.

Fig. 4: Moderately dipping lava flows of the Antrim Plateau Volcanics, southwestern margin of Hardman Basin (view looking south). The base of the formation is near the foot of the scarp on the right hand side of the photograph.

in the western part of the area where they have been downfolded and cut by numerous faults and lineaments that trend predominantly northwest to north-northwest. The increase in the degree of deformation correlates with the proximity of the Halls Creek Mobile Zone (Dow & Gemuts, 1969; Plumb & Derrick, in press; Fig. 1). The volcanics and overlying Negri Group have been downwarped and, in places, downfaulted to form the markedly asymmetrical Hardman, Rosewood, and Argyle Basins (Matheson & Teichert, 1948; Fig. 2). These basins probably formed as a result of isostatic movements and readjustment of basement blocks in the western part of the area after the outpouring of vast quantities of lava. Dips range from about 30^0 in volcanics bordering the western margin of the Hardman Basin to subhorizonal in volcanics around the southern and eastern margins of the basin, farther away from the mobile zone. Parts of the southeastern margin of the Hardman Basin and of the northern margin of the Rosewood Basin are bounded by steep monoclinal flexures in the Headleys Limestone. The underlying volcanics responded to the stresses responsible for the downfolding of the Headleys Limestone by fracturing or movement along pre-existing fault planes. Displacements on most faults in the volcanics have been small (less than 100 m) and mainly vertical.

The Helen Springs Volcanics, in the east, have not undergone any major tectonic deformation. However, small domes ranging from about 0.5 km to 2.5 km across have been produced in the volcanics by the upwarping of underlying siltstones (Randal & Brown, 1969). The basal lava flow dips outward from the domes at angles ranging from 5^0 to 30^0. The Nutwood Downs Volcanics are flat-lying or gently dipping (less than 5^0) in a westerly direction. The Peaker Piker and Colless Volcanics appear flat-lying.

FLOW CHARACTERISTICS
Internal fabric, extent, and thickness of individual flows

The general scarcity of phenocrysts, flow banding, and pronounced elongation of vesicles, indicates that most of the lavas of the Antrim Plateau Volcanics and correlative formations were highly fluid at the time of their extrusion, and that they crystallised under relatively static conditions. The lavas were extruded subaerially: flow tops are generally highly oxidised and pillow lavas and palagonite (or altered palagonite) have not been definitely identified. It has not been possible to determine the extent of individual lava flows, or to correlate sequences from different parts of the lava field, because of the lack of continuous outcrop, the similarity in hand-specimen appearance of most lava flows, and the general absence of persistent marker units.

Secondary alteration, mainly as a result of low-grade burial metamorphism, but also partly due to weathering and probably deuteric or hydrothermal processes, has affected all the primary phases in most flows from the Antrim Plateau Volcanics and their correlatives to varying degrees. At the base of most flows is a thin, slightly to moderately vesicular or amygdaloidal, extensively altered chilled zone, generally less than 1 m thick. This zone grades upwards into relatively fresh, fine- to coarse-grained massive basalt making up the bulk of most flows and in which columnar jointing is rarely well developed. Higher up is a transitional zone of altered, slightly to moderately vesicular or amygdaloidal basalt, commonly characterised by spheroidal weathering. At the top of each flow is a highly vesicular or amygdaloidal zone of intensely altered basalt comprising about 15-30 percent of the total flow thickness. The predominantly spheroidal or almond-shaped vesicles range in size from less than 1 mm to 20 mm across and are commonly filled with a variety of secondary minerals, the most common being quartz, prehnite, calcite, chalcedony, hematite, agate, pumpellyite, natrolite, analcime, and chlorite. The poorly exposed upper and lower surfaces of most flows appear to be planar; rubbly layers characteristically found in the upper and basal parts of aa and blocky lava flows (Macdonald, 1972) are rare. Individual thicknesses of 43 lava flows, penetrated in 9 stratigraphic holes drilled in essentially flat-lying sequences in the eastern and western belts of Antrim Plateau Volcanics, range from about 10 m to about 114 m, the average flow thickness being about 38 m.

Flow breccias or agglomerates (Traves, 1955; Sweet et al., 1974), or both, form extensive and prominent layers up to 40 m thick, at or near the top of the sequence in parts of the East Kimberley and Victoria River regions. The breccias consist of unsorted, closely packed, angular to subrounded fragments (up to about 60 cm across) of massive to highly vesicular basalt set in a matrix of intensely altered, fine-grained volcanic detritus and, locally, minor red-brown siltstone and sandstone, confined to the upper parts of the units. Some of the breccias may be autobrecciated parts of lava flows (see Macdonald, 1972); in at least three places around the margins of the Hardman Basin, breccia grades downwards into massive non-brecciated basalt. Elsewhere, the breccias appear to form discrete units; they may have resulted from violent explosive activity.

Mineralogy

Massive basalts from the Antrim Plateau Volcanics and correlative formations are typically pale to dark grey and fine- to medium-grained. The rocks commonly contain dark red-brown patches and veinlets rich in secondary iron oxides and small dark green clots of chlorite. A few flows are distinguished by small, prominent glomeroporphyritic aggregates of plagioclase phenocrysts.

The lavas are tholeiitic in character and consist essentially of plagioclase, clinopyroxene, opaque minerals, and a mesostasis of red-brown to almost colourless devitrified glass or a quartzo-feldspathic residuum (e.g., see Wilkinson, 1967). Quartz, hornblende, biotite, and apatite are common accessories. A few flows contain scattered chlorite or 'iddingsite' pseudomorphs after olivine. Intersertal textures are the most common. Porphyritic textures are rarely well developed, the phenocryst content generally being less than 10 percent. Subophitic and ophitic textures are poorly developed in coarse-grained rocks from the central and lower parts of a few thick flows.

Rapidly quenched upper and basal parts of flows intersected in stratigraphic holes consist predominantly of dark red-brown to almost black semi-opaque devitrified glass containing microlites of plagioclase, clinopyroxene, and opaque oxide, and, rarely, scarce euhedral laths (0.5 mm x 1.5 mm) of plagioclase and rectangular prisms (0.2 mm x 0.1 mm) of clinopyroxene. The devitrified glass is commonly partly replaced by chlorite, more rarely by prehnite, pumpellyite, calcite, and quartz.

In most massive basalts the groundmass laths and granules range from 0.1 mm to 0.3 mm in length. Most phenocrysts are less than 2 mm long. Phenocrysts rarely show flow alignment, and the groundmass grains are generally randomly orientated. Small pegmatoid veinlets are present in some samples. The well-formed constituent minerals are the same as those in the enclosing host rocks, but the pegmatoids are characterised by coarse grainsize and high contents (over 50 percent) of red-brown devitrified glass charged with opaque oxide granules.

Fresh and little-altered groundmass plagioclase grains in the massive basalts are about sodic labradorite in composition (An 50-63; determined optically by the maximum extinction angles method). In the highly crystalline rocks feldspar mainly forms small euhedral tabular grains showing normal zoning. In addition most massive basalts contain minor amounts of euhedral to subhedral plagioclase phenocrysts, mainly as scattered isolated laths, rarely as glomeroporphyritic aggregates. Some phenocrysts appear to be partly resorbed.

Plagioclase grains, particularly phenocrysts, are commonly partly replaced by sericite. In intensely altered rocks calcic plagioclase is extensively replaced by prehnite, pumpellyite, albite, and rare calcite.

Clinopyroxene forms mainly small subhedral blocky grains in the massive, highly crystalline basalts. Some medium- to coarse-grained rocks contain relatively large subhedral to anhedral grains (0.3-0.5 mm) and phenocrysts (0.5-2.0 mm) of clinopyroxene. The phenocrysts are commonly twinned and weakly zoned; many enclose fine submicroscopic exsolution lamellae parallel to the (001) plane of the host clinopyroxene. The proportion of clinopyroxene granules to feldspar laths decreases markedly in the fine-grained rapidly chilled rocks, and in some rapidly quenched basalts clinopyroxene is absent.

Augite appears to be the most common pyroxene, but pigeonite $(2V = 0^0\text{-}15^0)$ and subcalcic augite $(2V = 15^0\text{-}30^0)$, generally associated with a more calcic clinopyroxene $(2V$ exceeds $30^0)$, have been identified in many flows. The pyroxenes appear as colourless grains in thin sections and are indistinguishable in general appearance. Pigeonite, as small groundmass grains, is most abundant in the quartz-rich rocks. Massive basalts (specimens 70770872A, 70770874, Table 2) from the interiors of two particularly thick (more than 90 m) lava flows intersected in stratigraphic holes contain phenocrysts of pigeonite.

In most massive basalts clinopyroxene shows only minor replacement by green or yellowish-green chlorite or, rarely, by pale yellow-green to blue-green, slightly pleochroic, fibrous amphibole. In the more intensely altered amygdaloidal basalts from the upper and basal parts of flows, however, clinopyroxene is commonly completely replaced by amorphous brown to black secondary iron oxides.

Magnetite and subordinate ilmenite form mainly small, subhedral to anhedral, interstitial grains in massive highly crystalline rocks. Small euhedral and partly resorbed magnetite phenocrysts, commonly containing small inclusions of plagioclase and, rarely, clinopyroxene, are present in a few flows. Magnetite grains are commonly extensively replaced around boundaries and along fractures by secondary hematite and hydrated iron oxides (goethite or lepidocrosite, or both).

The bulk of the interstitial late-stage residuum crystallised as fine-grained micrographic intergrowths of quartz and alkali feldspar in many of the coarser grained, relatively slowly cooled basalts from the central and lower parts of thick flows. Rarely, groundmass plagioclase grains are mantled by narrow overgrowths of alkali feldspar. Many of the highly crystalline basalts from the interiors of thick flows also contain minor interstitial primary quartz and traces of primary poikilitic hornblende and biotite (Table 2). The hornblende is distinctly pleochroic from greenish yellow to pale pink and the biotite from dark brownish red to nearly colourless. Quartz, alkali feldspar, and primary hornblende and biotite have rarely been observed in flood-basalt flows, many of which are thinner and cooled more rapidly than the relatively thick lava flows of the Antrim Plateau Volcanics. Konda & Green (1974; p. 1192) reported alkali feldspar, quartz, and amphibole in the groundmasses of flows of 'basaltic andesite' and a dyke of 'andesite or trachyandesite' from the Keeweenawan lava sequence of Minnesota. Tholeiitic dolerites commonly contain a quartzo-feldspathic mesostasis and minor primary amphibole and biotite (Wilkinson, 1967).

Minor olivine crystallised as euhedral to subhedral grains (0.1 mm x 0.05 mm) and small phenocrysts (0.4 mm x 0.3 mm) in only a few flows, and in the thin sections examined by the writer has been completely replaced by chlorite or 'iddingsite' and opaque minerals. Glover (in Traves, 1955) described specimens of basalt from the Antrim Plateau Volcanics that contain abundant small phenocrysts of unaltered olivine.

Chemical composition

Chemical and modal analyses and CIPW norms of 18 samples (selected from a total of about 55 analysed specimens) are listed in Table 2. Only the least-altered samples from the massive central parts of flows were analysed. On the whole the lavas are remarkably uniform in composition, and most oxides display only very limited ranges in composition (Table 2; Figs 5a, b). For example, silica contents in the specimens analysed range from about 50 to about 56 percent, nearly all values falling between 52 and 54 percent (Fig. 5a).

Table 2. Chemical and modal analyses of selected volcanic rocks from the Antrim Plateau Volcanics and correlative formations (sample localities are listed in Table 4).

	7077-2102	6977-0107	6977-0118	7077-0719	*7077-6004	7077-2142	7077-2141	7077-0539
SiO_2	50.6	50.9	51.6	52.0	52.0	52.1	52.5	52.6
TiO_2	0.60	0.76	2.05	0.84	1.10	1.09	1.00	0.81
Al_2O_3	16.1	15.7	13.0	15.6	14.2	14.7	14.8	15.0
Fe_2O_3	1.69	1.15	4.10	1.55	6.60	2.25	6.25	1.55
FeO	6.05	6.75	10.70	7.70	3.90	7.85	4.75	7.50
MnO	0.13	0.15	0.20	0.17	0.16	0.17	0.15	0.15
MgO	8.80	8.70	3.90	7.20	6.15	7.00	5.70	6.85
CaO	11.10	8.85	7.90	10.60	5.25	10.30	9.50	10.40
Na_2O	2.25	3.30	2.50	2.20	3.20	2.10	2.55	2.35
K_2O	0.36	1.24	1.43	0.98	4.30	0.81	0.88	0.88
'P_2O_5	0.08	0.07	0.16	0.12	0.11	0.11	0.11	0.10
H_2O+	0.84	1.87	1.63	0.61	1.97	0.68	0.37	0.54
H_2O-	1.26	0.31	0.59	0.58	0.77	0.90	1.13	0.86
CO_2	0.12	0.03	0.05	0.05	0.05	0.08	0.08	0.11
Total	99.98	99.78	99.81	100.20	99.76	100.14	99.77	99.70
$\dfrac{100\,Mg}{Mg + Fe^{++}}$ (mol.)	72.1	69.7	38.2	56.1	59.5	61.5	55.5	61.9
CIPW norms								
Q	0.3	0.0	6.8	1.8	0.0	4.5	5.1	3.4
or	2.2	7.5	8.7	5.9	26.3	4.9	5.3	5.3
ab	19.4	28.6	21.7	18.8	28.0	18.0	22.0	20.2
an	33.5	25.0	20.5	30.1	12.1	28.7	26.9	28.3
di ⎰ wo	9.0	8.1	7.6	9.1	5.8	9.1	8.3	9.5
di ⎨ en	5.9	5.1	3.1	5.3	3.3	5.3	4.5	5.5
di ⎱ fs	2.4	2.5	4.6	3.5	2.2	3.4	3.6	3.7
hy ⎰ en	16.5	3.7	6.9	12.8	1.6	12.4	10.0	11.9
hy ⎱ fs	6.7	1.8	10.4	8.4	1.1	7.9	8.1	8.0
ol ⎰ fo	0.0	9.4	0.0	0.0	7.7	0.0	0.0	0.0
ol ⎱ fa	0.0	5.0	0.0	0.0	5.6	0.0	0.0	0.0
mt	2.5	1.7	5.3	2.3	3.9	3.3	3.7	2.3
il	1.2	1.5	4.0	1.6	2.2	2.1	1.9	1.6
ap	0.2	0.2	0.4	0.3	0.3	0.3	0.3	0.2
cc	0.3	0.1	0.1	0.1	0.1	0.2	0.2	0.3
$\dfrac{100\,an}{an + ab}$	63.3	46.6	48.7	61.6	30.2	61.0	55.0	58.3
Modal analyses (volume per cent)								
Plagioclase	40	38	31	33		38	45	40
Clinopyroxene	55	46	26	49		32	37	45
Opaque minerals	1	1	7	3		2	4	4
Olivine (altered)	—	<1	<1	—		—	—	—
Quartz	—	—	—	<1		—	—	2
Mesostasis	4	15	37	15		28	14	10
Hornblende	—	—	—	·<1		<1	—	<1
Mica	—	—	—	<1		—	—	<1

Specimens analysed at Australian Mineral Development Laboratories, Adealide.
Fe_2O_3 contents adjusted in all normative calculations according to the formula: percent Fe_2O_3 = percent TiO_2 + 1.5; 'excess, Fe_2O_3 being recalculated as FeO (see Irvine and Baragar, 1971).

Table 2 cont.

7077-0715	7077-2104	7077-2166	*7077-3000	7077-2106	7077-0863	7077-0872A	7077-0721	7077-0615	7077-0248
52.7	52.9	52.9	53.0	53.1	53.5	53.7	54.2	55.1	56.4
0.95	0.84	1.25	1.01	1.09	0.90	0.84	1.27	0.94	1.64
15.4	14.8	14.4	14.2	14.3	14.7	15.1	14.4	14.8	12.8
2.00	2.10	3.50	5.25	3.45	3.50	1.33	3.50	7.95	3.70
7.40	7.20	7.55	3.90	7.30	5.05	6.65	7.35	1.10	8.20
0.16	0.16	0.18	0.05	0.17	0.14	0.15	0.18	0.13	0.29
7.05	6.70	5.70	8.65	5.55	6.70	6.75	5.15	5.10	3.70
9.30	10.10	9.15	5.65	9.25	9.35	10.00	8.40	7.05	5.80
3.05	2.50	2.20	2.10	2.70	2.35	2.20	2.40	3.00	2.45
1.35	1.10	1.12	1.41	1.28	1.24	1.01	1.50	2.10	2.80
0.11	0.11	0.13	0.11	0.10	0.11	0.11	0.16	0.13	0.15
0.27	0.44	0.92	2.60	0.55	0.81	0.51	0.75	1.60	1.53
0.47	1.01	0.81	4.10	0.78	1.40	1.17	0.33	0.58	0.45
0.03	0.16	0.08	0.30	0.08	0.13	0.16	0.11	0.29	0.07
100.24	100.12	99.89	99.63	99.70	99.88	99.68	99.70	99.87	99.98
63.0	62.4	55.1	71.1	55.0	66.4	64.4	53.3	60.0	42.9
0.0	3.5	7.3	5.6	4.6	6.4	6.1	8.3	7.2	11.8
8.0	6.6	6.8	9.0	7.7	7.5	6.1	9.0	12.8	16.9
25.9	21.4	19.0	19.2	23.2	20.4	19.0	20.6	26.1	21.2
24.5	26.3	26.6	27.2	23.5	26.6	29.0	24.5	21.3	16.0
8.8	9.5	7.6	0.1	9.2	8.1	8.3	6.7	5.0	5.0
5.2	5.6	4.1	0.1	4.9	5.2	5.0	3.6	2.9	2.2
3.2	3.5	3.2	0.0	4.0	2.4	3.0	3.0	1.8	2.7
11.1	11.3	10.3	23.1	9.2	12.0	12.2	9.5	10.1	7.2
6.8	7.1	8.1	8.7	7.4	5.6	7.3	7.9	6.2	8.7
0.9	0.0	0.0	0.0	0.0	0.0	0.0	0.0	0.0	0.0
0.6	0.0	0.0	0.0	0.0	0.0	0.0	0.0	0.0	0.0
2.9	3.1	4.1	3.9	3.8	3.6	2.0	4.1	3.6	4.7
1.8	1.6	2.4	2.1	2.1	1.8	1.6	2.5	1.8	3.2
0.3	0.3	0.3	0.3	0.2	0.3	0.3	0.4	0.3	0.4
0.1	0.4	0.2	0.7	0.2	0.3	0.4	0.3	0.7	0.2
48.6	55.1	58.4	58.6	50.3	56.6	60.4	54.3	45.0	43.1
35	40	47		34	38	41	33	36	31
49	42	32		48	36	43	47	27	31
2	3	4		5	4	3	8	4	6
—	—	—	—	—	—	—	—	—	—
—	2	—		3	2	3	—	**1	**6
14	13	17		10	21	11	12	33	26
—	—	<1	—	<1	—	<1	—	—	—
—	<1	—	—	<1	—	—	—	—	—

* These are the least-altered specimens of Peaker Piker (70773000) and Colless (70776004) Volcanics collected; however, the thin sections and chemical analyses indicate that the rocks have been fairly extensively altered.
** Mainly secondary.

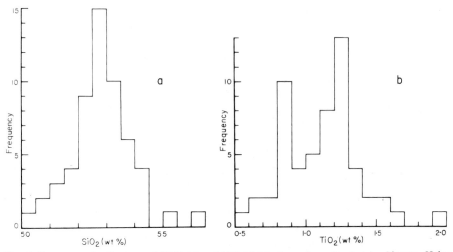

Fig. 5: Frequency distributions of SiO2 (a) and TiO2 (b) in lavas from the Antrim Plateau, Helen
Springs, Nutwood Downs, Peaker Piker, and Colless Volcanics.

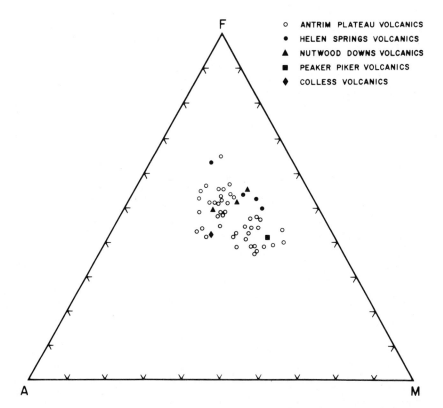

Fig. 6: AFM plot of chemical data from the Antrim Plateau, Helen Springs, Nutwood Downs, Peaker
Piker, and Colless Volcanics. A = Na2O+K2O, F = FeO + 0.8998 Fe2O3, M = MgO.

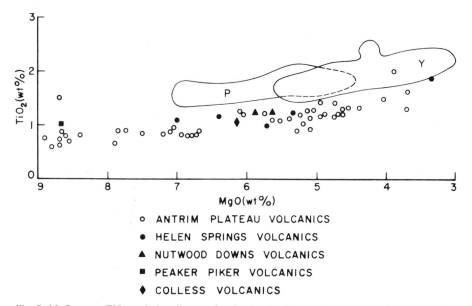

Fig. 7: MgO versus TiO2 variation diagram for the Antrim Plateau lavas and correlative formations : showing the fields of variation delineated by Wright et al. (1973) for the Picture Gorge Basalt and 'lower' basalt (P) and Lower Yakima Basalt (Y) of the Columbia River Group. All the other chemical types defined by Wright et al. from the group are characterised by higher TiO2 contents, and MgO contents ranging from about 6.5 percent to about 2.5 percent.

The majority of the lavas from the Antrim Plateau Volcanics and their correlatives are characterised by relatively high SiO_2 and MgO contents, and by low TiO_2 and total Fe (as FeO) contents compared with most flood-basalt sequences. The lavas contain normative hypersthene and many bear normative quartz; some have bulk chemical compositions more akin to andesites than basalts (Table 2). The lavas show marked enrichment in alkalis, particularly K_2O, and most have alumina contents ranging between 13.5 and 15 percent. Fe_2O_3 contents range from 1.15 to 7.95 percent. The high Fe_2O_3 values and correspondingly low FeO contents are mainly the result of secondary oxidation and alteration. The rocks display a tholeiitic trend of moderate iron enrichment on an AFM diagram (Fig. 6), but because of their relatively high alkali contents the northern Australian lavas plot closer to the alkalis corner than many other tholeiitic suites. The wide scatter of points is probably mainly a reflection of partial alteration of some of the samples. The only specimen from the Colless Volcanics, for example, is fairly extensively altered and has an anomously high K_2O content (Table 2).

TiO_2 contents in the Antrim Plateau Volcanics and their correlatives show a bimodal distribution with maxima between 0.8 and 0.9 percent and 1.2 and 1.3 percent (Fig. 5b). These values are significantly lower than TiO_2 contents of most lavas analysed from most of the better known flood-basalt sequences such as the Deccan Traps and Columbia River Group (Sukheswala & Poldervaart, 1958; Waters, 1961; Bose, 1972; Wright et al., 1973; Fig. 7). Karroo basalts from Lesotho are also characterised by relatively low TiO_2 contents (Cox & Hornung, 1966), but SiO_2 and K_2O contents in these basalts are generally lower than in most lavas from the Antrim Plateau Volcanics and correlatives.

Table 3. Comparison of analyses of little-altered massive specimens from flow interiors and extensively altered amygdaloidal and vesicular specimens from the upper and basal parts of flows intersected in stratigraphic holes drilled in the Antrim Plateau Volcanics (sample localities are listed in Table 4).

	Interior of flow			Upper part of flow 70773053	Basal part of flow 70770518	Central part of flow 70770872A	Upper part of flow 70773052	Basal part of flow 70772034	Central part of flow 70776088	Upper part of flow 70776087
	70770539	70772104	70770874							
SiO_2	52.6	52.9	53.0	51.5	51.3	53.7	50.9	48.8	51.5	48.4
TiO_2	0.81	0.84	0.85	0.86	0.89	0.84	0.93	0.88	0.89	0.93
Al_2O_3	15.0	14.8	14.9	14.0	15.2	15.1	14.7	14.9	15.4	14.6
Fe_2O_3	1.55	2.10	3.35	7.55	3.70	1.33	8.75	5.15	3.55	5.35
FeO	7.50	7.20	5.95	1.97	5.70	6.65	1.88	2.4	4.75	5.05
MnO	0.15	0.16	0.14	0.09	0.15	0.15	0.12	0.09	0.12	0.07
MgO	6.85	6.70	6.75	8.10	7.05	6.75	8.05	7.75	6.75	10.2
CaO	10.40	10.10	9.80	3.00	7.65	10.00	2.70	5.9	9.8	1.50
Na_2O	2.35	2.50	2.20	4.60	3.55	2.20	4.10	1.38	2.35	3.05
K_2O	0.88	1.10	1.04	2.40	1.56	1.01	2.00	5.5	1.20	3.0
P_2O_5	0.10	0.11	0.10	0.09	0.10	0.11	0.10	0.10	0.10	0.10
H_2O+	0.54	0.44	0.57	3.60	2.10	0.51	3.25	3.25	1.11	4.25
H_2O-	0.86	1.01	1.31	2.05	0.93	1.17	2.70	1.84	2.4	3.0
CO_2	0.11	0.16	0.17	0.28	0.08	0.16	0.08	2.35	0.15	0.15
Total	99.70	100.12	100.13	100.09	99.96	99.68	100.26	100.29	100.07	99.65
Total Fe (as FeO)	8.90	9.09	8.96	8.76	9.03	7.85	9.75	7.04	7.95	9.86

CIPW norms

Q	3.4	3.5	7.0	0.0	0.0	6.1	2.3	2.4	5.1	0.0
C	0.0	0.0	0.0	0.0	0.0	0.0	1.4	1.7	0.0	4.5
or	5.3	6.6	6.3	15.1	9.5	6.1	12.5	34.1	7.3	19.2
ab	20.2	21.4	18.9	41.2	31.0	19.0	36.8	12.3	20.6	27.9
an	28.3	26.3	28.2	11.1	21.6	28.0	13.0	14.5	⁄28.9	6.3
di ⎰ wo	9.5	9.5	8.2	0.9	6.8	8.3	0.0	0.0	8.3	0.0
di ⎨ en	5.5	5.6	5.3	0.8	4.7	5.0	0.0	0.0	5.9	0.0
di ⎱ fs	3.7	3.5	2.2	0.0	1.7	3.0	0.0	0.0	1.6	0.0
hy ⎰ en	11.9	11.3	11.8	15.4	10.5	12.2	21.3	20.3	11.5	27.2
hy ⎱ fs	8.0	7.1	4.9	0.0	3.7	7.3	0.0	0.0	3.1	3.7
ol ⎰ fo	0.0	0.0	0.0	3.6	2.1	0.0	0.0	0.0	0.0	0.2
ol ⎱ fa	0.0	0.0	0.0	0.0	0.8	0.0	0.0	0.0	0.0	0.03
mt	2.3	3.1	4.9	4.4	5.5	2.0	4.0	5.8	5.3	8.4
he	0.0	0.0	0.0	5.0	0.0	0.0	6.5	1.4	0.0	0.0
il	1.6	1.6	1.6	1.7	1.7	1.6	1.9	1.8	1.8	1.9
ap	0.2	0.3	0.2	0.2	0.2	0.3	0.3	0.3	0.3	0.3
cc	0.3	0.4	0.4	0.7	0.2	0.4	0.2	5.6	0.4	0.4
$\frac{100\ an}{ab + an}$	58.3	55.1	59.8	21.2	41.1	60.4	26.1	54.1	58.4	18.5

Specimens analysed at Australian Mineral Development Laboratories, Adelaide.

Modal and normative mineralogy indicate that the northern Australian lavas range from olivine tholeiite to tholeiitic andesite, quartz tholeiite being by far the most abundant rock type. Using a modal classification individual lava flows commonly show a range in rock type suggesting that some differentiation occurred during consolidation of the flows. However, most of the modal variations can be attributed to differing degrees of crystallinity. Chemical analyses of massive basalts collected from different levels in several relatively thick flows show no significant variation in composition within the individual flows. Analyses of rocks from the chilled vesicular upper and basal parts of flows, however, reflect the changes imposed upon them as a result of redistribution of elements during secondary alteration and depart significantly from the original compositions.

Table 4. Locality index of chemically analysed specimens from the Antrim Plateau, Helen Springs, Nutwood Downs, Peaker Piker, and Colless Volcanics.

Antrim Plateau Volcanics

69770107	Hilltop at 129°30'24"E, 16°25'30"S
69770118	Hilltop at 131°30'18"E, 16°17'00"S
70770248	Lava flow overlain by Headleys Limestone at 129°17'42"E, 17°32'48"S
70770518	Cuttings from 177-178 m interval, flow no. 3 (64-178 m), in stratigraphic hole Limbunya No. 2, sited at 130°00'06"E, 17°52'12"S
70770539	Cuttings from 303-305 m interval, flow no. 5 (195-305+ m), in stratigraphic hole Limbunya No. 1, sited at 129°22'36"E, 17°25'00"S
70770615	Basal flow at 128°06'24"E, 18°17'18"S
70770715	Second lowermost flow, western margin of Hardman Basin — at about 128°14'E, 17°39'S
70770719	Second lowermost flow, northwestern margin of Hardman Basin — at about 128°19'E, 17°22'S
70770721	Third lowermost flow, northwestern margin of Hardman Basin at about 128°19'E, 17°22'S
70770863	Core from depth of about 70 m, flow no. 2 (46-93 m), in stratigraphic hole Waterloo No. 1, at 129°05'06"E, 16°31'06"S
70770872A	Core from depth of about 138 m, flow no. 3 (64-178 m), in stratigraphic hole Limbunya No. 2, at 130°00'06"E, 17°52'12"S
70770874	Core of massive basalt from depth of about 237 m, flow no. 5 (195-305+ m), in stratigraphic hole Limbunya No. 1
70772034	Cuttings from 128-130 m interval, flow no. 4 (104-130 m), in stratigraphic hole Wave Hill No. 1, at 131°14'42"E, 17°21'48"S
70772102	Cuttings from 119-122 m interval, flow no. 2 (62-148 m) in stratigraphic hole Waterloo No.2, at 129°24'24"E, 16°25'00"S
70772104	Cuttings from 273-274 m interval, flow no. 5 (195-305+ m), in stratigraphic hole Limbunya No. 1
70772106	Cuttings from 50-52 m interval, flow no. 1 (0-56 m), in stratigraphic hole Waterloo No. 2
70773052	Cuttings from 66-67 m interval, flow no. 3 (64-178 m), in stratigraphic hole Limbunya No. 2
70773053	Cuttings from 198-200 m interval, flow no. 5, in stratigraphic hole Limbunya No. 1
70776087	Cuttings from 105-107 m interval, flow no. 4 (104-130 m), in stratigraphic hole Wave Hill No. 1
70776088	Cuttings from 63-64 m interval, flow no. 2 (62-148 m), in stratigraphic hole Waterloo No. 2

Helen Springs Volcanics

70772141	Massive basalt in hillside at about 133°54'06"E, 18°26'06"S
70772142	Small gully at about 133°52'42"E, 18°26'42"S

Nutwood Downs Volcanics

70772166	Lava flow at about 134°04'42"E, 15°36'54"S (Flicks Waterhole)

Peaker Piker Volcanics

70773000	Core sample from about 149-150 m interval in stratigraphic hole Alroy No. 2, at 136°17'30"E, 19°30'50"S

Colless Volcanics

70776004	Floater of massive basalt at about 138°19'E, 18°39"S

Alteration

The pattern of alteration in the lavas shows a persistent tendency for the degree of alteration to increase from the massive part of each flow to the highly amygdaloidal or vesicular top and base of each flow. Joints and fractures in the lavas and the vesicular parts of flows acted as channelways for migrating connate and meteoric fluids which produced the sequence of secondary phases within the flows. The oxides which show the greatest susceptibility to metasomatic redistribution are CaO, K_2O, Fe_2O_3, FeO, H_2O+, H_2O-, and Na_2O (Table 3).

Extensively altered amygdaloidal basalts from the basal and upper parts of most flows are significantly enriched in K_2O, Na_2O, H_2O+, and Fe_2O_3 and depleted in SiO_2, FeO, and CaO compared with relatively fresh, massive basalts from the central parts of the flows (Table 3). MgO and TiO_2 are slightly higher in the altered rocks but Al_2O_3 and total iron (as FeO) display little variation. Volatile contents are generally much higher in the altered amygdaloidal basalts. Similar trends have been described in tholeiitic basalts from the Portage Lake Series (Jolly & Smith, 1972) and in altered andesitic flows from a Cretaceous volcanic sequence, central Chile (Levi, 1969).

MODE AND LOCATION OF ERUPTION

The lavas are characterised by a general lack of recognisable eruptive centres. Small swarms of northwest-oriented dykes cut the volcanics around the south-southeastern margin of the Hardman Basin. Most of the dykes are completely brecciated and consist of angular fragments (up to 15 cm across) of intensely altered, mainly massive to slightly vesicular basalt in a finely comminuted, extensively chloritised basaltic matrix. Slickensides are common on the walls of the dykes. The dykes closely resemble the flow breccias exposed around the margin of the Hardman Basin. Only two, thin (less than 50cm wide), poorly exposed dykes of extensively altered massive basalt have been found.

Flood basalts are generally regarded as having been erupted relatively quietly, from extensive fissures developed in zones of crustal tension (e.g., Washington, 1922; Gibson, 1966; Clifford, 1968; Choubey, 1971; Macdonald, 1972). In many other flood-basalt sequences, dykes or other structures indicating sites of former vents have not been reported from large areas of the lava fields (e.g., Butler & Burbank, 1929; West, 1959; Swanson, 1967; Schmincke, 1967), and it is widely accepted that the featureless character of the final lava plains resulted from the burying of any vents by the superposition of thickening lava flows during the later stages of the eruption. The Antrim Plateau lavas mostly cover an area that has been relatively stable since at least the early Carpentarian (Plumb & Derrick, in press), and there is little evidence (except for the few dykes) of any tensional features which may have acted as loci for feeders to the flows. The northern Australian lavas are thickest in and near the Halls Creek Mobile Zone, but there is no evidence of major extrusion of lava along the length of the zone in the early Cambrian.

ACKNOWLEDGEMENTS

I wish to thank my wife, Joyce, for help in drawing the diagrams. Permission to publish the paper has been given by the Director, Bureau of Mineral Resources.

REFERENCES

Amaral, G., Cordani, U.G., Kawashita, K., & Reynolds, J.H., 1966: Potassium-argon dates of basaltic rocks from southern Brazil. *Geochim. et Cosmochim. Acta, 30*, pp. 159-189.

Baragar, W.R.A., 1969: The geochemistry of Coppermine River basalts. *Geol. Surv. Can. Pap. 69-44.*

Blake, D.H., in press: Birrindudu, Northern Territory — 1:250,000 Geological Series. *Bur. Miner. Resour. Aust. explan. Notes* SE/52-11.

Bofinger, V.M., 1967: Geochronology in the East Kimberley area of Western Australia. *Ph. D. Thesis, Australian National University* (unpubl.).

Bose, M.K., 1972: Deccan basalts. *Lithos, 5*, pp. 131-145.

Brooks, C.K., 1973: Tertiary of Greenland — a volcanic and plutonic record of continental break-up. *Amer. Assoc. Petrol. Geol. Mem. 19*, pp. 150-160.

Butler, B.S., & Burbank, W.S., 1929: The copper deposits of Michigan. *U.S. Geol. Surv. prof. Pap. 144.*

Carter, E.K., Brooks, J.H., & Walker, K.R., 1961: The Precambrian mineral belt of north-western Queensland. *Bur. Miner. Resour. Aust. Bull. 51.*

Choubey, V.D., 1971: Narmada-Son lineament, India. *Nature, 232*, pp. 38-40.

Clifford, P.M., 1968: Flood basalts, dike swarms and sub-crustal flow. *Can. J. Earth Sci., 5*, pp. 93-96.

Compston, W., 1974: The Table Hill Volcanics of the Officer Basin — Precambrian or Palaeozoic? *J. geol. Soc. Aust., 21*, pp. 403-411.

Cox, K.G., 1971: Karroo lavas and associated igneous rocks of southern Africa. *Bull. volcanol., 35*, pp. 867-886.

Cox, K.G., 1972: The Karroo volcanic cycle. *J. geol. Soc. Lond., 128*, pp. 311-336.

Cox, K.G., & Hornung, G., 1966: The petrology of the Karroo basalts of Basutoland. *Amer. Mineral., 51*, pp. 1414-1432.

Dow, D.B., 1965: Evidence of a late Pre-Cambrian glaciation in the Kimberley region of Western Australia. *Geol. Mag., 102*, pp. 407-414.

Dow, D.B., & Gemuts, I., 1969: Geology of the Kimberley region, Western Australia: The East Kimberley. *Bur. Miner. Resour. Aust. Bull. 106.*

Dunn, P.R., 1963a: Hodgson Downs, N.T. — 1:250,000 Geological Series. *Bur. Miner. Resour. Aust. explan. Notes* SD/53-14.

Dunn, P.R., 1963b: Urapunga, N.T. — 1:250,000 Geological Series. *Bur. Miner. Resour. Aust. explan. Notes* SD/53-10.

Dunn, P.R., & Brown, M.C., 1969: North Australian plateau volcanics. *Geol. Soc. Aust. Spec. Publ., 2*, pp. 117-122.

Edwards, A.B., & Clarke, E. de C., 1940: Some Cambrian basalts from the East Kimberley, Western Australia. *J. Proc. Roy. Soc. W. Aust., 26*, pp. 77-94.

Einarsson, Th., 1973: Geology of Iceland. *Amer. Assoc. Petrol. Geol. Mem. 19*, pp. 171-175.

Fawcett, J.J., Brooks, C.K., & Rucklidge, J.C., 1973: Chemical petrology of Tertiary flood basalts from the Scoresby Sund area. *Medd. om Gronland, 195*, pp. 1-54.

Gibson, I.L., 1966: Crustal flexures and flood basalts. *Tectonophysics, 3*, pp. 447-456.

Gemuts, I., & Smith, J.W., 1968: Gordon Downs, Western Australia — 1:250,000 Geological Series. *Bur. Miner. Resour. Aust. explan. Notes* SE/52-10.

Gregory, A.C., 1858: Journal of the north Australian exploring expedition under the command of Augustus C. Gregory. *J. Roy. geogr. Soc., 28*, pp. 1-137.

Hardman, E.T., 1885: Report on the geology of the Kimberley district, Western Australia. *W. Aust. parl. Pap. 34.*

Hays, J., 1967: Land surfaces and laterites in the north of the Northern Territory. *in* Jennings, J.N., & Mabbutt, J.A. (Eds), *Landform Studies from Australia and New Guinea*, pp. 182-210. Canberra, Australian National University Press.

Hodgson, I.M., in press a: Tanami, Northern Territory — 1:250,000 Geological Series. *Bur. Miner. Resour. Aust. explan. Notes* SE/52-15.

Hodgson, I.M., in press b: The Granites, Northern Territory — 1:250,000 Geological Series. *Bur. Miner. Resour. Aust. explan. Notes* SF/52-3.

Irvine, T.N., & Baragar, W.R.A., 1971: A guide to the chemical classification of the common volcanic rocks. *Can. J. Earth Sci., 8.*, pp. 523-548.

Jensen, H.I., 1915: Report on the geology of the country between Pine Creek and Tanami. *N. Terr. Bull., 14*, pp. 5-19.

Jolly, W.T., & Smith, R.E., 1972: Degradation and metamorphic differentiation of the Keweenawan tholeiitic lavas of northern Michigan, U.S.A. *J. Petrol., 13*, pp. 273-309.

Kaulback, J.A., & Veevers, J.J., 1969: Cambrian and Ordovician geology of the southern part of the Bonaparte Gulf Basin, Western Australia. *Bur. Miner. Resour. Aust. Rep. 109.*

Konda, T., & Green, J.C., 1974: Clinopyroxenes from the Keweenawan lavas of Minnesota. *Amer. Mineral., 59*, pp. 1190-1197.

Krieg, G.W., Jackson, M.J., & van de Graaff, W.J.E., in press: Officer Basin. *in* Knight, C.L. (Ed.), *The Economic Geology of Australia and Papua New Guinea*. Melbourne, Aust., Inst. Min. Metall.

Levi, B., 1969: Burial metamorphism of a Cretaceous volcanic sequence west from Santiago, Chile. *Contr. Mineral. & Petrol., 24*, pp. 30-49.

Lowry, D.C., Jackson, M.J., van de Graaff, W.J.E., & Kennewell, P.J., 1972: Preliminary results of geological mapping in the Officer Basin, Western Australia, 1971. *Geol. Surv. W. Aust. ann. Rep. for 1971.* pp. 50-56.

Macdonald, G.A., 1972: *Volcanoes.* Englewood Cliffs, Prentice-Hall Inc.

Matheson, R.S., & Teichert, C., 1948: Geological reconnaissance in the eastern portion of the Kimberley Division, Western Australia. *Geol. Surv. W. Aust. ann. Rep. for 1945,* pp. 27-41.

McDougall, I., & Rüegg, N.R., 1966: Potassium-argon dates on the Serra Geral Formation of South America. *Geochim. et Cosmochim. Acta, 30,* pp. 191-195.

Nalivkin, D.V., 1973: *Geology of the U.S.S.R.* (Translated from the Russian by N. Rast and edited by N. Rast and T.S. Westoll). Edinburgh, Oliver and Boyd.

Öpik, A.A., 1967: The Ordian Stage of the Cambrian and its Australian *Metadoxididae. Bur. Miner. Resour. Aust. Bull., 92,* pp. 133-169.

Peers, R., 1969: A comparison of some volcanic rocks of uncertain age in the Warburton Range area. *Geol. Surv. W. Aust. ann. Rep. for 1968,* pp. 57-61.

Plumb, K.A., & Derrick, G.M., in press: Geology of the Proterozoic rocks of northern Australia. *in* Knight, C.L. (Ed.), *The Economic Geology of Australia and Papua New Guinea.* Melbourne, Aust. Inst. Min. Metall.

Randal, M.A., 1963: Katherine, N.T. — 1:250,000 Geological Series. *Bur. Miner. Resour. Aust. explan. Notes* SD/53-9.

Randal, M.A., 1969: Larrimah, Northern Territory — 1:250,000 Geological Series. *Bur. Miner. Resour. Aust. explan. Notes* SD/53-13.

Randal, M.A., & Brown, M.C., 1969: Helen Springs, N.T. — 1:250,000 Geological Series. *Bur. Miner. Resour. Aust. explan. Notes* SE/53-10.

Roberts, H.G., Halligan, R., & Playford, P.E., 1968: Mount Ramsay, Western Australia — 1:250,000 Geological Series. *Bur. Miner. Resour. Aust. explan. Notes* SE/52-9.

Roberts, H.G., Gemuts, I., & Halligan, R., 1972: Adelaidean and Cambrian stratigraphy of the Mount Ramsay 1:250,000 Sheet area, Kimberley region, Western Australia. *Bur. Miner. Resour. Aust. Rep. 150.*

Schmincke, H.U., 1967: Stratigraphy and petrography of four upper Yakima Basalt flows in south-central Washington. *Geol. Soc. Amer. Bull., 78,* pp. 1385-1422.

Siedner, J.A., & Miller, J.A., 1968: K-Ar age determinations on basaltic rocks from South-West Africa and their bearing on continental drift. *Earth & Planet. Sci. Letters, 4,* pp. 451-458.

Smith, J.W., & Roberts, H.G., 1963: Mount Drummond, N.T. — 1:250,000 Geological Series. *Bur. Miner. Resour. Aust. explan. Notes* SE/53-12.

Snavely, P.D., MacLeod, N.S., & Wagner, H.C. 1973: Miocene tholeiitic basalts of coastal Oregon and Washington and their relations to coeval basalts of the Columbia Plateau. *Geol, Soc. Amer. Bull., 84,* pp. 387-424.

Sukheswala, R.N., & Poldervaart, A., 1958: Deccan Traps of the Bombay area, India. *Geol. Soc. Amer. Bull., 69,* pp. 1475-1494.

Swanson, D.A., 1967: Yakima Basalt of the Tieton River area, south-central Washington. *Geol. Soc. Amer. Bull., 78,* pp. 1077-1110.

Sweet, I.P., Mendum, J.R., Bultitude, R.J., & Morgan, C.M., 1974: The geology of the southern Victoria River region, Northern Territory. *Bur. Miner. Resour. Aust. Rep. 167.*

Traves, D.M., 1955: The geology of the Ord-Victoria region, northern Australia. *Bur. Miner. Resour. Aust. Bull. 27.*

Walter, M.R., 1972: Stromatolites and the biostratigraphy of the Australian Precambrian and Cambrian. *Spec. Pap. Palaeontology, 11.*

Washington, H.S., 1922: Deccan traps and other plateau basalts. *Geol. Soc. Amer. Bull., 33,* pp. 765-804.

Waters, A.C., 1961: Stratigraphic and lithologic variations in the Columbia River basalt. *Amer. J. Sci., 259,* pp. 583-611.

West, W.D., 1959: The source of the Deccan Trap flows. *J. geol. Soc. India, 1,* pp. 44-52.

White, W.S., 1960: The Keweenawan lavas of Lake Superior, an example of flood basalts. *Amer. J. Sci., 258A,* pp. 367-374.

Wilkinson, J.F.G., 1967: The petrography of basaltic rocks. *in* Hess, H.H. & Poldervaart, A, (Eds), *Basalts: The Poldervaart Treatise on Rocks of Basaltic Composition, 1,* pp. 163-214. New York, John Wiley & Sons, New York.

Wright, T.L., Grolier, M.J., & Swanson, D.A., 1973: Chemical variation related to the stratigraphy of the Columbia River basalt. *Geol. Soc. Amer. Bull., 84,* pp. 371-386.

REVIEW OF MINERALOGY AND CHEMISTRY OF TERTIARY CENTRAL VOLCANIC COMPLEXES IN SOUTHEAST QUEENSLAND AND NORTHEAST NEW SOUTH WALES

A. EWART, A. MATEEN[1], and J. A. ROSS

Department of Geology and Mineralogy, University of Queensland, Brisbane, QLD 4067

ABSTRACT

The Late Oligocene to Early Miocene volcanics of southeast Queensland and northeast New South Wales are dominated by mafic lavas (Albert, Beechmont, and Hobwee Basalt groups) of predominantly alkali olivine basalt to hawaiite compositions; additional lava types include: mugearite; tholeiitic andesite and dacite; metaluminous trachyte (1- and 2-feldspar types); rhyolite (1- and 2-feldspar potassic types); and peralkaline trachyte and comendite, plus plutonic equivalents within the Mount Warning and Focal Peak complexes. There is complete continuity of mineral compositions from basalt (and gabbro) through to rhyolite, with extreme Fe-enriched compositions (ferrohedenbergite and fayalite) developed in the rhyolites. Spinel is absent from most lavas, but an almost stoichiometric ilmenite occurs in nearly all lavas. Iron-titanium oxide equilibration temperatures deduced for some trachytes, dacite, and rhyolite lie between 885 and 970°C. These silicic lavas are deduced to have been relatively anhydrous, and crystallisation occurred between QFM and WM buffer conditions. Na enrichment of pyroxenes occurs only after extreme Fe-enrichment is attained (reflecting interplay of fo_2, and a_{Al2O3}^{liquid}). The rhyolite, comendite, and to lesser extents the trachyte and dacite, exhibit strong evidence of extreme crystal fractionation, e.g. strong depletion of Eu, Sr, Ba, V, Ni, Cr, Mg, and variable enrichment of Rb, Zr, Zn, Nb, Be, U, Th. The potassic nature of the rhyolite and trachyte, however, suggests the role of crustal assimilation (or anatexis) in their genesis, and is supported by Pb and Sr isotope compositions. The peralkaline lavas are interpreted as the end product of a long line of liquid descent; certain evidence suggests that the comendite could represent the end product of continued crystal fractionation, under quartz-feldspar ternary minima restrictions, from rhyolitic magma.

INTRODUCTION

Southeast Queensland and northeast New South Wales are characterised by extensive Late Oligocene to Early Miocene basalt-trachyte-rhyolite-comendite volcanism emanating from several major centres. This region includes several of at least 50 Cainozoic provinces that extend roughly along the eastern margin of Australia. Wellman & McDougall (1974a) distinguished two major volcanic groups — central volcano provinces and lava field provinces — and have demonstrated a southward migration of volcanism of the central volcano provinces at a rate of 66 ± 5 mm per year. Volcanism commenced some 70 m.y. ago.

This paper is concerned with a summary of current unpublished work on some of the central volcano provinces and smaller volcanic centres. The following is a brief outline of the geological setting of those described in this paper (Fig. 1).

Mount Warning and northern flanks of Tweed shield volcano (McTaggart, 1961; Solomon, 1964; Wilkinson, 1968; Green, 1969; Stevens, 1971; Ewart et al., 1971). The Mount Warning plutonic complex (22.4-23.1 m.y., Webb et al., 1967)

[1]Permanent address: Pakistan Atomic Energy Commission, Lahore, Pakistan.

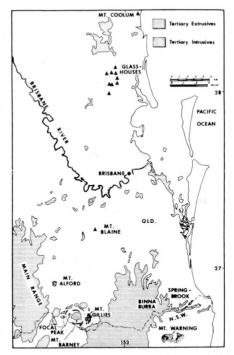

Fig. 1: Distribution of main volcanic centres discussed in this paper.

represents the eroded remnant of the core of the Tweed shield volcano. It is a ring structure, elliptical in plan, and between 5.5 and 8 km in diameter. Gabbros (showing, in part, inward dipping lamination and layering) form an outer belt enclosing a central syenite, itself enclosing a core of trachyandesite. A complex basalt-trachyte-syenite ring dyke, and a monzonite, cut the complex, together with minor dykes including comendites.

The following simplified volcanic succession (age 20.0-21.8 m.y., Wellman & McDougall, 1974b) is recognised on the northern flank of the Tweed volcano (excluding those lavas not emanating from the Tweed volcano):

	Comendite intrusion?	
Lamington volcanics	Hobwee basalt group	600 m
	Binna Burra rhyolite group	240 m
	Beechmont basalt group	270 m

The Hobwee and Beechmont Groups are composed of numerous flows which show much petrological overlap and thus are treated as one group in this paper. Current work has shown that the Binna Burra Rhyolite consists of at least three distinct phases; the Binna Burra Rhyolite sensu stricto (composed of several flows and pyroclastics); the Springbrook Rhyolite; and a xenolithic dacite lava flow within the Binna Burra region. In this paper the term Binna Burra Rhyolite is used in accordance with this modified subdivision.

Focal Peak shield volcano (Stephenson, 1956, 1959; Ross, 1974). Focal Peak lies

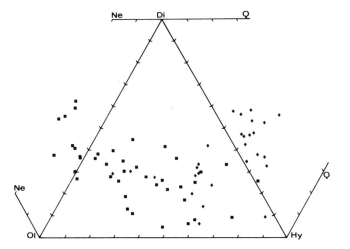

Fig. 2: Normative (CIPW) Ne-Ol-Di-Hy-Q components of analysed mafic lavas (basalts, hawaiites, mugearites). Squares are Albert Basalts; diamonds are Beechmont and Hobwee Basalts. Constant Fe_2O_3 of 1.5 percent assumed.

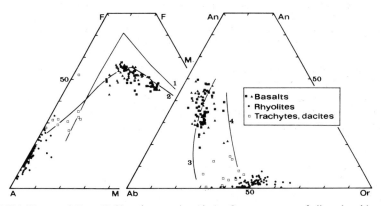

Fig. 3: AFM (F = total Fe as FeO) and normative Ab-An-Or components of all analysed lavas from region. Squares are Albert Basalts; triangles are Beechmont and Hobwee Basalts. Curve 1 is Hawaiian tholeiitic trend; curves 2 and 3 are Hawaiian alkalic trend; curve 4 is Gough Island trend (Coombs & Wilkinson 1969; note that curves 1 and 2 are based on F = Fe_2O_3 + FeO).

within a complex ring structure, about 14 km in diameter, marked by more-or-less peripheral rhyolite intrusions (including Mount Gillies), which also include the massive Mount Barney granophyre, several syenitic intrusives, the gabbro-syenite intrusive of Focal Peak, doleritic cone sheets, and minor trachytic intrusives. Two distinct volcanic units are associated with this centre: the Mount Gillies Rhyolite Formation and the older Albert Basalt Formation. Focal Peak is now interpreted as the major source of the Albert Basalt, although some contributions were made by adventive vents. The rhyolite has been dated at 22.7-23.6 m.y. (Webb et al., 1967; Ross, 1974) and was largely fed by the numerous dykes and subordinate plugs of the Mount Gillies area, in which subsidence occurred.

Mount Alford Ring-Complex (Stevens, 1959, 1962). This complex consists of a

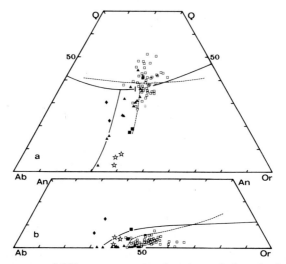

Fig. 4: Normative quartz and feldspar components of trachytes, dacites, rhyolites, and comendites. Hollow squares are rhyolites; filled triangles are comendites and peralkaline trachyte; filled diamonds are dacites (excluding hybrid, xenolithic dacites); filled squares are 2-feldspar trachytes; stars represent 1-feldspar trachytes. Solid lines in (a) are boundary curve and thermal valley for 1000 bars confining pressure (Tuttle & Bowen 1958). Dotted curves in (a) are from James & Hamilton (1969) for join [Ab-Or-Q]97An3, and in (b) is 2-feldspar boundary curve (quartz saturated), also from the same authors.

central plug of microdiorite and subordinate granophyre, intruded by tholeiitic andesite (non-calcalkaline type). The intrusion (3-4 km in diameter) is emplaced into sandstones which have been steeply upturned. Later dykes and breccias are found; the dykes occur as a semicircular zone of rhyolitic and trachytic ring-dykes, and a second, non-arcuate, set of cross-cutting rhyolitic, comenditic, and basaltic dykes.

Mount Blaine is one of the Flinders Group of volcanic plugs, forming an isolated group of small trachytic, and less abundant rhyolitic and basaltic intrusive plugs, emplaced in Early Jurassic sandstones of the Marburg Formation.

Glass House Mountains (Stevens, 1971; Bryan & Stevens, 1973) constitute a spectacular group of domes and conical peaks, each representing an eroded centre of eruption, dated at 24.8 m.y. (Webb et al., 1967; Wellman & McDougall, 1974). Basalts to the north, in the Maleny region, are of similar age. The peaks comprise comendite and slightly less abundant trachyte.

CHEMISTRY AND NOMENCLATURE OF LAVAS
Mafic lavas (Albert, Beechmont, and Hobwee Basalt Groups)

Chemical analyses show that these lavas in fact range from alkali olivine basalt through hawaiite to mugearite, using the normative-chemical nomenclature of Coombs & Wilkinson (1969). Hawaiite is the predominant type, but a complete mineralogical and chemical gradation exists through the series. Thus, the term 'basalt', as used in the stratigraphic nomenclature, is somewhat misleading. A more convenient two-fold classification, which can be correlated with mineralogy, is into (a) nepheline-normative and olivine-normative lavas, and (b) quartz-normative lavas, as shown in Figure 2, but based on an assumed constant value of 1.5% Fe_2O_3 (as most samples studied show some evidence of secondary hydration and oxidation).

Figure 2 emphasises not only the gradational variation of the basalt compositions, but also their transitional affinities (Green, 1968, 1969; Wilkinson, 1968). The Albert Basalt Group is nepheline-normative to transitional, whereas the Beechmont/Hobwee Groups are transitional to quartz-normative. These mafic lavas show no tendency for strong Fe-enrichment (Fig. 3); as a suite, they are relatively enriched in potash, TiO_2, and P_2O_5, when compared to tholeiitic sequences (cf. Jamieson and Clarke 1970).

Silicic and intermediate lavas

These include tholeiitic andesite and dacite; oversaturated metaluminous and peralkaline trachyte; metaluminous rhyolite; and comendite. They are plotted in Figure 4 in the quartz-alkali feldspar and feldspar systems. The comendites (which are in reality distinguished by their sodic mineralogy as subsequently described) most commonly contain normative *ac* and *ns*; occasional samples show traces of normative *an*, but these also have evidence of incipient weathering effects.

The rhyolites are classed as potassic, 2-feldspar rhyolites which are relatively low in MgO and CaO. Although a number of analysed rocks do contain normative *C,* those with significant normative *C* (e.g. more than 0.5%) are either strongly devitrified or exhibit evidence of alkali leaching, with the notable exception of the Springbrook rhyolites. Thus, the original compositions of the rhyolitic magmas (except Springbrook) are presumed to have been metaluminous. In Figure 4a, the scatter of points around the ternary minimum probably reflects devitrification and oxidation of certain samples, as all available data are plotted.

Discussion of the dacitic and trachytic compositions is deferred to the next section.

MINERALOGY

Basalt, hawaiite, mugearite

Plagioclase, augite, olivine, ilmenite, and apatite are present in all samples so far examined.

Albert Basalt Group. These lavas typically contain phenocryst olivine (zoned to Fe-rich rims, total compositional range Fo40-77, and plagioclase (labradorite, Fig. 5,1P). Only rarely, however, is phenocrystic plagioclase a conspicuous phase. Groundmass minerals include (Figs 5,6) clinopyroxene (salite and salitic augite), olivine (Fo50), ilmenite and titanomagnetite (the two latter minerals not always in equilibrium, see Fig. 7), and feldspar which shows continuous compositional variation from labradorite through andesine and calcic anorthoclase to sanidine (Fig. 5,1G), the most potassic feldspar being interstitial. Interstitial chlorophaeite is common.

Beechmont and Hobwee Basalt Groups are of two types:
(i) Olivine-normative group. Mineralogically, these are similar to the Albert

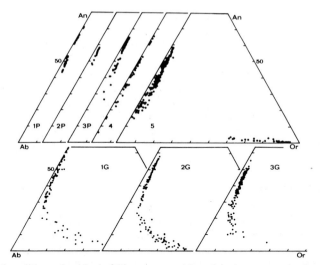

Fig. 5: Basaltic, hawaiitic, and gabbroic feldspar compositions, based on composite data, in terms of Ab, An, and Or (mol. %). P refers to phenocrysts, G to groundmass. 1, Albert Basalt (representing basalts and hawaiites); 2, Beechmont olivine-normative hawaiites; 3, Beechmont and Hobwee quartz-normative hawaiites; 4, Focal Peak gabbro and monzonite; 5, Mount Warning gabbros.

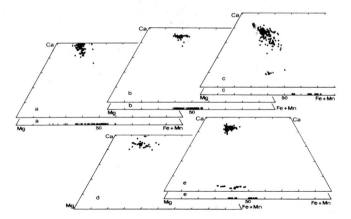

Fig. 6: Basaltic, hawaiitic, and gabbroic pyroxene and olivine compositions, based on composite data, in terms of Ca, Mg, and Fe + Mn (mol. %). a, Albert Basalt (basalt and hawaiite); b, Beechmont olivine-normative hawaiites; c, Beechmont and Hobwee quartz-normative hawaiites; d, Focal Peal gabbro and monzonite; e, Mount Warning gabbros. For a, b, and c, filled circles and squares are groundmass pyroxenes and olivines respectively; open circles and squares are phenocrystic pyroxenes and olivines.

Basalt, except phenocrystic olivine and plagioclase are less abundant and the groundmass clinopyroxene is rather less calcic (Fig. 6b).

(ii) Quartz-normative group. These are commonly strongly porphyritic, with plagioclase the dominant phase (labradorite, zoned abruptly to andesine on the rims), and minor phenocryst diopsidic augite (Figs 5, 6). Groundmass phases consist of: augite ranging to subcalcic augite (most grains showing slight Fe-enrichment towards their rims); pigeonite (not all samples), which is commonly rimmed by subcalcic and low-Ca augite; olivine (Fe-rich, Fo_{27-46}) which is present

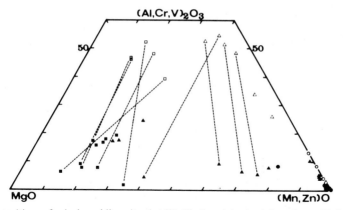

Fig. 7: Compositions of spinels and ilmenites (wt %). Tie-lines join coexisting phases. Hollow and filled squares are spinels and ilmenites respectively from basalt, hawaiite, and mugearite; hollow and filled triangles are rhyolitic, trachytic, and dacitic spinels and ilmenites respectively; hollow and filled circles are comenditic spinels and ilmenites respectively.

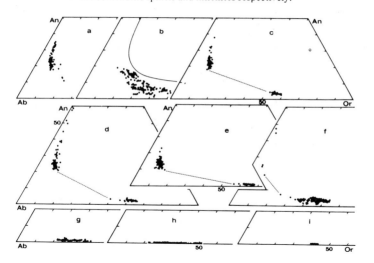

Fig. 8: Feldspar phenocryst compositions, based on composite data, of silicic and intermediate lavas in terms of Ab, An, and Or (mol. %). a, represents tholeiitic andesites and dacites; b, one-feldspar trachytes; c, two-feldspar trachytes; d, Springbrook rhyolites; e, Binna Burra rhyolites; f, Mount Gillies rhyolites; g, peralkaline trachyte; h, Glass House comendites; i, Binna Burra comendite.

in all samples examined, but only as sporadic small interstitial grains; feldspar, ranging from labradorite to calcic anorthoclase and significantly not as potash-enriched as in the olivine-normative basalts; ilmenite. Titanomagnetite occurs in only some samples. A fine-grained siliceous residue and an interstitial Fe-rich glassy (partly devitrified) residuum are present.

It is clear that the basalts show a systematic correlation between mineral chemistry and bulk composition, which is considered significant despite the hydrated nature of the basalts (Table 1). Thus, hydration has evidently not changed the essential chemical characteristics of the lavas.

EWART AND OTHERS

Table 1. Averaged analyses of some representative comendites, rhyolites, trachytes, and 'basalts' from Tertiary volcanic centres, S.E. Queensland.

| | COMENDITE | RHYOLITES | | | TRACHYTE | 'BASALTS' | | |
	Average Glass Houses Comendite	Binna Burra	Springbrook	MtGillies	Glass Houses (average 1-feldspar type)	Beechmont and Hobwee ol-normative	qtz-normative	Albert
No. of analyses	6	10	4	30	3	3	5	41
SiO_2	72.23	74.42	74.04	76.23	62.64	48.51	53.96	48.46
TiO_2	0.16	0.07	0.37	0.15	0.53	2.61	2.22	2.27
Al_2O_3	12.58	12.19	12.12	11.42	16.77	15.19	14.60	15.81
Fe_2O_3	2.05	0.75	1.17	0.58	2.82	3.89	3.31	3.98
FeO	1.22	0.45	0.95	0.61	1.49	8.02	6.82	7.36
MnO	0.06	0.01	0.03	0.02	0.09	0.15	0.15	0.15
MgO	0.03	0.04	0.33	0.12	0.54	4.88	3.83	5.75
CaO	0.13	0.42	0.46	0.48	1.47	6.75	7.24	7.42
Na_2O	5.98	3.49	3.33	3.27	5.81	3.83	3.91	3.78
K_2O	4.54	5.15	4.63	4.98	5.50	1.62	1.56	1.46
P_2O_5	0.01	0.01	0.06	0.02	0.15	0.83	0.46	0.69
H_2O+	0.47	2.39	2.14	—	0.61	2.38	0.78	—
H_2O-	0.23	0.38	0.23	—	1.20	1.28	1.01	—
Loss on ignition	—	—	—	2.08	—	—	—	2.88
*Trace elements (p.p.m.)**								
Rb	337	471	206	157	102	24	29	25
Sr	1.3	1.0	67.	9	117	468	452	635
Ba	20	12	259	79	820	415	430	376
V	n.d.	1.2	ca.4	ca.2	ca.2	135	165	175
Cr	n.d.	n.d.	n.d.	ca.2	n.d.	35	71	103
Ni	4	ca.2	ca.4	6.8	4.3	45	23	81
Co	n.d.	n.d.	—	ca.1	ca.2	35	28	ca. 38
Cu	6	11	ca.7	9.6	9	34	30	43
Zn	315	147	51	115	102	211	177	129
Zr	1450	137	282	330	872	264	238	238
Pb	44	45	20	21	15	2.7	3.1	3.8
Be	ca.16	ca.15	ca.4	—	4.6	—	—	—
Ga	52	ca.40	ca.24	—	39	—	—	—
B	ca.20	ca.50	ca.19	—	n.d.	—	—	—
Sc	ca.1	ca.5	ca.37	—	7.2	—	—	—
Nb	220	58	14	34	52	26	19	28
Y	164	87	33	71	53	34	29	28
La	115	27	36	67	106	27	22	25.5
Ce	220	61	75	154	212	61	52	56
Nd	104	31	32	75	81	32	24	27.6
Σ REE+Y	732	267	213	456	530	192	159	170
Th	41.4	45	18.8	20	9	1.8	2.4	3.1
U	8.1	9.7	4.2	ca.3.7	2.3	0.65	0.78	ca.0.88
K/Rb	112	91	187	263	461	581	447	498
Th/U	5.1	4.6	4.5	ca.5.4	3.9	2.8	3.1	ca.3.0

n.d. = not detected
— = not determined
ca. = average of less No. of samples

*U by delayed neutron counting. Be, Ga, B, Sc by optical spectrography.
 Remaining elements by X-ray fluorescence.

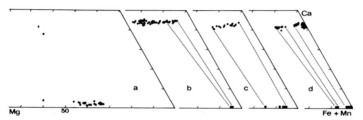

Fig. 9: Pyroxene and olivine phenocryst compositions, based on composite data, of silicic and intermediate lavas, in terms of Ca, Mg, and Fe + Mn (mol. %). a, represents Springbrook rhyolites; b, trachytes; c, tholeiitic andesite and dacite; d, Mount Gillies rhyolites. Tie-lines join coexisting compositions. Filled circles are pyroxenes; squares are olivines.

Also shown in Figures 5 and 6 are the comparative data for pyroxene and feldspar compositions of the Focal Peak and Mount Warning gabbros. Both sets of gabbros contain a plagioclase and a potash-rich feldspar (not analysed in Focal Peak rocks), the latter replacing the equivalent high-temperature anorthoclase of the mafic lavas. The augites and olivines have similar compositional fields to the mafic lavas, the only difference being the orthopyroxene occurrence in certain of the Mount Warning gabbros.

Silicic and intermediate lavas

These may be conveniently divided into the following four types, based mainly on phenocryst assemblages (Figs 8,9):

(i) *Tholeiitic andesite and dacite* (not of island-arc type): Characterised by plagioclase (zoned from andesine through oligoclase to calcic anorthoclase), clinopyroxene (ferroaugite-ferrohedenbergite), ilmenite, $\pm$ titanomagnetite, $\pm$ fayalitic olivine. These lavas plot within the plagioclase field of the feldspar and quartz-feldspar systems (Fig. 4). An unusual hybrid dacite from Binna Burra contains ferropigeonite.

(ii) *Trachyte:* Three distinct types are recognised:
(a) One-feldspar trachytes (metaluminous) containing a calcic anorthoclase - calcic sanidine feldspar, calcic ferroaugite-ferrohedenbergite, $\pm$ fayalitic olivine, and titanomagnetite. In Figure 4 these project towards the termination of the 2-feldspar boundary curve, and presumably define the thermal valley (Fig. 4a) appropriate to these lavas.
(b) Two-feldspar trachytes (metaluminous) with plagioclase (andesine - potash oligoclase), sanidine, calcic ferroaugite - ferrohedenbergite, fayalitic olivine, ilmenite, and titanomagnetite. In Figure 4 these project within the potash-feldspar field and close to the 2-feldspar boundary curve.
(c) Peralkaline trachyte, with low-Ca anorthoclase, traces of interstitial quartz, arfvedsonite, aegirine, and ilmenite (all strictly groundmass phases).

(iii) *Rhyolite:* Three distinct types are recognised, although compositionally similar:

(a) Binna Burra type, characterised by potassic oligoclase (commonly showing resorption), sanidine, quartz, and ilmenite. An extremely Fe-rich fluor-biotite

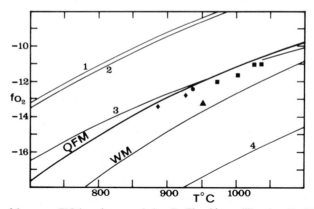

rig. 10: Log fo_2 versus T°C based on coexisting Fe-Ti oxide equilibration (Buddington & Lindsley, 1964). Filled squares are basalt, hawaiites, and mugearite; filled circle is rhyolite (Mount Gillies); filled diamonds are trachytes; filled triangle is dacite. QFM (1 bar) from Wones & Gilbert (1969); WM (1 bar) from Eugster & Wones (1962). Curve 1 represents reaction: 4 diopside + 4 ilmenite + O_2(gas) = 2 hematite + 4 sphene + 4 enstatite; curve 2 represents: 4 ilmenite + O_2(gas) = 4 rutile + 2 hematite; curve 3 represents: ilmenite + hedenbergite + $\frac{1}{2}O_2$(gas) = 2/3 magnetite + sphene + quartz; curve 4 represents: hedenbergite+ ilmenite = 2Fe (α) + rutile + wollastonite + quartz + O_2 (gas).

(Fe/Fe + Mg = 0.82-0.90) and traces of allanite are present in one flow, while a small number of rather spongy-looking grains of an almost pure ulvospinel occur in heavy-mineral concentrates from another flow. No other ferromagnesians occur. Lavas have low phenocryst contents (less than 5%).

(b) Springbrook type, characterised by plagioclase (andesine - potash oligoclase), sanidine, ± quartz, ferrohypersthene, and ilmenite. Phenocryst contents are less than 5 percent, and basaltic xenoliths are common.

(c) Mount Gillies type, characterised by sanidine, quartz, ferrohedenbergite, fayalite, ilmenite, traces of zircon, and in certain samples by rare ferrohypersthene and ferropigeonite. Titanomagnetite occurs in only one specimen examined. Plagioclase is absent in most samples, but in others, occasional partially resorbed crystals occur enclosed in sanidine; it thus seems likely that plagioclase has crystallised but has subsequently been resorbed in most of the rhyolites (Rahman & MacKenzie, 1969). Phenocryst contents average 15-25 per cent. A rare accessory, tentatively identified as belonging to the chevkinite group, is present.

(iv) *Comendite:* These are generally phenocryst-poor with anorthoclase-sanidine and quartz phenocrysts. Groundmass phases include Ca-poor anorthoclase-sanidine, quartz, fluor-arfvedsonite, ± aenigmatite, ± aegirine, ± ilmenite. In the most strongly peralkaline samples, arfvedsonite is the main, if not only, Fe-silicate phase. The compositional range of the alkali feldspar phenocrysts is variable; in the Glass House comendites the range is rather wide, Or23-52 (cf. Nicholls & Carmichael, 1969).

Iron-Titanium Oxides

An important feature of most of the rocks examined is the absence of a spinel phase and the presence of an almost stoichiometric ilmenite (FeTiO3). Figure 10 shows the equilibration T-fo_2 data deduced from coexisting Fe-Ti oxides in those samples in which they do coexist (see also Fig. 7). From Figure 10, equilibration temperatures of between 885 and 950°C are indicated for the trachyte, dacite, and rhyolite samples, at an oxygen potential below that defined by the QFM buffer and approaching the WM buffer. The basalt, hawaiite, and mugearite equilibration temperatures, all from groundmass oxides, should represent solidus temperatures.

Consideration of reactions such as:

(a) $FeTiO_3$ + $CaFeSi_2O_6$ + $1/3O_2 \rightleftharpoons 2/3Fe_3O_4$ + $CaTiSiO_5$ + SiO_2
     ilmenite   hedenbergite   gas   magnetite   sphene   quartz

(b) $4FeTiO_3$ + O_2 $\rightleftharpoons$ $4TiO_2$ + $2Fe_2O_3$
     ilmenite   gas   rutile   hematite

(c) $2/3\,Fe_3O_4$ + SiO_2 $\rightleftharpoons$ Fe_2SiO_4 + $1/3O_2$
     magnetite   quartz   fayalite   gas

indicates that the presence of ilmenite without spinel is unlikely to occur at oxygen potentials above that defined by QFM, and in fact it is deduced that it indicates oxygen potentials lower than the lowest T-fo_2 equilibration data in Figure 10. By analogy with the Skaergaard intrusion, crystallisation is thus likely to have occurred very close to WM buffer conditions (Lindsley et al., 1969). Calculations of reaction (c) (using data of Carmichael et al., 1974), in terms of $a\ ^{spinel}_{Fe_3O_4}$ with fo_2 defined by the WM buffer (Eugster & Wones, 1962), indicate that a nearly pure ulvöspinel would be in equilibrium with fayalite and quartz. The occurrence of rare spinel grains (Usp. 95-99) in the Binna Burra rhyolite is thus significant.

Reference to Figure 7 illustrates the strong relative enrichment of (Mn, Zn)O, and depletion of MgO, in the Fe-Ti oxides of the comendites and rhyolites. Certain of the comenditic ilmenites contain over 0.9 per cent ZnO.

TRACE ELEMENT GEOCHEMISTRY

Of special significance are the trace element abundance patterns of the silicic lavas, especially the rhyolites and comendites.

The Binna Burra rhyolites exhibit abnormal enrichment of Pb, Rb, U, Th, and Be; more significantly, they exhibit very strong depletion of Sr, Ba, V, Cr, Ni, and Mg. The Mount Gillies and Springbrook rhyolites show these features to a progressively lesser degree, respectively. The Rb versus K/Rb plot (Fig. 11) is especially significant, as there is a logarithmic relation between Rb and K/Rb, the Binna Burra rhyolites lying at the extreme end of the fractionation trend. Thus, the criterion of peralkalinity does not necessarily imply a highly fractionated geochemistry, as seen, for example, in abnormally low K/Rb ratios. The REE data are also significant (Fig. 12), being dominated by intense Eu depletion, which again is most pronounced in the Binna Burra rhyolite and least in the Springbrook rhyolite. The trace element evidence clearly points towards crystal fractionation processes in controlling the development of the rhyolites. The relatively more highly-fractionated trace element chemistry of the Binna Burra lavas can be explained in terms of extended crystal fractionation of these rhyolitic magmas under quartz-feldspar ternary minima restrictions.

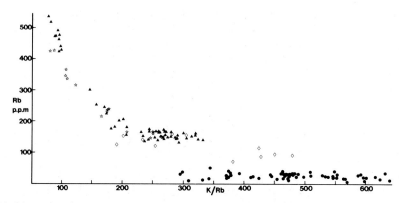

Fig. 11: Rb versus K/Rb. Filled triangles are rhyolites; stars represent comendites; hollow diamonds represent trachytes, tholeiitic andesites, and dacites; filled circles are basalts, hawaiites, and mugearites.

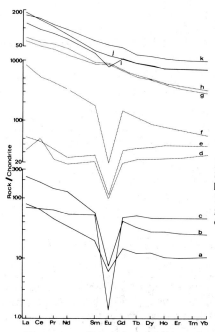

Fig. 12: Chondrite-normalised REE data. a, Springbrook rhyolite; b, Mount Gillies rhyolite; c, Binna Burra rhyolite; d, Binna Burra comendite; e Mount Alford comendite; f, Glass House comendite; g, Hobwee quartz-normative hawaiite; h, Beechmont olivine-normative hawaiite; i, dacite; j, trachyte (1-feldspar); k, tholeiitic andesite.

The comendites also exhibit strong depletion of Sr, Ba, V, Ni, and Cr, but their main trace element characteristics are the enrichments in Zr, Nb, Y, Zn, Σ REE, and Ga which always accompany the development of peralkalinity. Again, a pronounced Eu depletion occurs in their REE patterns (Fig. 12). Zr/Hf ratios range from 59-67 in the comendites, compared with 24-59 in the rhyolites.

PETROGENESIS OF SILICIC AND INTERMEDIATE LAVAS

Equilibration temperatures

Fe-Ti oxide equilibration temperatures of two trachytes, a dacite, and the Mount Gillies rhyolite lie between 885 and 950°C (Fig. 10). Additional equilibration temperatures of ferrohedenbergite-fayalite pairs from the Mount

Gillies rhyolites have been calculated (using the clinopyroxene-olivine exchange reaction of Powell & Powell, 1974), which give 1 bar temperatures between 890 and 970°C (average 925°C). Crude temperature estimates for the Springbrook rhyolite of 900-1000°C have been obtained from the orthopyroxene-ilmenite exchange reaction*:

$$MgSiO_3 \ + \ FeTiO_3 \ \rightleftharpoons \ FeSiO_3 \ + \ MgTiO_3$$

enstatite   ilmenite   ferrosilite   geikielite

This has been calibrated using New Zealand rhyolite data (Ewart, et al., 1975) for which temperatures and pressures are independently available.

These relatively high equilibration temperatures, together with the absence of hydrous phases, the general paucity of pyroclastic deposits associated with the Springbrook and Mount Gillies rhyolites, and the inferred WM T-fo_2 buffer conditions, all indicate that these rhyolitic magmas were relatively dry. This implies viscosities of $10^{7.5}$-$10^{8.5}$ poise at 900-950°C (Shaw, 1972).

Slightly lower temperature estimates of about 800-900°C are obtained for the Binna Burra rhyolites by means of the plagioclase thermometer (Kudo & Weill, 1970), which are consistent with the occurrence of fluor-biotite and a greater pyroclastic component associated with the Binna Burra lavas.

Equilibration pressures

P $total$ for phenocryst equilibration has been estimated for three lavas (quartz-free) by combining the following reactions, and using T-fo_2 from Fe-Ti oxides $[\Delta G_r^\circ / 2.303RT + \Delta V_r^\circ (P-1)/2.303RT = 9689/T - 2.298 - 0.10752 (P-1)/T]$:

$$SiO_2 \ + \ CaAl_2SiO_6 \ \rightleftharpoons \ CaAl_2Si_2O_8$$
glass   Ca-Tschermak's   anorthite
                pyroxene

$$2/3 \ Fe_3O_4 \ + \ SiO_2 \rightleftharpoons Fe_2SiO_4 \ + \ 1/3 O_2$$
magnetite   glass   fayalite   gas

Results for the Mount Blaine trachyte give P $total$ = 6.4 kb (about 23 km); a Tweed volcano dacite gives 1.4 kb (about 5 km); the third sample, a Mount Gillies rhyolite, gives a value between 2.5 and 5.6 kb (9-20 km) depending on the assumptions made concerning $a_{SiO_2}^{liquid}$ (plagioclase being absent). Calculations using coexisting ferrohypersthene and fayalite in these rhyolites indicated disequilibrium between these two phases.

Evidence for crystal fractionation and crustal assimilation

(i) Data presented in Figure 13a for the silicic and intermediate lavas (excluding peralkaline types) show an almost complete continuity of phenocryst clinopyroxene and olivine compositions to the Fe-enriched end-members, the latter occurring in

---

*$\Delta G_r^\circ / 2.303RT$ = 1169/T-0.049 for reaction with components in standard state (Robie & Waldbaum, 1968; Ewart et al., 1975, orthoferrosilite data) and 1169/T + 0.083 using New Zealand rhyolite data, and assuming ideality for the pyroxenes and ilmenites.

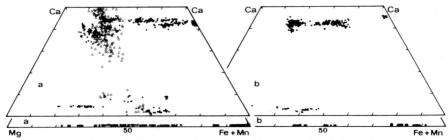

Fig. 13: Summary of all available pyroxene (excluding sodic pyroxenes) and olivine data from (a) S.E. Queensland volcanic rocks; and (b) Mount Warning plutonic igneous complex. In (a), open and filled triangles refer to groundmass pyroxenes and olivines respectively, and filled circles and squares to phenocryst pyroxenes and olivines.

the Mount Gillies rhyolites. The continuity to Mg-enriched compositions is complete if the Mount Warning rock data are included (Figure 13b). These mineral data imply a continuous liquid line of descent, and obvious comparison can be made with the Skaergaard trends (Brown & Vincent, 1963).

(ii) Trace element abundances provide convincing evidence of the role of crystal fractionation in the development of the rhyolitic and comenditic lavas, for example, the depletion of Sr, Ba, V, Ni, Cr, and variable enrichment of such elements as Rb, U, Th, Zr, Nb, and Zn. The Eu anomalies provide strong evidence for feldspar fractionation. Similar comments apply to lesser extents to the trachytes and dacites (e.g., Fig. 12).

Reference to Figure 11 indicates that the basalts and hawaiites do not form a continuum, with respect to Rb and K, with the more silicic lavas; this does not necessarily preclude a crystal fractionation derivation of the silicic lavas, but simply that the exposed basalts and hawaiites are not part of the postulated fractionation scheme, as is usually supposed.

(iii) Isotopic evidence is available for the Lamington lavas (Ewart et al., in prep.). Pb isotopes indicate at least two distinct groups within these hawaiites: certain olivine-normative types less radiogenic than modern mid-ocean ridge basalt, and a second more radiogenic group including both olivine- and quartz-normative hawaiites. Pb isotopic compositions of the Binna Burra rhyolites, the comendite, and the Springbrook rhyolites are all distinctive, and all more radiogenic than the hawaiites. Initial Sr^{87}/Sr^{86} ratios of the Springbrook rhyolite (0.7049) are also more radiogenic than the hawaiites (0.7038-0.7043); the same is probably true of the Binna Burra rhyolites and comendite (initial Sr^{87}/Sr^{86} greater than 0.71), but their extremely high Rb/Sr ratios make estimation of initial Sr isotopic compositions rather uncertain. These data suggest crustal assimilation during the development of these acidic magmas. Assimilation is further suggested by the correlations, in the Springbrook and Binna Burra rhyolites, between Σ Pb and Σ Sr, and their isotopic compositions. Thus, relative to the Springbrook rhyolite, the Binna Burra rhyolites contain less radiogenic $Pb^{206}/^{204}$ and $Pb^{208}/^{204}$ but higher Σ Pb, and apparently more radiogenic Sr, but highly depleted Σ Sr.

(iv) The rhyolites are classed as two-feldspar potassic rhyolites, which project into the sanidine fields in Figure 4 and are modally dominated by phenocrystic sanidine. Carmichael (1963) pointed out that such rhyolites are unlikely to be simple

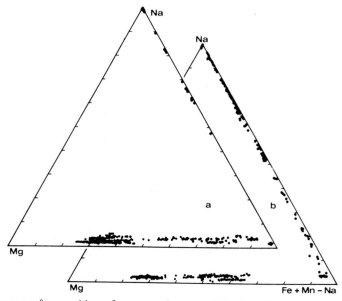

Fig. 14: Summary of compositions of pyroxenes, in terms of Mg, Fe + Mn — Na, and Na (mol. %) in (a) volcanic rocks of S.E. Queensland; and (b) Mount Warning plutonic igneous complex.

differentiates from a basaltic precursor, which would produce one-feldspar sodic liquids; the tholeiitic andesites and dacites, in fact, seem to correspond to such derivative liquids (note the lower P_{total} estimate for the Tweed dacite). Carmichael (1963) concluded that potassic rhyolites result from sialic fusion or fractional crystallisation of tholeiitic magma modified by sialic contamination. As these Queensland rhyolites are deduced to have been relatively dry, this implies an anhydrous crustal material, such as granulite or charnockite. Allowance for the effect of normative *an* in the rhyolites (James & Hamilton, 1969) does not seem to change this conclusion (Fig. 4a); it should be noted that the [Ab-Or-Q]97An3 thermal valley shown in Fig. 4a is slightly too far shifted towards the Q-Or sideline, judging from the projections of the trachyte compositions.

It should also be noted that comparable rhyolites are evidently unknown in oceanic regions, but do occur in the western U.S.A. (Christiansen & Lipman, 1972).

The problem of Na-enrichment

One problem within the region is the contemporaneous occurrence, within the one province, of metaluminous, peraluminous, and peralkaline silicic lavas. Although it was initially assumed that comendites and rhyolites were separately derived, certain syenites of Mount Warning provide a possible link between the two. The syenites contain perthite, quartz, ilmenite, fayalite, ferrohedenbergite zoned continuously to aegirine, and an arfvedsonitic amphibole. The mineralogy indicates a late-stage change from a metaluminous (cf. Mount Gillies Rhyolite) to peralkaline mineralogy. The Na-enrichment in the volcanic and plutonic pyroxenes (the latter from Mount Warning) are summarised in Figure 14. Clearly, Na-

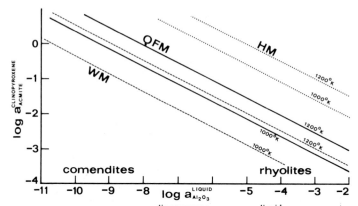

Fig. 15: Schematic relation between log $a^{\text{clinopyroxene}}_{\text{acmite}}$ and log $a^{\text{liquid}}_{\text{Al}_2\text{O}_3}$ calculated according to the reaction (1 bar): $\frac{1}{2}\text{Al}_2\text{O}_3(\text{liq})$ + acmite + $1\frac{1}{2}$ quartz = albite + $\frac{1}{4}\text{O}_2$ + $\frac{1}{2}$ fayalite, using three T-f_{O_2} buffers (WM, QFM, HM, using data from Eugster & Wones 1962, and Wones & Gilbert 1968) each at two temperatures (1000 and 1200°K). In the calculations $a^{\text{quartz}}_{\text{SiO}_2}$, $a^{\text{olivine}}_{\text{fayalite}}$, and $a^{\text{feldspar}}_{\text{albite}}$ assumed constant at 1.0, 1.0, and 0.75, respectively. The estimated approximate $a^{\text{liquid}}_{\text{Al}_2\text{O}_3}$ of selected S.E. Queensland comendites and Mount Gillies metaluminous rhyolites are indicated. See text for further explanation.

enrichment occurs only after extreme Fe-enrichment (occurring only in the peralkaline magmas), thus contrasting with the Sakhalin (Yagi, 1953) and Shonkin Sag (Nash & Wilkinson, 1970) pyroxenes.

A possible explanation, with special reference to the Mount Warning syenite, is illustrated in Figure 15, based on the following reaction at 1 bar ($\Delta G^{\circ}_r/2.303RT$ = 2356/T-2.740 for components in their standard states, using Robie & Waldbaum 1968, Stull & Prophet 1971):

$$\underset{\text{liquid}}{\frac{1}{2}\text{Al}_2\text{O}_3} + \underset{\text{acmite}}{\text{NaFe}^{3+}\text{Si}_2\text{O}_6} + \underset{\text{quartz}}{1\frac{1}{2}\text{SiO}_2} \rightleftharpoons \underset{\text{albite}}{\text{NaAlSi}_3\text{O}_8} + \underset{\text{gas}}{\frac{1}{4}\text{O}_2} + \underset{\text{fayalite}}{\frac{1}{2}\text{Fe}_2\text{SiO}_4}$$

Approximate estimates of $a^{\text{liquid}}_{\text{Al}_2\text{O}_3}$ have been obtained, for the Mount Gillies and comenditic assemblages, from the following reactions:

$$\underset{\text{quartz}}{\frac{1}{2}\text{SiO}_2} + \underset{\text{hedenbergite}}{\text{CaFeSi}_2\text{O}_6} + \underset{\text{liquid}}{\text{Al}_2\text{O}_3} \rightleftharpoons \underset{\text{anorthite}}{\text{CaAl}_2\text{Si}_2\text{O}_8} + \underset{\text{fayalite}}{\frac{1}{2}\text{Fe}_2\text{SiO}_4}$$

$$\underset{\text{acmite}}{\text{NaFeSi}_2\text{O}_6} + \underset{\text{liquid}}{\frac{1}{2}\text{Al}_2\text{O}_3} + \underset{\text{quartz}}{\text{SiO}_2} \rightleftharpoons \underset{\text{albite}}{\text{NaAlSi}_3\text{O}_8} + \underset{\text{magnetite}}{\frac{1}{3}\text{Fe}_3\text{O}_4} + \underset{\text{gas}}{1/12\,\text{O}_2}$$

$$\underset{\text{hedenbergite}}{\text{CaFeSi}_2\text{O}_6} + \underset{\text{liquid}}{\text{Al}_2\text{O}_3} \rightleftharpoons \underset{\substack{\text{Ca-Tschermak's}\\\text{pyroxene}}}{\text{CaAl}_2\text{SiO}_6} + \underset{\text{quartz}}{\frac{1}{2}\text{SiO}_2} + \underset{\text{fayalite}}{\frac{1}{2}\text{Fe}_2\text{SiO}_4}$$

Figure 15 illustrates that increasing oxygen potential, for a given $a^{\text{liquid}}_{\text{Al}_2\text{O}_3}$, favours increasing $a^{\text{clinopyroxene}}_{\text{acmite}}$. The low oxygen potential (WM buffer) deduced for the Queensland silicic lavas is thus consistent with a late stage of acmite development in these pyroxene solid solutions which coincides with the development of strong peralkalinity. The interaction between fo_2, temperature, and varying a $^{liquid}_{Al_2O_3}$ thus gives considerable scope for variation in soda enrichment in igneous pyroxenes at differing stages of crystallisation. Figure 15 implies high oxygen potential within the Sakhalin rocks, which is supported by the occurrence of hematite and sphene in these rocks (Yagi, 1953).

SUMMARY

(i) The mafic lavas are dominated by alkali olivine basalt and hawaiite, and are volumetrically dominant in the field over the more silicic lavas (approximately in ratio 4:1 to 10:1, respectively, in areas studied).

(ii) Crystallisation of the trachytic, dacitic, and rhyolitic lavas is deduced to have occurred under T-fo_2 buffer conditions below QFM, and probably close to WM buffer conditions for the non-spinel-bearing lavas. The magmas are deduced to have been essentially anhydrous, although F was clearly a significant volatile component in the comendites and the Binna Burra rhyolite.

(iii) The rhyolite and metaluminous trachyte show strong evidence of derivation by crystal fractionation; two possibilities exist for the rhyolite: (a) derivation by crystal fractionation from a basaltic precursor, modified, however, by (possibly lower) crustal assimilation — this would also apply to the trachyte, and is preferred; or (b) the rhyolite represents localised lower crustal melts, modified by subsequent crystal fractionation.

(iv) The tholeiitic andesite and dacite are interpreted to represent products of fractionation from essentially uncontaminated basaltic precursors (excepting the hybrid dacite of Binna Burra).

(v) The comendites also exhibit strong evidence of crystal fractionation control, and may represent (a) end products of a long liquid line of descent, from basalts through trachytic derivatives; or (b) very late stage differentieates from rhyolitic magmas, i.e. by continued crystal fractionation under quartz-feldspar ternary minima restrictions. For the Lamington comendite, this receives some additional support from Pb and Sr isotopic data, and least-squares linear mixing calculations.

ACKNOWLEDGEMENTS

Analytical work was supported by A.R.G.C. grant (B67/16723) to A.E.; most of the probe analyses were carried out (A.E.) at the University of California, Berkeley, under N.S.F. grant (GA43177X) to I.S.E. Carmichael, the remaining probe analyses being undertaken (J.A.R.) at the Research School of Earth Science, A.N.U. S.R. Taylor is acknowledged for providing mass spectrograph facilities for REE determination (A.M.) at A.N.U. A.S. Bagley and G. Langthaler, University of Queensland, have assisted greatly in other aspects of the analytical work.

REFERENCES

Brown G.M,. and Vincent, E.A., 1963: Pyroxenes from the late stages of fractionation of the Skaergaard Intrusion, East Greenland. *J. Petrology, 4*, pp. 175-197.

Bryan, W.B., and Stevens, N.C., 1973: Holocrystalline pantellerite from Mt Ngun-ngun, Glass House
    Mountains, Queensland, Australia. *Amer. J. Sci., 273,* pp. 947-957.
Buddington, A.F., and Lindsley, D.H., 1964: Iron-titanium oxide minerals and synthetic equivalents. *J.
    Petrology, 5,* pp. 310-357.
Carmichael, I.S.E., 1963: The crystallization of feldspar in volcanic acid liquids. *Quart. J. Geol. Soc.
    Lond., 119,* pp. 95-113.
Carmichael, I.S.E., Turner, F.J., and Verhoogen, J., 1974: *Igneous Petrology.* McGraw-Hill Book
    Company, 739 pp.
Christiansen, R.L., & Lipman, P.W., 1972: Cenozoic volcanism and plate-tectonic evolution of the
    Western United States. II. Late Cenozoic. *Phil. Trans. R. Soc. Lond., A. 271,* pp. 249-284.
Coombs, D.S., & Wilkinson, J.F.G., 1969: Lineages and fractionation trends in undersaturated volcanic
    rocks from the East Otago Volcanic Province (New Zealand) and related rocks. *J. Petrology, 10,*
    pp. 440-501.
Eugster, H.P., & Wones, D.R., 1962: Stability relations of the ferruginous biotite, annite. *J. Petrology,
    3,* pp. 82-125.
Ewart, A., Hildreth, W., & Carmichael, I.S.E., 1975: Quaternary acid magma in New Zealand. *Contr.
    Mineral. and Petrol. 51,* pp. 1-27.
Ewart, A., Oversby, V.M., & Mateen, A., in prep. Petrology and isotope geochemistry of Tertiary lavas
    from the northern flank of the Tweed volcano, southeastern Queensland.
Ewart, A., Paterson, H.L, Smart, P.G., & Stevens, N.C., 1971: Binna Burra, Mount Warning.
    Geological Excursions Handbook (G. Playford, Ed.), *A.N.Z.A.A.S. 43rd Congress & Geol.
    Soc. Australia Inc., Qld Division, Brisbane.* pp. 63-78.
Green, D.C., 1968: Further evidence for a continuum of basaltic compositions from southeast
    Queensland. *J. Geol. Soc. Aust., 15,* pp. 159.
Green, D.C., 1969: Transitional basalts from the Eastern Australian Tertiary province. *Bull.
    Volcanologique, 33,* pp. 930-941.
James, R.S., & Hamilton, D.L., 1969: Phase relations in the system NaAlSi3O8 — KAlSi3O8 —
    CaAl2Si2O8 — SiO2 at 1 kilobar water vapour pressure. *Contr. Mineral. and Petrol., 21,* pp.
    111-141.
Jamieson, B.G., & Clarke, D.B., 1970: Potassium and associated elements in tholeiitic basalts. *J.
    Petrology, 11,* pp. 183-204.
Kudo, A.M., & Weill, D.F., 1970: An igneous plagioclase thermometer. *Contr. Mineral. and Petrol.,
    25,* pp. 52-65.
Lindsley, D.H., Brown, G.M., & Muir, I.D., 1969: Conditions of the ferrowollastonite-
    ferrohedenbergite inversion in the Skaergaard intrusion, East Greenland. *in* Pyroxenes and
    Amphiboles: Crystal Chemistry and Phase Petrology. *Mineralogical Soc. Amer., Spec. Publ.
    No. 2,* pp. 193-201.
McTaggart, N.R., 1961: The sequence of Tertiary volcanic and sedimentary rocks of the Mount
    Warning volcanic shield. *Trans Roy. Soc. New South Wales, 95,* pp. 135-144.
Nash, W.P., & Wilkinson, J.F.G., 1970: Shonkin Sag Laccolith, Montana. I. Mafic minerals and
    estimates of temperature, pressure, oxygen fugacity and silica activity. *Contr. Mineral. and
    Petrol., 25,* pp.241-269.
Nicholls, J., & Carmichael, I.S.E., 1969: Peralkaline acid liquids: A petrological study. *Contr. Mineral.
    and Petrol., 20,* pp. 268-294.
Powell, M., & Powell, R., 1974: An olivine-clinopyroxene geothermometer. *Contr. Mineral. and
    Petrol., 48,* pp. 249-263.
Rahman, S., & MacKenzie, W.S., 1969: The crystallization of ternary feldspars: A study from natural
    rocks. *Amer. J. Sci., 267-A,* pp. 391-406.
Robie, R.A., & Waldbaum, D.R., 1968: Thermodynamic properties of minerals and related substances
    at 298.15°K (25.0°C) and one atmosphere (1.013 bars) pressure and at higher temperatures. *U.S.
    Geol. Surv. Bull. 1259,* 256 pp.
Ross, J.A., 1974: The Focal Peak shield volcano, southeast Queensland — evidence from its eastern
    flank. *Proc. R. Soc. Qld., 85,* pp. 111-117.
Shaw, H.R., 1972: Viscosities of magmatic silicate liquids: An empirical method of prediction. *Amer. J.
    Sci., 272,* pp. 870-893.
Solomon, P.J., 1964: The Mount Warning shield volcano. A general geological and geomorphological
    study of a dissected shield. *Pap. Dep. Geol. Univ. Qld, 5,* pp. 1-12.
Stephenson, P.J., 1956: The geology of the Mt Barney Central Complex. *Unpubl. Ph. D. Thesis, Univ.
    London.*
Stephenson, P.J., 1959: The Mt Barney central complex, S.E. Queensland. *Geol. Mag., 96,* pp. 125-136.

Stevens, N.C., 1959: Ring-structures of the Mt Alford district, South-east Queensland. *J. Geol. Soc. Aust., 6,* pp. 37-50.

Stevens, N.C., 1962: The petrology of the Mt Alford ring-complex, S.E. Queensland. *Geol. Mag., 99,* pp. 501-515.

Stevens, N.C., 1970: Miocene lava flows and eruptive centres near Brisbane, Australia. *Bull. Volcanologique, 34,* pp. 353-371.

Stevens, N.C., 1971: The Glass Houses. Geological Excursions Handbook (G. Playford, Ed.), *A.N.Z.A.A.S. 43rd Congress & Geol. Soc. Aust. Inc., Qld Division, Brisbane.* pp. 153-158.

Stull, P.R., & Prophet, H., 1971: JANAF thermochemical tables (2nd ed.), *Nat. Stand. Ref. Data Ser, U.S. Nat. Bur. Standl, 37.*

Tuttle, O.F., & Bowen, N.L., 1958: Origin of granite in the light of experimental studies in the system $NaAlSi_3O_8 - KAlSi_3O_8 - SiO_2 - H_2O$. *Geol. Soc. Amer. Mem., 74,* 153 pp.

Webb, A.W., Stevens, N.C., & McDougall, I., 1967: Isotopic age determinations on Tertiary volcanic rocks and intrusives of south-eastern Queensland. *Proc. Roy. Soc. Qld, 79,* pp. 79-92.

Wellman, P., & McDougall, I., 1974a: Cainozoic igneous activity in eastern Australia. *Tectonophysics, 23,* pp. 49-65.

Wellman, P., & McDougall, I., 1974b: Potassium-argon ages on the Cainozoic volcanic rocks of New South Wales. *J. geol. Soc. Aust., 21,* pp. 247-272.

Wilkinson, J.F.G., 1968: The magmatic affinities of some volcanic rocks from the Tweed Shield Volcano, S.E. Queensland-N.E. New South Wales. *Geol. Mag., 105,* pp. 275-289.

Wones, D.R., & Gilbert, M.E., 1969: The fayalite-magnetite-quartz assemblage between 600°C and 800°C. *Amer. J. Sci., 276-A,* pp. 480-488.

Yagi, K., 1953: Petrochemical studies on the alkalic rocks of the Morotu District, Sakhalin. *Bull. Geol. Soc. Amer., 64,* pp. 769-810.

SOME LONG BASALTIC LAVA FLOWS IN NORTH QUEENSLAND

P. J. STEPHENSON and T. J. GRIFFIN

Department of Geology, James Cook University of North Queensland, Townsville, QLD 4811

ABSTRACT

Each of eight basaltic flows in north Queensland extends more than 80 km from its source. Petrography indicates the lavas were erupted with phenocrysts of olivine varying in amount up to 10 percent. Groundmass textures indicate that crystallisation began effectively only after the lavas had ceased flowing. Chemically, the rocks are hawaiite, or are transitional to alkali olivine basalts, containing moderate to low amounts of normative nepheline, or low amounts of normative hypersthene. The eight flows are chemically similar, but show conspicuous differences in potash contents. Within any one flow, chemical variations are smaller. The chemistry does not suggest unusually low viscosity, and it is concluded that the main factors in producing the long flows were: high, continued effusion rates; favourable topography which channelled the flows; and in some cases, the existence of lava tubes.

INTRODUCTION

Five large areas of Pliocene to Recent basaltic rocks exist in north Queensland, comprising a total area of more than 20,000 km² (Fig. 1). More than 300 vents have been identified, but few of these have acted as major eruptive centres, to judge from the small size of the volcanoes and the extent of their lava fields.

While it is possible to identify the sources of some lavas, it is considerably more difficult to trace and delineate individual flows. This is due to age and disguising effects of weathering and erosion, and to the rather close similarity of many lavas, especially those erupted from the same source in succession.

In some instances individual flows can be outlined, and some have considerable lengths. This paper describes eight examples longer than 80 km, and considers reasons for their great length. The location of these flows and their outlines are shown in Figures 3-5, and 7, and some dimensions are given in Table 1. T. J. Griffin is responsible for Undara and other McBride province information. P. J. Stephenson contributed the material covering other provinces.

CENTRES OF ERUPTION AND MORPHOLOGY OF THE FLOWS

The volcanoes which erupted the long flows vary in state of preservation in accordance with their age. The youngest, Toomba volcano, has almost perfectly preserved details, whereas the oldest, Marabon, has been considerably denuded. Five centres still have craters. Only the Toomba flow has its original pahoehoe surface preserved. The nature of the other flows, pahoehoe or aa, has not been confirmed, but they are thought likely to have been dominantly pahoehoe also.

Undara crater (Figs 2, 3) is an impressive steep-sided depression 340 m across and 48 m deep with an unbreached rim rising 20 m above the surrounding lava field. The cone has only minor pyroclastic deposits, but near-horizontal lavas, from 10 cm to more than 1 m thick, crop out in the crater wall. Several long flows issued

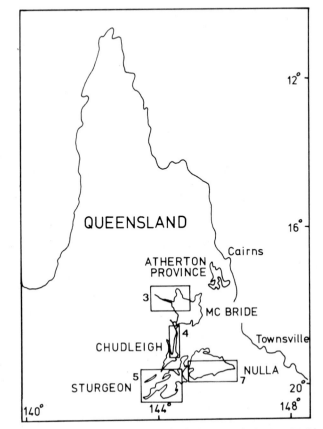

Fig. 1: Cainozoic basaltic provinces in north Queensland. Areas marked 3-5, and 7, locate Figures 3, 5, and 7, respectively.

Table 1. Some dimensions of long lava flows

	Length	Area	Thickness[1]	Volume	Average Slope	Age
UNDARA	160 km	700-1550 km²	15 m	10-23 km³	0.3°	0.19 my[2]
BARKER	100-130	350-400	10	3-4	0.3-0.2°	? 0.2[5]
TWINS	120	250	(10)	2	0.4°	? 0.5[5]
BARABON	90	>700	15	>10	0.15°	? 3-4[5]
BIRDBUSH	110	480	(10)	5	0.2°	1.3[3]
KANGERONG	100	600	(10)	6	0.15°	2.4[3]
HANN CREEK	115	>500	(10)	5	0.2°	1.3[3]
TOOMBA	120	700	5-10	3-6	0.2°	? 0.01[4]

[1] Relative estimate based on edge relief, available topographic contours, and outcrop evidence. Bracketed figures are nominal, in the absence of contour data.
[2] Griffin & McDougall (1975).
[3] Wyatt & Webb (1970).
[4] Polach et al. (in prep.).
[5] Estimated from relative erosion, compared with other volcanoes in the region of known age.

from it (Fig. 2); one, approximately 90 km long, entered the Lynd River to the north, and another longer one moved northwest and then followed the precursors of Cassidy Creek and Einasleigh River to attain a total length of 160 km. These two

Fig. 2: Aerial view of Undara crater. A lava tube commences northwest of the crater, evident as a narrow zone of dark vegetation running towards the upper right edge. The major flows occurred in that direction. Older volcanoes are visible towards the skyline, including Racecourse Knob, a shield volcano with minor pyroclastic cones.

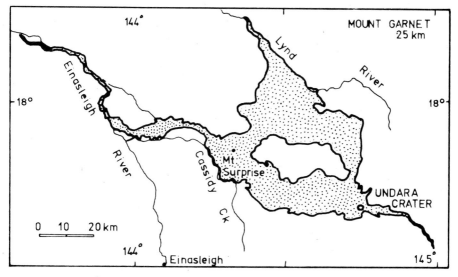

Fig. 3: Undara lava field, McBride province.

long flows may be branches of the same flow. Lava tubes are widely developed and in the longest flow a tube system extends for perhaps more than 100 km (Atkinson et al., in prep.). The lava surface is now composed of boulders, but is extremely rough with pressure ridges up to 4 m high, generally parallel to the flow direction and locally with more irregular humps and depressions. The average slope from

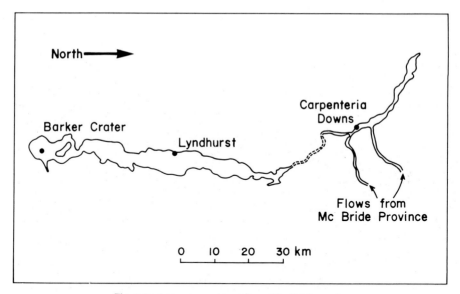

Fig. 4: Barker crater and lava flow, Chudleigh province.

crater to terminus is 0.3°, but the section from 39 to 74 km slopes at only 0.09°. This section contains the long narrow ridge feature known as the 'wall', which may contain an undrained lava tube.

Barker Crater (Fig. 4) is 700 m across with a flat floor 40 m below the rim. The cone is partly pyroclastic, rises to 50 m above the surrounding lava field, and is breached to the south by a narrow valley leading out into a lava ridge with five collapses along a short lava tube. This ridge fans out into the lava field around the volcano and appears to have fed the long flow to the north which reached and followed the Einasleigh River (Fig. 4). It filled the main river channel and flooded out to occupy most of the wide valley (it is 6 km across at Lyndhurst, 40 km from the crater). The end of the flow has not been found, but it appears to continue almost to Carpentaria Downs. Lavas crop out 30 km farther down the river, but some consist of flows from the McBride province.

The Twins crater (Fig. 5) occupies a relatively steep lava cone which rises 60 to 80 m above its lava field. The crater is 500 m across, with low walls up to 20 m above the flat, soil-covered floor. There appear to be only minor pyroclastic deposits. The crater is breached to the south and the long flow (Fig. 5) can be traced across the tableland by its low, weathered pressure ridges. It descended from the tableland 50 km from the volcano down a likely precursor of Whitecliff Gorge Creek and reached Galah Creek; its final 25 km now stands as a low, meandering narrow ridge above the plain. The flow is the narrowest of the eight examples and this may reflect its average gradient, 0.4°, the steepest found.

North of Barabon (Fig. 5) a long mesa capped by basaltic rocks trends back northeast towards the closest known volcanoes, Bald Mountain and Mount Stewart, one of which was probably the source. Although possibly as old as 3 or 4 m.y. (estimated from relative erosion in comparison with other, dated basaltic

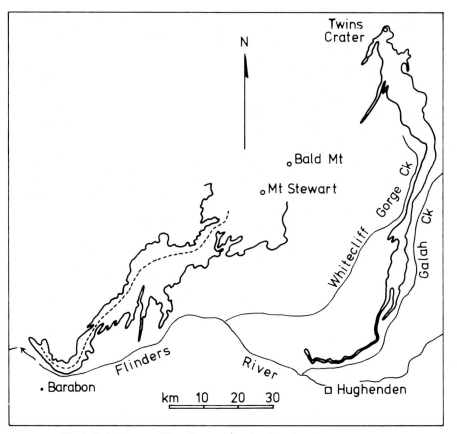

Fig. 5: Twins crater and lava flow, and Barabon lava flow, Sturgeon province.

lavas of similar age in the region), the mesa carries irregular hummock ridges in a narrow axial zone (Fig. 6), suggesting flow down a former valley of the Flinders River system having very subdued relief. Today the mesa stands 50 m above the plain through erosion inversion.

In the Nulla province, several long flows were first delineated by Wyatt & Webb (1970) who confirmed ages by K:Ar dating. These flows include the Birdbush, Kangerong, Hann Creek, and Toomba flows (Fig. 7). The Toomba flow is very young (probably Recent, Polach et al., in prep.) and has no soil cover. It is pahoehoe, with an abundance of complex ridge and depression patterns, and it can be traced west to its source, a low-profile, broad lava cone. Near the crest of this cone, lava-draped scarps suggest northeast fissures with three major zones of eruption. However, the lavas forming the cone are petrographically distinct from the slightly earlier main Toomba flow, edges of which can be found beneath the later lavas. East of the cone, the main flow has some elongate narrow depressions which may be collapsed lava tubes.

The source of other long flows in the Nulla province cannot be recognised with certainty. The Birdbush flow probably issued from the denuded cone south of Nulla

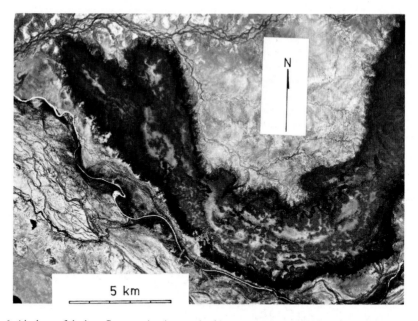

Fig. 6: Airphoto of the long flow termination north of Barabon (Fig. 5). The flat mesa has scrub-covered edges and has relic flow ridges parallel to axis. (Richmond Run 6, CAB 309-168).

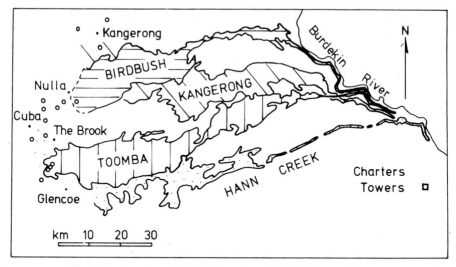

Fig. 7: Long lava flows in the Nulla province. Small circles are volcanoes.

which has a small crater preserved, but flow continuity is difficult to demonstrate and petrography and geochemistry do not resolve the question. The Kangerong flow may have been erupted from Black Hill, 6 km south of Kangerong, and a possible source for the Hann Creek flow may have been the large crater south of Cuba. Despite uncertainties, it can be seen from Figure 7 (showing all known eruption centres) that each of the flows must be more than 100 km long.

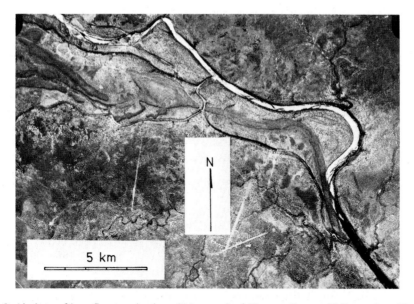

Fig. 8: Airphoto of long flow terminations, 20 km north of Charters Towers, Nulla province (Fig. 7).
West of the present Burdekin River (white sandy course) several lava ridges can be seen which formed as
flows downs previous Burdekin courses. (Townsville Run 9, CAB 198-5088).

The four long flows in the Nulla province all flowed east, down precursors of
the Basalt River, Lolworth Creek, or Hann Creek (Burdekin River tributaries).
They reached the former Burdekin River and followed it (Fig. 8).

Comparative dimensions for the eight flows given in Table 1 show Undara to
be the largest. However, accepting much uncertainty in nominating thicknesses, the
magnitudes for volume appear to be similar. The large volume tabled for Undara is
an estimate of its total lava field, which could involve several separate flows. The
average slopes are all very low (0.15° to 0.4°), and gradients well away from the
volcanoes are even more gentle (0.09° for Undara).

PETROLOGY

Forty-two rocks have been examined in thin section and 36 have been
chemically analysed. Each flow has been found to show relatively narrow
compositional variation, and the different flows show many similarities.

All the lavas carry subhedral to euhedral olivine crystals up to 1 mm and
rarely 3 mm, in amounts ranging from less than 1 percent up to approximately 10
percent. Some olivines tend to be glomeroporphyritic and show varying degrees of
iddingsite alteration. Their optic sign is neutral to negative with high axial angle,
suggesting compositions Fo 80-90. They do not show strain bands characteristic of
xenocrysts from peridotite nodules. They appear to be phenocrysts, but to test this,
microprobe analyses for olivine cores in a specimen from Undara were compared
with crystal equilibrium compositions predicted from the chemical rock analysis using the relationships established by Roeder & Emslie (1970). For prediction, the

analysis was first adjusted to our routine oxidation value, $Fe^{3+}/Fe^{2+} = 0.2$, in recognition of the common oxidation weathering of most specimens, especially older lavas. The calculated olivine composition was then Fo 82, which compared very well with the microprobe result, Fo 81.5. Although this closeness of agreement may be fortuitous, it does support the crystals' being true phenocrysts.

Most of the rocks are holocrystalline and the groundmass contains titanaugite, plagioclase, olivine, and opaque oxides. Some specimens also contain very fine-grained interstitial material or brown glass. In different rocks plagioclase cores range from andesine to labradorite (An 45-65), measured by twin extinction angles. Secondary calcite and zeolite occasionally fill vesicles. Texturally, the pyroxene is mostly intergranular in finer rocks, but in coarser specimens it ranges through subophitic to ophitic in relation to plagioclase; coarse ophitic pyroxene crystals can attain 2 mm in length.

The lavas are vesicular, even more than 100 km from their source. A characteristic of the more coarsely-grained specimens is the presence of irregular intersertal cavities, many triangular in appearance. The rocks show little flow orientation of the groundmass plagioclase laths.

From their porphyritic nature, the uniform groundmass textures, and the nature of vesicular cavities, it is concluded that very little crystallisation must have taken place in the lavas during their flow, and that most occurred only when flow movement had slowed considerably, close to arrest of the flow.

Chemically, the rocks are hawaiite, some transitional to alkali olivine basalt, according to the classification of Coombs & Wilkinson (1969). Representative analyses from each flow are listed in Table 2. They are chemically similar, but show some differences in alkalis, especially potash. The range in chemistry is, however, considerably narrower than the range found among other basaltic rocks from any one of the five major provinces in north Queensland.

Attempts to correlate some of the long flows with one of several possible source volcanoes, using petrological criteria, have not been successful. Diagnostic petrographic or chemical features are hard to establish for individual flows. Another problem in correlation is the common occurrence of different lavas at the same eruption centre. At the Toomba cone, for example, the rocks are porphyritic in plagioclase, sector-zoned augite, and olivine, and were erupted after the long flow. Similar porphyritic lavas occur around the Mount Stewart centre and several of the possible vents for the Kangerong, Birdbush, and Hann Creek Flows. In those places too, the porphyritic lavas could be later than the main flows.

TEMPERATURES AND VISCOSITIES

Lava viscosity may be a significant factor influencing the length achieved by flows. For this reason estimates of viscosity have been calculated from the chemical analyses. All the rocks are porphyritic only in olivine, suggesting they were erupted at temperatures just below the olivine liquidus. Liquidus temperatures have been estimated from the chemistry, using the findings of Roeder & Emslie (1970). Using these temperatures, viscosities were calculated (Table 2) by the method of Shaw

Table 2. Representative chemical analyses of long lava flows, and estimated liquidus temperatures and calculated anhydrous viscosities.

	Undara	Barker	Twins	Barabon	Birdbush	Kangerong	Hann Creek	Toomba
SiO_2	49.5	49.5	49.5	46.6	49.8	49.2	50.7	47.5
TiO_2	1.78	2.00	2.05	2.12	1.72	1.85	1.78	2.00
Al_2O_3	16.2	15.4	14.9	14.3	15.6	15.25	15.7	14.9
Fe_2O_3	2.67	3.15	4.06	2.98	1.89	1.91	2.93	1.63
FeO	7.79	7.50	7.01	8.79	9.01	9.02	8.12	8.50
MnO	0.16	0.16	0.16	0.17	0.15	0.15	0.15	0.16
MgO	7.62	8.55	8.15	9.20	8.33	8.40	7.85	8.55
CaO	8.07	8.47	8.55	7.55	8.30	8.25	7.48	8.30
Na_2O	4.08	3.50	2.98	4.21	3.62	3.44	3.92	3.85
K_2O	1.75	1.70	2.23	1.99	1.30	1.32	1.49	2.15
H_2O+	nd	nd	nd	nd	nd	nd	nd	1.03
P_2O_5	0.52	0.44	0.56	0.68	0.48	0.47	0.52	0.54
Total	100.14	100.37	100.15	98.59	100.20	99.26	100.64	99.11

nd: not determined

	Undara	Barker	Twins	Barabon	Birdbush	Kangerong	Hann Creek	Toomba
Olivine liquidus temperature	1200°	1230°	1215°	1245°	1220°	1215°	1210°	1230°
Calculated viscosity (Pa s)	25.9	16.1	18.9	7.3	17.1	16.4	19.7	5.8

	Einasleigh	Clarke River	Hughenden	Richmond	Townsville	Townsville	Townsville	Townsville
Grid Reference	198100 716600	211900 591600	195400 401900	679400 399500	414300 513500	411600 512500	409900 507300	413800 513800

Analyses: Toomba, by Australian Mineral Development Laboratories, Adelaide; Undara, by T. J. Griffin (Atomic Absorption); other analyses, by P. J. Stephenson and others (Atomic Absorption). Atomic Absorption: HF-boric acid digestion; P, spectrophotometric; Fe^{2+} by titration; except for Fe^{2+}, samples dried at 110°C.

(1972). These are minimum viscosity estimates, as actual temperatures are likely to have been somewhat lower.

The temperatures range from 1200 to 1245°C, and the corresponding viscosities from 6 to 26 Pa s. These are anhydrous values which can be compared with those quoted by Bottinga & Weill (1972) for various other lavas. The long-flow lavas do not appear to have unusually low viscosities. The presence of water in lavas lowers viscosity, but there are no indications that these Queensland lavas were unusually hydrous. It is concluded that exceptionally low viscosity was not a factor in the development of the long flows.

DISCUSSION

Walker (1973) discussed the development of very long basaltic lava flows. After considering a number of possible factors, he concluded that their formation is characterised by high rates of effusion. The characteristics of the long flows in north Queensland appear to confirm his conclusions.

Relatively mobile lava was erupted in considerable volume from vents at which pyroclastic activity was usually minor (the pyroclastic cone at Barker Crater may have developed at a later stage). According to Walker (1973), the high rate of effusion maintains lava advance at the toe of the flow; slower or discontinuous eruption builds successive overriding flows and a steeper cone. Extrapolating Walker's relationships (Walker, 1973, p. 112), the average effusion rate for the Undara flow must have exceeded 1000 m^3s^{-1}. However, effusion rates could have been much greater, as the estimated flow volume using a rate of 1000 m^3s^{-1} would have taken longer than three months to erupt. We envisage relatively rapid formation of the main flow, from a sustained discharge, perhaps within only a week; slower discharge over a longer period would have formed successive shorter flows since long periods of feebler eruption or pauses would occur.

The low average slopes of the flows are noteworthy (all less than 0.5°). For lavas to continue flowing a great distance it is probably essential for the flow to remain confined and not seriously diminished by spreading out or branching. These requirements are likely to be even more critical for low gradients. Each of the north Queensland flows was confined by a narrow stream course. These inland stream courses in north Queensland consist of dry stretches with occasional waterholes, for most of the year.

CONCLUSIONS

Apart from a large volume of lava the most important factor favouring the development of long flows in north Queensland is believed to have been high rates of effusion. Topographic channelling into dry stream courses, confining the width, was probably another contributing influence. Several of the flows are known to have had lava tubes operating in them (Undara, Barker, and probably Toomba) and this may have added to possible length through improved flow efficiency. Chemical evidence suggests that the flows were not charaterised by unusually low viscosity.

ACKNOWLEDGEMENTS

This paper is dedicated to the memory of Tony Taylor, who held a keen interest in some of the lavas he first saw near Charters Towers as a serviceman during the last war, before he studied geology.

Funding for continuing basaltic research has been provided by the Australian Research Grants Committee (to P. J. Stephenson) and James Cook University (to T. J. Griffin). Laboratory work was carried out at James Cook University, with some at the Australian National University. At James Cook, P. J. Stephenson gratefully acknowledges the careful work of S. Cumming, G. O'Donnell, J. B. Gallo, and D. Tam who helped him in basalt analysis. At ANU, T. J. Griffin is grateful for help and advice from N. Ware in the use of microprobe equipment.

REFERENCES

Atkinson, A., Griffin, T. J. & Stephenson, P. J. (in prep.): A major lava tube system from Undara Volcano, north Queensland.
Bottinga, Y. & Weill, D. F. 1972: The viscosity of magmatic silicate liquids: a model for calculation. *Am.J.Sci., 272(5)*, pp. 438-75.
Coombs, D. A. & Wilkinson, J. F. G. 1969: Lineages and frationation trends in undersaturated volcanic rocks from the East Otago Volcanic Province (New Zealand) and related rocks. *J. Petrology, 10(8)*, pp. 440-501.

Griffin, T. J., & McDougall, I. 1975: Geochronology of the Cainozoic McBride Volcanic Province, northern Queensland. *J. Geol. Soc. Aust.* 22.

Polach, H., Wyatt, D. H. & Stephenson, P. J., (in prep.): The age of the Toomba volcano, Nulla province, north Queensland.

Roeder, P. L. & Emslie, R. F., 1970: Olivine-liquid equilibrium. *Contr. Mineral. & Petrol., 29*, 275-89.

Shaw, H. B., 1972: Viscosities of magmatic silicate liquids: an empirical method of prediction. *Am. J. Sci., 272(9)*, pp. 870-93.

Walker, G.P.L., 1973: Lengths of lava flows. *Phil. Trans. R. Soc. Lond. A., 274*, pp. 107-18.

Wyatt, D.H. & Webb, A.W., 1970: Potassium-argon ages of some northern Queensland basalts and an interpretation of late Cainozoic history. *J. Geol. Soc. Aust., 17(1)*, pp. 39-52.

GEOLOGICAL INTERPRETATION OF THE GRAVITY ANOMALY AT MOUNT PORNDON VOLCANO, VICTORIA, AUSTRALIA

C.O. McKEE[1] and L. THOMAS

Department of Geology, University of Melbourne, Melbourne, VIC 3052

ABSTRACT

The Mount Porndon Volcanic Complex covers an area of about 200 km^2 and is composed of basaltic lava flows, tuff deposits, and scoria hills. Seven gravity traverses radiating from the centre of the Complex indicate a broad gravity maximum bounded by steep flanks and flattened over the centre of the Complex. The positive gravity feature is intepreted as being caused by an outcropping disc of basalt, with a less dense region near the centre in the vicinity of vents which produced the known tuff and scoria deposits. It is suggested that the disc of basalt was formed by the infilling of a caldera.

INTRODUCTION

The Mount Porndon Volcanic Complex (lat. 38°19'S, long. 143°17'E) forms part of the Newer Volcanic Suite (Singleton & Joyce, 1969) of southeastern Australia (Fig. 1) and lies 2.4 km south of the Princes Highway at Pomborneit (Fig. 2). In Victoria the Newer Volcanics extend west from Melbourne almost to the South Australian border, and are bounded to the north by the Victorian Highlands and to the south by the Otway Ranges (Fig. 1). The Volcanics consist of lava flows and minor pyroclastic deposits and contain between 350 and 400 small vents (Singleton & Joyce, 1969) ranging in age from Upper Pliocene to Holocene (McDougall et al., 1966). Underlying the Newer Volcanics are Cainozoic sediments which in turn overlie Mesozoic sediments of the Otway Group; Palaeozoic sediments and granitic intrusions form the basement (Geological Map of Victoria, 1:1,000,000, 1963, Geological Survey of Victoria).

A gravity survey was carried out on Mount Porndon between February and September 1972, following geological mapping (Fig. 2) in the previous year by D.J. Ellis, A.K. Ferguson, and T.J. Nolan. The authors' interest in Mount Porndon arose out of findings of bedded tuff east and north of the mount and the location of Mount Porndon within a belt of maars (Singleton & Joyce, 1969). Because the central part of the Complex is covered by scoria deposits and lava flows, other tuff deposits are not known; it was hoped that a gravity study would elucidate the structure beneath the central pyroclastic deposits and lava flows.

GEOLOGY

Formation of the Mount Porndon Volcanic Complex commenced with the eruption of basaltic lavas onto peneplanated Tertiary sediments (Skeats & James, 1937). These first extensive lavas, 'Hawaiite I' (Fig. 2) formed the Stony Rises basalt field (Skeats & James, 1937).

In a new effusive phase, 'Hawaiite II' lavas (Fig. 2) of similar composition flowed about 1.5 km in all directions from the central vent of Mount Porndon,

---

[1] Present address: Geological Survey of Papua New Guinea, Volcanological Observatory, P.O. Box 386, Rabaul, Papua New Guinea.

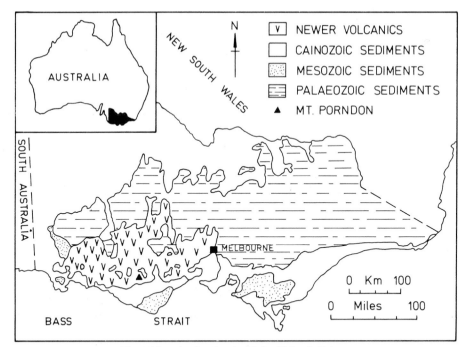

Fig. 1: Victoria, showing the Newer Volcanic Suite.

according to Skeats & James (1937), producing a 'pancake' of lava. The edge of this lava body, called the 'Ring Barrier' by Skeats & James, is an almost continuous ridge up to 25 m higher than the Stony Rises surface but higher by only 10 m or less than the level of lava immediately within the Ring Barrier. At many places around the outside base of the Ring Barrier the Hawaiite I surface is depressed and slopes down towards the Ring Barrier. A curious feature of the surface of the Hawaiite II sheet is that in the southern half, where it is not covered by later pyroclastic deposits and lava flows, an inwards slope from the Ring Barrier is found (Skeats & James, 1937; Fig. 4, topographic profile 3). Skeats & James considered that this inwards slope and the Ring Barrier may have been produced when lava was withdrawn down the vent or vents up which it had risen. In an inspection of the southern boundary of the Hawaiite II lava body by one of the authors (McKee), two circular lava cones about 15 to 20 m high, a deep crater-like depression bordered by a low rim of lava, and a few poorly defined flow margins were found. The slope of the Hawaiite II lava surface towards the centre of the Complex and the attitudes of the flow margins indicate that at least in the southern part of the Hawaiite II body lava flowed inwards from the Ring Barrier.

The next and final phase of activity was both explosive and effusive producing deposits of tuff and scoria and flows of basanitoid lava from central vents. Mount Porndon itself is the largest of several craterless scoria hills which surround a small cratered scoria cone near the centre of the Complex. Most of the scoria and tuff is east and north of the centre of the Complex, presumably because prevailing southwesterly winds deposited it there. Thick deposits of tuff exposed in quarries directly overlie the Hawaiite II lava. A thick basanitoid flow in the northern part of

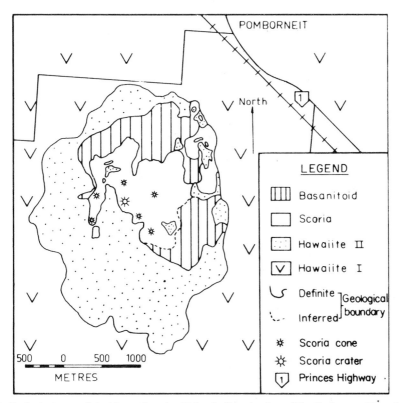

Fig. 2: Geological map of the Mount Porndon Volcanic Complex from 'The Geology and Petrology of the Mount Porndon Volcanic Complex' (separate unpublished Honours reports, 1971, University of Melbourne), courtesy of D.J. Ellis, A.K. Ferguson, and T.J. Nolan.

the Complex underlies scoria deposits whereas a thin flow to the east overlies scoria and was, perhaps, the concluding episode of the volcanic activity at the Mount Porndon Complex. Figure 2 illustrates the geology within the Ring Barrier.

GRAVITY DATA COLLECTION AND REDUCTION

Six gravity traverses were laid out in a radial pattern about the centre of the Complex (Fig. 3), and a seventh shorter traverse was added later to give additional detail. The traverses were positioned to avoid the roughest topography (scoria mounds, quarries, etc.), so reducing terrain corrections and simplifying levelling. Locations of the traverse lines are shown in Figure 3, but station spacings of 100 ft (30 m) and 200 ft (61 m) are too small to permit individual stations to be shown at this scale. In total, 250 gravity stations were occupied.

The visible structure of Mount Porndon may be regarded as essentially an upper level of scoria overlying a lower level of lava, the two levels being separated by a horizontal plane. Fifty-two typical specimens were collected from the two levels for density measurements: the scoria has a mean density of 1.63 g cm^{-3} (standard deviation, 0.28) and the lava, 2.59 g cm^{-3} (standard deviation, 0.15). This strong density contrast and the horizontal layering suggested that the separation

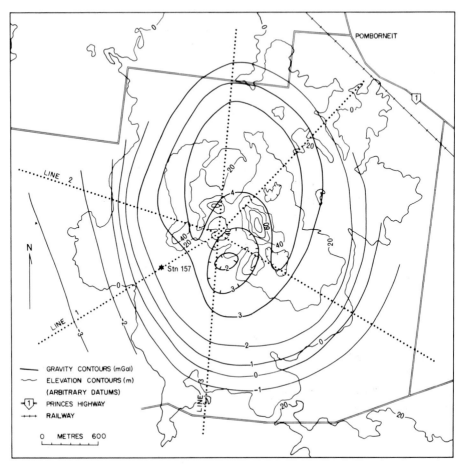

Fig. 3: Topographic and Bouguer gravity anomaly contour map of the Mount Porndon Volcanic Complex.

plane would be a convenient datum level for reduction of gravity measurements to Bouguer anomalies; station 157 was assumed to lie on the datum plane.

The Bouguer plate corrections were calculated using densities of 1.63 and 2.59 g cm^{-3} for rocks above and below the datum, respectively. Terrain corrections were included in the calculation of Bouguer anomalies.

The errors associated with the various stages of data collection and reduction, i.e. gravity meter reading, levelling, latitude correction, terrain correction, were examined. The largest source of error (up to 0.1 mGal) probably lies in the terrain corrections; the errors result from deficiencies in the topographic information near the station. The Bouguer gravity values are believed to be reliable to within 0.15 mGal, and on each traverse were smoothed with a three-point running average method.

The smoothed gravity values and station elevations are shown as profiles in

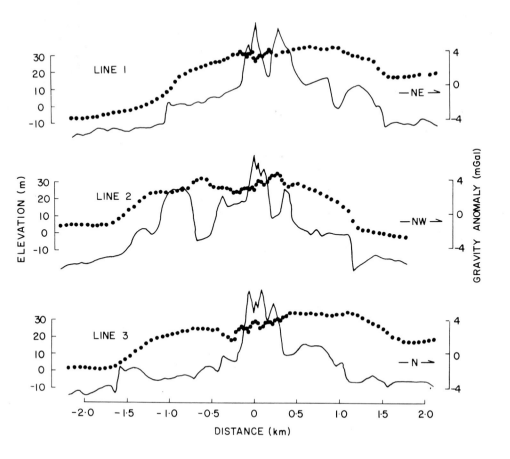

Fig. 4: Gravity (solid circles) and elevation profiles along traverse lines.

Figure 4 and as contours in Figure 3. Apparent correlation between the gravity anomaly and elevation is evident at some locations. The anomaly smoothing process will only reduce the effect of bodies which are small (100 m) compared with the size of the Complex, and may act to reduce random errors between stations.

INTERPRETATION

The principal Bouguer anomaly feature at Mount Porndon is slightly elliptical, and has dimensions of about 3.1 km by 2.6 km, its long axis trending northwards (Fig. 3). The anomaly is positive, approximately 4 mGal in amplitude, and has steep gradients about its edges and a broad flattened central region containing a minor relative minimum (Fig. 3) in the vicinity of the scoria hills. The edges of this main gravity feature correspond with the Ring Barrier.

The principal anomaly is superimposed on a regional variation which, judged from the values of the Bouguer anomalies at the ends of the lines, is of the form of a

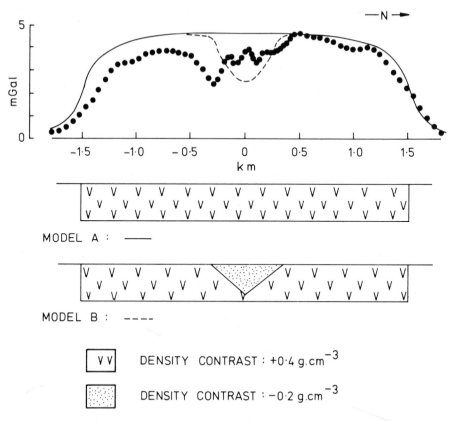

Fig. 5: Gravity profiles computed for two simple models having axial symmetry. Solid circles are smoothed Bouguer gravity values along Line 3.

gravity trough plunging to the south. This agrees with the regional field determined by Schoenharting & Brunt (1971).

Short-wavelength anomalies are superimposed on the main anomaly. The most significant of these are gravity minima which coincide with positive topographic features, scoria hills, whose roots lie below the datum plane. Other short-wavelength anomalies are probably associated with density variations just below the surface.

Because the steep flanks of the main anomaly and its simple shape suggest a shallow disturbing body of simple form, it was concluded that modelling with vertical cylinders would be most appropriate. The best fit of computed anomalies to the observed curves was obtained with models of discs.

The basic model is an elliptical disc with its upper surface at the top of the Hawaiite II lava body (Fig. 5, model A). The horizontal dimensions of the disc are 3.05 km (north-south) by 2.56 km (east-west). A density contrast with the surrounding rocks of 0.4 g cm⁻³ was taken, on the basis of the measured densities

(2.6 g cm^{-3}) for the dense volcanic rocks and the densities (2.2 g cm^{-3}) for Tertiary rocks quoted by Schoenharting & Brunt (1971).

The thickness of the disc was adjusted so that the computed gravity anomaly would give the best match to the observed anomaly, particularly near the flanks of the anomaly, as it was expected that complicating factors would modify the response near the central region. The thickness finally used was 330 m although thinning of the disc towards the south is indicated by the smaller anomaly amplitude in the southern half of line 3 (Fig. 5).

This simple disc model produces computed anomalies with larger amplitudes in the central region than are observed. The difference between the observed and calculated anomalies represents a negative gravity anomaly which was modelled by a cone of radius 330 m and depth 380 m (Fig. 5, model B), representing a scoria and tuff 'plug'. The density assumed was 2 g cm^{-3} and may be a lower limit if the plug materials have been compacted.

The computed gravity profiles for these two models, and the residual gravity values on line 3, are shown in Figure 5. The profiles show that the longer-wavelength variations in the gravity field observed in the central region of the Complex could well be explained by plugs of less dense materials, but the centroid of these features would need to be displaced towards the south by about 200 to 300 m (see Fig. 3).

The gradient at the southern end of line 3 (Fig. 5) is not well matched by Model A, and on some traverses (line 1 NE end, line 2 NW end, line 3 S end) the 'foot' of the anomaly has a sharper curvature than the 'shoulder'. Taken together with the fact that the foot coincides with a topographical cliff (the Ring Barrier), an 'overthrust' configuration is suggested. This could well be expected if the lava is filling a depression with sloping walls.

DISCUSSION

The dimensions of the proposed lava disc suggest that the gravity anomaly might be due to the infilling of a caldera by lava.

The mechanism of cauldron subsidence (e.g. Bailey & Maufe, 1960; Williams, 1941) may have been responsible for caldera formation at Mount Porndon. Surficial volcanic activity involving the eruption of lava flows, mainly from ring fractures, onto basins formed by cauldron subsidence is believed to have occurred at Glen Coe, Scotland (Clough et al., 1909), Oslo graben cauldrons, Norway (Oftedahl, 1960), and Ossipee Mountains, New Hampshire, USA (Kingsley, 1931). Stages in the development of the disc of lava at Mount Porndon according to this model are presented diagramatically in Figure 6.

The Newer Volcanic activity in southeastern Australia was of the areal type (Joyce, 1975, in press). Shirinian (1968) characterised areal volcanism by 'large areal development, multi-orifice character or mass nature of eruptions, monogenetic nature of the centres represented by both vent and fissure types and relative independence of separate volcanoes or their groups connected with near-surface secondary foci' whose depth ranges from 0.5 km to 3 km. Joyce (1975, in

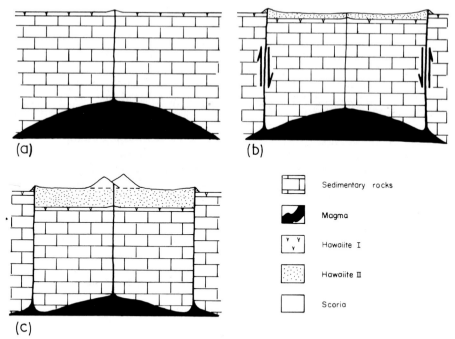

Fig. 6: Development of the Mount Porndon Volcanic Complex according to caldera hypothesis: (a) eruption of lava flows to form Stony Rises — updoming also possible; (b) waning of eruption and partial withdrawal of magma in chamber, beginning of cauldron subsidence, and new eruptions of lava flows from ring fracture; (c) further eruptions of lava flows and continued subsidence producing lava disc — final phase of basanitoid lava extrusion and scoria and tuff eruption from central vents.

press) believes that intermediate magma chambers may have formed at the contact between the folded Palaeozoic sediments and the overlying basin deposits, at a depth of about 0.8 km below Mount Porndon (unpublished data, Frome-Broken Hill Co. Pty Ltd.). Such a high-level magma chamber would greatly facilitate the formation of ring fractures and the operation of the cauldron subsidence mechanism to produce a lava infilled caldera. The inwards slope of the surface of the Hawaiite II lava body is explained if lava welled up around the edge of a sinking plate of country rock and flowed inwards from the ring fracture.

Consideration has been given to a second hypothesis to account for the proposed disc of lava below the Mount Porndon Complex. The location of the Complex within a belt of maars and the discovery by the gravity method (W.S. Fischer, unpublished report*) of a slab of dense material up to 270 m thick within the maar crater (diameter 2 km) and beneath the scoria deposits at Mount Noorat, some 50 km west from Mount Porndon, suggested the possible existence of a lava infilled maar beneath Mount Porndon.

Singleton & Joyce (1969) hypothesised that in the formation of a maar, violent phreatomagmatic eruptions produce a conical throat of shattered rock beneath the crater of the volcano. Modelling of conical bodies at the base of the Mount

* 'A Gravity Study of Tertiary Volcanic Structures.' by W.S. Fischer (unpublished Honours report, 1973, University of Melbourne).

Porndon lava disc, having radius equal to that of the disc, walls dipping at 30^0-60^0, and a density of 2 g cm^{-3}, was tried, but the anomaly amplitudes of these composite models were considerably smaller than that of the observed gravity anomaly. It was concluded that the most reasonable way to improve the fit was by increasing the density in the conical throat. Compaction of material in a maar throat (Lorenz, 1973) and the presence of igneous intrusions would both serve to increase the density of the throat rocks. Fischer (op. cit.) found that modelling of a basalt slab alone satisfactorily accounted for the main gravity anomaly at Mount Noorat, suggesting that if a maar throat exists there its density is not significantly anomalous.

Whereas Mount Noorat is known to be a buried maar (Singleton & Joyce, 1969) there is no geological or geomorphological evidence for the existence of a tuff ring beneath the Mount Porndon Complex. If there were a tuff ring below the Stony Rises lava flows one might expect an annular gravity minimum surrounding the principal anomaly, unless compaction of the tuff reduced the density contrast to insignificance.

If a maar crater at Mount Porndon were infilled with lava, in order that the gravity anomaly retain its simple shape, the Hawaiite II lava would have had to cease advancing over the Stony Rises lava at the ground surface projection of the edge of the maar crater (assuming eruption from a central vent). Such a coincidence seems improbable, but may have been possible if the Hawaiite I lava filling the crater sank slightly (due to compaction of the maar throat rocks) before Hawaiite II lava was erupted, and the new lava was constrained by the walls of this depression.

CONCLUSIONS

The gravity study at the Mount Porndon Volcanic Complex leads to the following conclusions:

1) Below the surface flows of basanitoid lava, the scoria cones, and the tuff deposits of the Complex, is a large disc of basalt whose lower surface extends below the average upper level of the Tertiary sedimentary rocks surrounding the Complex.

2) The disc of basalt has the horizontal dimensions of 2.6 km by 3.1 km and a thickness of 330 m.

3) The phase of activity which produced the known tuff and scoria deposits which overlie Hawaiite II lava also produced a concealed body of scoria and tuff which is responsible for a small gravity depression.

4) Two hypotheses for the formation of the basalt disc are discussed, but on the available evidence, and because it is the simpler mechanism, infilling of a caldera is favoured.

ACKNOWLEDGEMENTS
We gratefully acknowledge the assistance of G.W. Boyd, R.J.S. Cooke, C. Kerr Grant, V. Hodor, E.B. Joyce, O.P. Singleton, P. Wood, Shell Development

(Aust.) Pty Ltd, and Frome-Broken Hill Co. Pty. Ltd., at various stages of this work. Adrian G. Power wrote the computer programs used. D. Campbell and L. Hodgson drafted the illustrations.

REFERENCES

Bailey, E.B., & Maufe, H.B. 1960: The geology of Ben Nevis and Glen Coe and the surrounding country, 2nd ed. *Geol. Surv. Scotland Mem.*
Clough, C.T., Maufe, H.B., & Bailey, E.B., 1909: The cauldron subsidence of Glen Coe and the associated igneous phenomena. *Geol. Soc. London Quart. J., 65,* pp. 611-676.
Joyce, E.B., 1975: Quaternary volcanism and tectonics in southeastern Australia. *Roy. Soc. N.Z. Bull. 13 (in press).*
Kingsley, L., 1931: Cauldron subsidence of the Ossipee Mountains. *Amer. J. Sci., 22,* pp.139-168.
Lorenz, V., 1973: On the formation of maars. *Bull. Volc., 37,* pp.183-204.
McDougall, I., Allsop, H.L., & Chamalaun, F.H., 1966: Isotopic dating of the Newer Volcanics of Victoria, Australia, and geomagnetic polarity epochs. *J. geophys. Res., 71,* pp.6107-6118.
Oftedahl, C., 1960: Permian igneous rocks of the Oslo graben, Norway. *21st Intern. Geol. Congr., Norway.* Guide to excursions no. A11 and C7.
Schoenharting, G., & Brunt, P.G., 1971: Colac-Geelong gravity survey. *Shell Development (Aust.) Pty Ltd* (unpubl.).
Shirinian, K.G., 1968: Endogenetic conditions of areal volcanism (on the example of Armenia). *Bull. Volc, 32,* pp.283-295.
Singleton, O.P., & Joyce, E.B., 1969: Cainozoic volcanicity in Victoria. *Geol. Soc. Aust., Spec. Pub. 2,* pp.145-154.
Skeats, E.W., & James, A.V.G., 1937: Basaltic barriers and other surface features of the Newer Basalts of Western Victoria. *Proc. Roy. Soc. Vic., 49 (2),* pp.245-278.
Williams, H., 1941: Calderas and their origin. *Univ. Calif. Berkeley Pub. Geol. Sci., v. 25.*

POTASSIUM VARIATION IN LAVAS ACROSS THE SUNDA ARC IN JAVA AND BALI

D. J. WHITFORD and I. A. NICHOLLS[1]

Research School of Earth Sciences, Australian National University, P.O. Box 4, Canberra, A.C.T. 2600

ABSTRACT

The Sunda arc of Indonesia has been used by several authors as an example of a volcanic arc where the K_2O content of rocks with the same SiO_2 content increases regularly with increasing depth to an underlying Benioff seismic zone. Lavas ranging from those showing affinities with the island-arc tholeiitic association through a broad spectrum of calcalkaline to high-K alkaline compositions have been analysed, in order to examine the variation of K_2O at 55 percent SiO_2 (K_{55}) in different volcanoes as a function of depth to the Benioff zone (h). K_{55} versus h for all volcanoes shows a moderately good correlation (r = 0.83). However, if only those volcanoes whose lavas (1) show a correlation between SiO_2 and K_2O which is significant at the 99.9 percent confidence level, and (2) exhibit relatively low $^{87}Sr/^{86}Sr$ ratios, are accepted, the correlation can be improved (r = 0.95). The results of the present work diverge significantly from some earlier investigations. The high-K alkaline lavas from Muriah can be divided into two series. A 'wet series' plots close to the K_{55}-h trend defined by the tholeiitic and calcalkaline rocks, which suggests a genetic relationship. In contrast, the 'dry series' rocks, which appear more abundant, apparently show no relationship at all with this trend. K_{55}-h variation in lavas across the Sunda arc is consistent with recent models of island-arc petrogenesis, whereby magmas are derived dominantly from the mantle wedge above the Benioff zone, the mantle having been modified by the addition of water and/or a melt component from the subducted oceanic crust.

INTRODUCTION

Variation in total alkali content of volcanic rocks across island arcs has long been recognised and' has been examined by Rittman (1953), Kuno (1959), and Sugimura (1960), among others. In recent years, a large body of data has accumulated dealing with the variation in the K_2O content of lavas erupted from volcanoes at different distances above an underlying Benioff seismic zone (e.g. Dickinson & Hatherton, 1967; Hatherton & Dickinson, 1969; Ninkovich & Hays, 1972; Nielson & Stoiber, 1973).

Since Dickinson & Hatherton (1967) established a close correlation between increasing K_2O contents of erupted lavas at a constant level of SiO_2 and increasing depth to the Benioff zone, the inverse problem has also been studied — that is, given an ancient island arc and a spatial variation of K_2O within a suite of lavas of the same SiO_2 content, the polarity and depths to a pre-existing Benioff zone may be inferred. This has been discussed by Dickinson (1972) and used extensively by Lipman (1971).

The Sunda arc of Indonesia has been used by various authors as an example of a volcanic arc where K_2O, at a constant level of SiO_2 in the volcanic rocks, increases regularly with increasing depth to the Benioff zone. Much of the analytical data for these studies has been extracted from Neumann van Padang (1951), and has been used to arrive at some surprisingly varied results. It is the purpose of this paper to critically re-examine relationships between K_2O, SiO_2, and depth to Benioff zone across the Sunda arc in Java and Bali, using data from

[1] Present address: Department of Earth Sciences, Monash University, Melbourne, VIC 3168

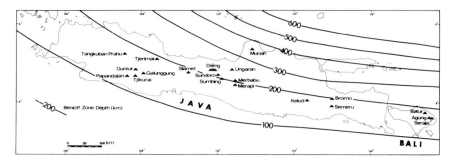

Fig. 1: Location of volcanoes sampled in Java and Bali for this study. The position of the underlying Benioff seismic zone is from Hamilton (1972).

Neumann van Padang (1951), together with recent analytical data which is supported by trace element and Sr isotopic data.

Java and Bali are attractive regions for a study of this kind because active volcanism is widespread, a wide variety of lavas has been erupted, corresponding to a considerable range of Benioff zone depths, and the position and attitude of the Benioff zone has been well defined (Fitch, 1970, 1972; Fitch & Molnar, 1970; Hamilton, 1972, 1973). Depths to the Benioff zone beneath individual volcanoes have been interpolated from the data presented by Hamilton (1972). It is obviously an oversimplification to suggest that the Benioff zone is perfectly planar, having no thickness, but such an approximation is adequate for the purposes of this paper. Sampling localities together with Benioff zone contours from Hamilton (1972) are shown in Figure 1.

All samples (excluding those from Neumann van Padang, 1951) have been analysed either by X-ray fluorescence spectrometry (Norrish & Hutton, 1969), or by electron probe microanalysis (Nicholls, 1974a).

Only samples free from macroscopic alteration were selected, to avoid ambiguities caused by alkali mobilisation during secondary alteration. For uniformity, all analyses have been recalculated on a volatile-free basis with total Fe calculated as FeO. Analyses from Neumann van Padang (1951) were recalculated in the same way, although all analyses with total H_2O greater than 2.5 percent, SO_3 greater than 0.5 percent, or with TiO_2 not detected, were rejected. By comparison with recent analytical data, analyses from Neumann van Padang (1951) are inferior, but K_2O/SiO_2 relationships are still meaningful, if somewhat imprecise.

VOLCANIC ROCKS OF JAVA AND BALI

Lavas ranging from those showing affinities with the island-arc tholeiitic association (Jakeš & Gill, 1970) through a broad spectrum of calcalkaline to high-K calcalkaline compositions were analysed. (For a more detailed discussion see

Fig. 2 (opposite): Variation diagrams of K_2O versus SiO_2 for the volcanoes of Java and Bali, showing least-squares linear regression lines, the interpolated value of K_2O at 55% SiO_2(K_{55}), and the correlation coefficient r. Dots refer to data from Neumann van Padang (1951) and squares refer to new data. The regression lines marked (1) for Guntur and Muriah have been constructed using all data. Regression lines (2) and (3) for Guntur refer to the low- and high-K series respectively. Regression lines (2) and (3) for Muriah refer to the 'wet' and 'dry' series respectively (see discussion).

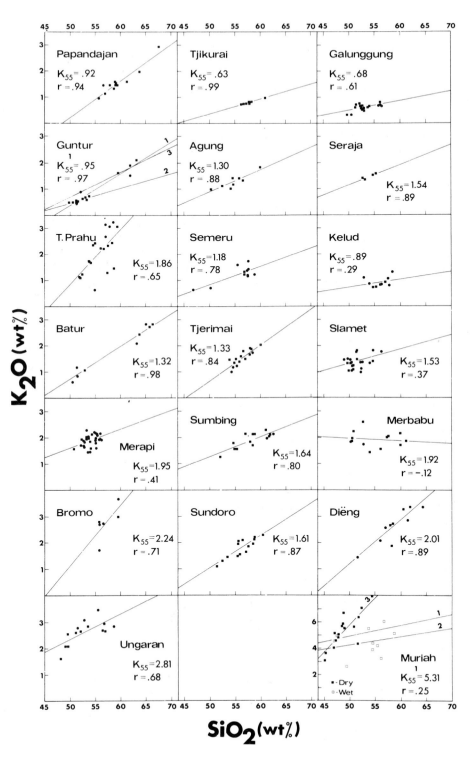

Whitford, 1975). In addition, rocks from the extinct high-K alkaline volcano Muriah were analysed in order to examine the relationship between these rocks and the more typical island-arc lavas. All samples except those from Tjikurai, Muriah, and Seraja are from presently active volcanoes, although all samples are thought to be early Pleistocene or younger (van Bemmelen, 1949).

RESULTS

Following the method of Dickinson & Hatherton (1967), variation diagrams of K_2O versus SiO_2 have been constructed for the rocks from individual volcanoes and a least-squares linear regression performed to find the line of best fit. Using this line the interpolated value of K_2O at 55% SiO_2 (K_{55}) has been calculated; K_{55} being preferred to K_{60} because of the predominance of basaltic andesites (52%-57% SiO_2) relative to andesites (57%-63% SiO_2) in analysed lava samples. In some cases the data would obviously be better fitted to a simple non-linear regression, but because of its simplicity and ease of interpretation and comparison, the linear regression has been preferred. Individual variation diagrams with regression lines, K_{55} values, and the correlation coefficients r are shown in Figure 2. Individual volcanoes, Benioff zone depths, number of analyses, K_{55} values, correlation coefficients, and the significance of r relative to the 99.9 per cent confidence limits are listed in Table 1. More than one regression line have been calculated for the volcanoes Guntur and Muriah. Lines marked (1) are the simple regression lines for all the data; the other regressions will be discussed below.

Table 1. Statistics

Volcano	Depth to Benioff Zone	Number of analyses	K55	r	Significance level (99.9%)
Papandajan	130	12	0.92	0.941	Yes
Tjikurai	140	8	0.63	0.990	Yes
Galunggung	150	18	0.68	0.610	No
Guntur	150	18(1)	0.95	0.970	Yes
		13(2)	0.73	0.861	Yes
		5(3)	1.14	0.923	No
Agung	160	8	1.30	0.877	Yes
Seraja	160	4	1.54	0.889	No
Tangkuban Prahu	170	20	1.86	0.647	No
Semeru	170	11	1.18	0.783	No
Kelud	180	10	0.89	0.291	No
Batur	180	9	1.32	0.982	Yes
Tjerimai	190	15	1.33	0.837	Yes
Slamet	190	22	1.53	0.366	No
Merapi	190	30	1.95	0.409	No
Sumbing	200	14	1.64	0.804	Yes
Merbabu	200	13	1.92	-0.121	No
Bromo	200	6	2.24	0.705	No
Sundoro	210	14	1.61	0.867	Yes
Diëng Complex	220	11	2.01	0.893	Yes
Ungaran	240	14	2.81	0.679	No
Muriah	360	27(1)	5.31	0.247	No
		10(2)	4.56	0.241	No
		17(3)	7.73	0.818	Yes
		Σ 284			

Guntur	(1) All data	Muriah	(1) All data
	(2) Low-K series		(2) 'Wet series'
	(3) High-K series		(3) 'Dry series'

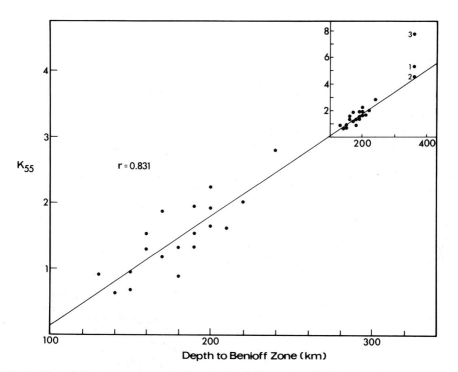

Fig. 3: Plot of K55 values versus depth to the Benioff zone for all data from the tholeiitic and calcalkaline volcanoes. Three points correspond to Muriah (h = 360 km): (1) all data, (2) 'wet series' rocks, (3) 'dry series' rocks (see discussion). A least-squares linear regression line for all data from all volcanoes, except Muriah, is also plotted with the correlation coefficient r.

Figure 3 shows K55 for each volcano against depth to the Benioff zone in the manner outlined by Dickinson & Hatherton (1967). These parameters have been plotted for all volcanoes irrespective of the significance of the correlation between SiO2 and K2O, and irrespective of any other geochemical consideration. Because of its clearly anomalous character, the value for Muriah (regression 1), has not been included in the regression. The three points corresponding to Muriah correspond to the three regression lines in Figure 2 (see below). The correlation shown in Figure 3 is significant at the 99.99 percent level. However the data are sufficiently scattered that when considered alone, volcanoes at Benioff zone depths from 160 to 220km show a weak positive correlation which is significant only at the 95 percent confidence limit. On a smaller scale, volcanoes over Benioff depths from 160 to 180km show a weak negative correlation; that is, K55 decreases with increasing depth to the Benioff zone. In fact, in Figure 3 the correlation is largely controlled by the end-points — the high-K calcalkaline rocks from Ungaran, the calcalkaline ones from Papandajan, and the tholeiitic rocks from Tjikurai, Guntur, and Galunggung.

The approach used to construct Figure 3 is unsatisfactory. For several volcanoes such as Merapi, Merbabu, Slamet and Tangkuban Prahu the correlation between SiO2 and K2O is poor, owing to a lack of dispersion in SiO2, a wide variation in K2O, or both. Hence, any K55 value calculated by a simple regression may have a large error.

The volcano Guntur shows two apparently distinct trends of K2O versus SiO2 variation. If the five samples showing distinctly higher K2O contents are considered separately, the value of K55 for the low-K2O rocks drops to 0.73 (regression 2), in much closer agreement with the nearby tholeiitic volcano Galunggung. The high-K2O rocks have a K55 value of 1.14 (regression 3) although this correlation is not significant at the 99.9 per cent level. This subdivision into high- and low-K series can be justified on the basis of major and trace element geochemistry — the low-K rocks are distinctly tholeiitic whereas the high-K rocks have a calcalkaline chemistry. The lavas from Tangkuban Prahu can also be divided into a high-K series and a low-K series, although the correlation coefficients for the subgroups are still not significant at the 99.9 percent level. Merapi shows a similar but less well defined division.

The above examples show that many of the K2O versus SiO2 variation diagrams are open to interpretation. It is obviously difficult to interpret the data objectively, but by applying two simple constraints the correlation in Figure 3 can be improved and made more meaningful.

1. By accepting only volcanoes whose K2O to SiO2 correlation is significant at the 99.9 percent level, the volcanoes Galunggung, Seraja, Tangkuban Prahu, Semeru, Kelud, Slamet, Merapi, Merbabu, Bromo, and Ungaran are removed from the data set. Within the group of volcanoes there are two sets. The first set, including Galunggung, Tangkuban Prahu, Slamet, Merapi, and Merbabu, shows a poor correlation between K2O and SiO2 because of either a lack of dispersion in SiO2 and/or a large scatter in K2O. The second set, including Seraja, Semeru, Kelud, Bromo, and Ungaran, shows a poor correlation between K2O and SiO2 largely because of the lack of sufficient precise analyses. Future work, however, may well result in significant K2O to SiO2 correlations for some of these volcanoes.

2. Sr isotopic studies on volcanic rocks from the Sunda arc (Whitford, 1975) show that there is a group of calcalkaline rocks characterised by relatively high $^{87}Sr/^{86}Sr$ ratios (0.7055-0.7059). These ratios are best explained in terms of magmatic contamination either from a radiogenic component on the downgoing slab, or from within the crust beneath the volcanoes. Because this type of contamination is believed to be secondary to the primary processes of magma genesis in island arcs in general, it can be argued that these volcanoes should be rejected. The volcanoes Papandajan, Slamet, Merapi and Merbabu have erupted lavas, some of which at least show high $^{87}Sr/^{86}Sr$ ratios and thus can be deleted.

Geochemical evidence suggests that there have been at least two types of contamination. The volcano Papandajan is characterised by abundant andesite and dacite, and the abundance of SiO2, K2O, Rb, and Cs, for example, suggests that some more basic magma has been contaminated by a sialic component enriched in all these elements. This has resulted in essentially linear elemental variation patterns which probably reflects simple mixing.

On the other hand, lavas from the other 'contaminated volcanoes' show a more complicated geochemistry. Basalts are common and although higher K2O contents generally correlate with the higher $^{87}Sr/^{86}Sr$ ratios, the pattern is not simple. The nature of the contaminant and the actual contamination process is

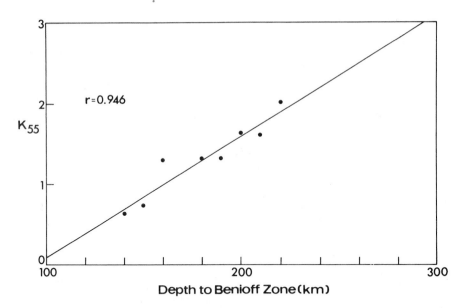

Fig. 4: Plot of K55 versus depth to the Benioff zone for 'preferred data' (see text for details), showing least-squares linear regression line, and correlation coefficient r.

unclear. However, not surprisingly, these volcanoes can be rejected on both criteria.

By applying the above criteria only eight of the original nineteen volcanoes remain. These acceptable data are plotted as K55 against depth to the Benioff zone in Figure 4. The data point at h = 150 km corresponds to the low-K rocks from Guntur. The correlation shown in Figure 4 is good (r = 0.946) and significant at the 99.99 percent confidence level.

DISCUSSION
Comparison with other data from Java and Bali and with data from other arcs.

As mentioned above, several other studies have used K55 versus depth to the Benioff zone data from the Sunda arc with analyses largely derived from Neumann van Padang (1951). K55 versus depth to the Benioff zone trend lines derived from this paper and elsewhere are shown in Figure 5. Data for Java from Hatherton & Dickinson (1969), Nielson & Stoiber (1973), and this work, which also includes Bali, have been plotted. The data of Hatherton & Dickinson (1969) show a considerable divergence from the present work in terms of Benioff zone depths. For example they quote h = 290 km for Ungaran whereas in this paper, using the data of Hamilton (1972), h = 240 km for the same volcano. Consequently their K55 data have been plotted in two ways: against the Benioff zone depths used in their paper, and also against the Benioff zone depths used in this paper.

From Figure 5 it is difficult to reconcile the data of Nielson & Stoiber (1973) with any of the other data. Their value of K55 of about 1.5 at h = 100 km (Nielson & Stoiber, 1973, Fig. 2) seems unrealistically high, and is in marked disagreement with the data presented in this paper. It should be noted that Nielson & Stoiber

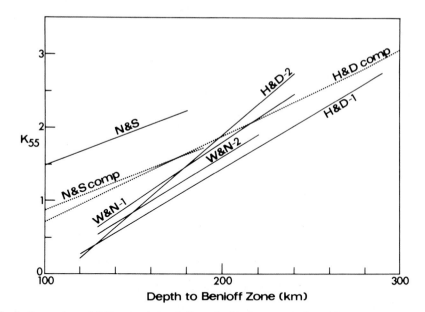

Fig. 5: Comparison of K55 versus h trends from the Sunda arc, together with two composite trends, representing averaged results for a number of island arcs: N & S — Nielson & Stoiber (6973) — Java; N & S comp. — Nielson & Stoiber (1973) — Composite; H & D 1 — Hatherton & Dickinson (1969) — Java, using their h values; H & D 2 — Hatherton & Dickinson (1969) — Java, using h values of Hamiton (1972); H & D comp. — Hatherton & Dickinson (1969) — Composite; W & N 1 — this paper — all data (Fig. 3); W & N 2 — this paper 'preferred data' (Fig. 4).

calculated their line in a slightly different manner, using a three-dimensional regression of K2O, SiO2, and depth to the Benioff zone from which they have derived their line by calculating the intersection of the regression plane with the plane corresponding to SiO2 = 55%. There is a better correspondence between the results presented by Hatherton & Dickinson (1969) and those presented here, despite the divergence in Benioff zone depths and the limited data available to Hatherton & Dickinson.

Also plotted in Figure 5 are two composite K55 versus h trends from Hatherton & Dickinson (1969) and Nielson & Stoiber (1973) which are estimates of average K55 versus h trends derived by averaging data from numerous island arcs around the world. It is interesting to note that the data from Java and Bali are somewhat different from the average trends of Hatherton & Dickinson (1969) and Nielson & Stoiber (1973), which themselves are quite similar. Whether or not the difference is significant is debatable, but of some importance in terms of models of magma genesis. Nielson & Stoiber (1973) concluded that there were significant differences in K versus h trends between different island arcs, although their data from Java, at least, are anomalous.

High-K alkaline rocks

Hatherton & Dickinson (1969) did not consider that the high-K alkaline rocks such as those found, for example, at Muriah and in the Roman province (Appleton, 1972) as being true 'island-arc' volcanics. On the other hand, Ninkovich & Hays

(1972) included this suite of rocks in their discussion, although they noted that on a K_2O versus SiO_2 variation diagram the points for the alkaline rocks scattered widely above a somewhat arbitrary curve corresponding to $h = 300$ km.

As shown in Figure 2, the K_2O versus SiO_2 data for Muriah are widely scattered and show no overall trend. The overall correlation (regression line 1) of K_2O with SiO_2 is not significant, and the calculated value of K_{55} does not plot near the trend defined by the tholeiitic and calcalkaline rocks in Figure 3. However, on petrographic criteria, two distinct suites of rocks can be recognised from Muriah (Nicholls & Whitford, in preparation).

A 'wet series' is characterised by abundant hydrous phenocrysts such as amphibole and biotite set in an apparently deuterically altered groundmass which is often characterised by abundant, pervasive calcite. Chemically these rocks tend to have lower K_2O/SiO_2 ratios due to generally lower K_2O and higher SiO_2. The K_{55} value (4.56) of the 'wet series' rocks plots very close to the trend defined by the more typical island-arc volcanic rocks (see Fig. 3), although the correlation between K_2O and SiO_2 is not statistically significant (see Fig. 2).

A 'dry series' is characterised by more mafic rocks, dominated by anhydrous phases, especially Ca-rich clinopyroxene in a fresh unaltered feldspathoid-rich groundmass. Rocks of this series tend to be less siliceous and richer in K_2O. The correlation between K_2O and SiO_2 is significant at the 99.9% confidence level and the interpolated K_{55} value (7.73) lies away from the trend defined in Figure 3.

Thus, whereas the 'wet series' rocks plot very close to the K_{55} versus h trend defined by the tholeiitic and calcalkaline rocks, which suggests some genetic relationship, the 'dry series' rocks apparently show no relation at all with this trend.

Petrogenetic significance of K55-h plots

A close correlation between the K_2O content at a constant level of SiO_2 in erupted lavas and the depth to the Benioff zone beneath island arcs implies a direct relationship between the Benioff zone and processes of magma genesis. On this basis, Dickinson & Hatherton (1967) suggested that in fact 'andesitic magmas are generated at Benioff zones'.

K_{55} versus h plots of the type developed by Dickinson & Hatherton (1967) imply that the K_{55} versus h relationship is the same for all island-arc systems. Such a relationship is compatible with these authors' view that the K_2O contents of island-arc magmas are determined primarily by the pressure under which magma production occurs at the Benioff zone. However, more detailed investigation appears to indicate that the K_{55} versus h relationship varies considerably between arcs (see e.g. Nielson & Stoiber, 1973). Such a variation may have important implications for the understanding of island-arc magma genesis. For example, demonstration of a dependence of the K_{55} versus h relationship on parameters such as the rate of subduction (and hence, perhaps, dip of the Benioff zone - Luyendyk, 1970) which affect the temperature distribution in the vicinity of the downgoing slab (e.g. Toksöz, et al., 1971) could imply a dependence of K_2O/SiO_2 on temperature within the subducted oceanic crust.

Attempts to explain the K2O-SiO2-h relationship for island-arc lavas (Jakeš & White, 1970; Marsh and Carmichael, 1974; Ringwood, 1974; Nicholls, 1974b) have stressed the importance of subducted oceanic crust as a source for potassium in island arc magmas. Green & Ringwood (1968) and Green (1972) proposed that andesitic magmas were produced directly by partial melting of quartz eclogite. On this basis, Jakeš & White (1970) proposed that K2O/SiO2 in andesitic magmas increased as a function of depth of partial melting, owing to the progressive reduction in the stability of phlogopite with increasing depth. Marsh & Carmichael (1974) have proposed a similar model in which the potassium-bearing phase involved is sanidine.

Ringwood (1974) and Nicholls & Ringwood (1973) proposed a more complex model to overcome a major objection to these earlier models — that liquids produced by partial melting of garnet-bearing assemblages are depleted in the heavy rare earth elements (Gill, 1974). According to this model the main source of island-arc magmas is the mantle wedge overlying the Benioff zone, but much of the K in these magmas and their mantle source is derived from the melting of the underlying oceanic crust. Modreski & Boettcher (1972, 1973) and Boettcher (1973) ascribed the K2O/SiO2-h variations to a systematic variation in the K2O content of a peridotitic source, but did not specify the source of this K.

In the model of Ringwood & Nicholls the first stage in the production of island-arc magmas is the subsolidus dehydration of basaltic oceanic crust at depths of 60 to 100 km to produce eclogite. The released water enters the overlying mantle, reduces its density and viscosity, and triggers diapiric uprise. Nicholls & Ringwood (1972, 1973) showed that about 20 percent partial melting of a diapir of water-saturated 'normal' peridotitic mantle could produce K-poor quartz tholeiite magmas similar to those of the island-arc tholeiitic series.

In the second stage of the model at depths greater than 100 km, partial melting of eclogite occurs. The first liquid to form under hydrous conditions is likely to be rhyodacitic or rhyolitic (Green & Ringwood, 1968; Stern & Wyllie, 1973), and rich in both Na and K. With increasing depth of melting, and decreasing availability of water, the first melt will probably become poorer in silica (approaching an andesitic to dacitic composition — Stern & Wyllie, 1973) and K/Na is likely to rise owing to the retention of Na in residual clinopyroxene and the progressive breakdown of phengitic mica in the eclogitic component (K. L. Harris, personal communication), and phlogopitic mica in the ultramafic component (Modreski & Boettcher, 1973; Boettcher, 1973; Wyllie, 1973).

Liquids produced in this way will tend to move upward into the overlying mantle. A large proportion is likely to react with the mantle, converting some of the olivine to pyroxene and garnet and enriching the mantle in K and related trace elements. If this hydrous geochemically modified mantle subsequently undergoes diapiric uprise, partial melting at depths less than 70 km is likely to produce basaltic to basaltic andesite magmas with K/Na increasing with increasing distance from the Benioff zone. These are the parental magmas of the calcalkaline suite (Nicholls, 1974b).

In this model, the K2O-SiO2-h relationship depends not only on the depth of eclogite melting at or near the Benioff zone, but also on the proportion of slab melt component added to the overlying peridotitic mantle, and probably the depth at which final separation of magma from residual modified peridotitic mantle takes place. The interplay of these factors may explain features such as variation in the K55-h relationship from arc to arc, major variations in the K_2O/SO_2 ratio of lavas within a single volcanic centre (e.g. Guntur — see above discussion) and apparent reversals in the increase in K_2O/SiO_2 and K_2O/Na_2O with depth to the Benioff zone, such as those observed in lavas of the New Britain area (R. W. Johnson, in preparation).

CONCLUSIONS

By applying two simple constraints to K_2O and SiO_2 data for a number of volcanoes across the Sunda arc in Java and Bali, a moderately good correlation between K55 and depth to the Benioff zone can be improved. It appears likely that additional data could result in improvements to this correlation. This study highlights several factors which should be considered in relation to K_2O/SiO_2/depth-to-Benioff-zone studies in island arcs.

1. There is a need for uniformity in assigning Benioff zone depths below particular volcanoes. It would be preferable if the same techniques for recalculating earthquake epicentre data were used on all island arcs so that meaningful comparisons can be made between them.

2. K versus h plots should be constructed within the framework of additional geochemical and geological data. This reduces the possibility of including volcanoes whose K_2O versus SiO_2 relationships have been disturbed by secondary processes such as contamination. If this is not done, solutions to the inverse problem of deducing depths to pre-existing Benioff zones are at best tenuous and may be misleading. This approach is also limited by the question of whether a 'universal' K55 versus h relationship exists for all arcs (see conclusion 3).

3. More rigorous examination of K55 versus h relationships in other island arcs would be most useful in placing additional constraints on processes of magma genesis. An important question now is (see conclusion 2) — do all island arcs display the same K55 versus h trend, or do they vary as a function of any other parameter, such as subduction rate or Benioff zone dip?

4. Recent work in experimental petrology and trace element and isotope geochemistry have highlighted the complex nature of processes of magma genesis in island arcs. Differences in the nature of the subducted oceanic crust, processes of contamination, temperature distributions in downgoing slabs, and many other difficulties suggest that great care is required in the interpretation of K_2O variation in lavas across island arcs.

ACKNOWLEDGEMENTS

Staff of the Bandung Institute of Technology (M. T. Zen), the Direktorat Geologi Indonesia — Volcanology Section (K. Kusumadinata), Universitas Padjadjaran, Bandung (M. Koesmono) and Universitas Gadjah Mada, Jogjakarta (I. Soemarinda) kindly provided assistance with fieldwork and also donated a number of samples. P. H. Beasley, E. H. Pederson and N. G. Ware provided technical assistance with the analytical work. S. R. Taylor, K. L. Harris, and W. B. Nance are thanked for critically reading the manuscript.

REFERENCES

Appleton, J. D., 1972: Petrogenesis of potassium-rich lavas from the Roccamonfina volcano, Roman Region, Italy. *J. Petrology, 13*, pp. 425-456.

Boettcher, A. L., 1973: Volcanism and orogenic belts — the origin of andesites. *Tectonophysics, 17*, pp. 223-240.

Dickinson, W. R., 1972: Evidence for plate tectonic regimes in the rock record. *Am. J. Sci., 272*, pp. 551-576.

Dickinson, W. R. & Hatherton, T., 1967: Andesitic volcanism and seismicity around the Pacific. *Science, 157*, pp. 801-803.

Fitch, T. J., 1970: Earthquake mechanisms and island arc tectonics in the Indonesian-Philippine region. *Bull. seism. Soc. Am., 60*, pp. 565-591.

Fitch, T. J., 1972: Plate convergence, transcurrent faults and internal deformation adjacent to Southeast Asia and the Western Pacific. *J. geophys. Res., 77*, pp. 4432-4460.

Fitch, T. J. & Molnar, P., 1970: Focal mechanisms along inclined earthquake zones in the Indonesia-Philippine region. *J. geophys. Res., 75*, pp. 1431-1444.

Gill, J. B., 1974: Role of underthrust oceanic crust in the genesis of a Fijian calc-alkaline suite. *Contr. Miner. Petrol., 43*, pp. 29-45.

Green, T. H., 1972: Crystallization of calc-alkaline andesite under controlled high-pressure hydrous conditions. *Contr. Miner. Petrol, 34*, pp. 150-166.

Green, T. H. & Ringwood, A. E., 1968: Genesis of the calc-alkaline igneous rock suite. *Contr. Miner. Petrol., 18*, pp. 105-162.

Hamilton, W., 1972: Preliminary tectonic map of the Indonesian region. *U.S. Geol. Surv. Open File Report*, 3 sheets.

Hamilton, W., 1973: Tectonics of the Indonesian region. *Bull. geol. Soc. Malaysia, 6*, pp. 3-10.

Hatherton, T. & Dickinson, W. R., 1969: The relationship between andesitic volcanism and seismicity in Indonesia, the Lesser Antilles and other island arcs. *J. geophys. Res., 74*, pp. 5301-5310.

Jakeš, P. & Gill, J., 1970: Rare earth elements and the island arc tholeiitic series. *Earth planet. Sci. Lett., 9*, pp. 17-28.

Jakeš, P & White, A. J. R., 1970: K/Rb ratios of rocks from island arcs. *Geochim. cosmochim. Acta, 34*, pp. 849-856.

Johnson, R. W. in prep: Late Cainozoic volcanoes at the southern margin of the Bismarck Sea, Papua New Guinea. Part 1. Distribution and major element chemistry.

Kuno, H., 1959: Origin of Cenozoic petrographic provinces of Japan and surrounding areas. *Bull. volcan, II, 20*, pp. 37-76.

Lipman, P. W., Protska, H. J. & Christiansen, R. L., 1971: Evolving subduction zones in the western United States as interpreted from igneous rocks. *Science, 174*, pp. 821-825.

Luyendyk, B. P., 1970: Dip of the downgoing lithospheric plate beneath island arcs. *Bull. geol. Soc. Am., 81*, pp. 3411-3416.

Marsh, B. D. & Carmichael, I. S. E., 1974: Benioff zone magmatism. *J. geophys. Res., 79*, pp. 1196-1206.

Modreski, P. J. & Boettcher, A. L., 1972: The stability of phlogopite + enstatite at high pressures: a model for micas in the interior of the earth. *Am. J. Sci., 272*, pp. 852-869.

Modreski, P. J. & Boettcher, A. L., 1973: Phase relationships of phlogopite in the system K_2O-MgO-CaO-Al_2O_3-SiO_2-H_2O to 35 kilobars: A better model for micas in the interior of the earth. *Am. J. Sci., 273*, pp. 385-414.

Neumann van Padang, M., 1951: *Catalogue of the active volcanoes of the world. Part 1, Indonesia.* Int. volcan. Ass. Italy.

Nicholls, I. A., 1974a: A direct fusion method of preparing silicate rock glasses for energy-dispersive electron microprobe analysis. *Chem. Geol. 14*, pp. 151-157.

Nicholls, I. A., 1974b: Liquids in equilibrium with peridotitic mineral assemblages at high water pressures. *Contr. Miner. Petrol, 45*, pp. 289-316.

Nicholls, I. A. & Ringwood, A. E., 1972: Production of silica saturated tholeiitic magmas in island arcs. *Earth Planet. Sci. Lett., 17*, pp. 243-246.

Nicholls, I. A. & Ringwood, A. E., 1973: Effect of water on olivine stability in tholeiites and the production of silica saturated magmas in the island arc environment. *J. Geol., 81*, pp. 285-300.

Nicholls, I. A. & Whitford, D. J., in prep: Petrology and chemistry of high-K lavas from Muriah, Java, Indonesia.

Nielson, D. R. & Stoiber, R. E., 1973: Relationship of potassium content in andesitic lavas and depth to the seismic zone. *J. geophys. Res., 78,* pp. 6887-6892.

Ninkovich, D. & Hays, J. D., 1972: Mediterranean island arcs and origin of high potash volcanoes. *Earth planet. Sci. Lett., 16,* pp. 331-345.

Norrish, K. & Hutton, J. T., 1969: An accurate X-ray spectrographic method for the analysis of a wide range of geological samples. *Geochim. cosmochim. Acta, 33,* pp. 431-454.

Ringwood, A. E., 1974: The petrological evolution of island arc systems. *J. geol. Soc. Lond., 130,* pp. 183-204.

Rittmann, A., 1953: Magmatic character and tectonic position of the Indonesian volcanoes. *Bull. volcan, II, 14,* pp. 45-58.

Stern, C. R. & Wyllie, P. J., 1973: Melting relations of basalt-andesite-rhyolite-H2O and a pelagic red clay at 30 kb. *Contr. Miner. Petrol., 42,* pp. 313-323.

Sugimura, A., 1960: Zonal arrangement of some geophysical and petrological features in Japan and its environs. *J. Fac. Sci. Tokyo. Univ. II, 12,* pp. 133-153.

Toksöz, M. N., Minear, J. W. & Julian, B. R., 1971: Temperature field and geophysical effects of a downgoing slab. *J. geophys. Res., 76,* pp. 1113-1138.

Van Bemmelen, R. W., 1949: *The Geology of Indonesia.* The Hague, Martinus Nijhoff.

Whitford, D. J. 1975: Strontium isotopic studies of the volcanic rocks of the Sunda arc, Indonesia, and their petrogenetic implications. *Geochim. Cosmochim. Acta. 39,* pp. 1287-1302.

Wyllie, P. J., 1973: Experimental petrology and global tectonics — a preview. *Tectonophysics, 17,* pp. 189-209.

PRIMARY MAGMAS ASSOCIATED WITH QUATERNARY VOLCANISM IN THE WESTERN SUNDA ARC, INDONESIA

I. A. NICHOLLS[1] and D. J. WHITFORD

Research School of Earth Sciences, Australian National University, P.O. Box 4, Canberra, A.C.T. 2600

ABSTRACT

Basaltic andesite and andesite are the dominant types amongst analysed Pleistocene to Recent lavas in the western Sunda arc (in the islands of Java and Bali). The overall silica mode is approximately 55%. Values of $Mg/Mg+Fe^{2+}$(approximate mode 0.48) and contents of Ni and Cr are in most cases too low for the lavas to have been derived directly from primary magmas produced by melting of peridotitic mantle. However, other trace element characteristics, especially abundances of heavy rare earth elements, are not those expected in primary magmas produced by partial melting of eclogite in subducted basaltic crust. Basaltic lavas are present in most of the volcanoes studied. It is proposed that these, and associated basaltic andesites, were derived from more mafic primary magmas by fractionation, mainly of olivine. To test this model, 'primary magma' compositions capable of equilibrium with olivine ($Mg/Mg+Fe^{2+}$ = 0.84-0.88) of a peridotitic mantle source, were calculated by a procedure involving incremental addition of olivine to analyses of basalt and basaltic andesite lavas. The results indicated that removal of less than 15 wt % olivine from primary olivine tholeiite to magnesian basaltic andesite magmas could produce most of the observed range of basalts and basaltic andesites. The production of more silicic lavas has also been investigated. Graphical representation of the compositional trends produced by removal from basalt and basaltic andesite of phases (olivine, orthopyroxene, clinopyroxene, amphibole, calcic plagioclase, and titanomagnetite) which crystallise at intermediate to low pressures (less than 10 kb) indicates that fractionation of these phases can produce the observed range of andesites and dacites.

INTRODUCTION

This paper assesses the relevance of theories of magma genesis in island arcs and continental margins to the Pleistocene-Recent lavas of the Sunda arc, in the islands of Java and Bali. The compositions and relative abundances of basaltic, andesitic, and dacitic lavas in active volcanoes are examined, in an attempt to determine the nature of primary magmas. The paper also considers the nature of processes by which, if these primary magmas are of basalt to basaltic andesite composition, the range of observed lava types may be generated.

COMPOSITIONAL FEATURES AND RELATIVE ABUNDANCE OF LAVA TYPES

Variation in the compositions of lavas from Java and Bali in terms of two parameters — SiO_2 content and atomic ratio $Mg/Mg+Fe^{2+}$ is summarised in Figures 1-3. Most of the analyses have been performed by instrumental methods (Whitford & Nicholls, this volume) but analyses from Neumann van Padang (1951) with Fe_2O_3 less than FeO, and H_2O (total) less than 2.5 wt. percent, have also been included. Data from all volcanoes for which instrumental analyses of lavas are available (those of West and Central Java, and Bali — Whitford & Nicholls, this volume, Fig. 1) are included in Figures 1-3. However, only volcanoes for which 10 or more instrumental analyses are available are considered in later fractionation calculations (Tables 1 & 2; Fig. 5).

[1] Present address: Department of Earth Sciences, Monash University, Melbourne, VIC. 3168.

Before calculation of $Mg/Mg+Fe^{2+}$, ferrous iron in all analyses was adjusted to FeO = 0.85 Σ FeO*, a relationship similar to that determined for basaltic to andesitic liquids equilibrated at oxygen fugacities of the wüstite-magnetite buffer (Fudali, 1965). Similar oxygen fugacities are believed to prevail in the upper mantle (Boettcher, 1973). While normalisation to a constant oxidation state, and an implied narrow range of mantle oxygen fugacities, probably represents an over-simplification for island arc regions, where mantle water contents may vary widely, the procedure is considered to be satisfactory for the purposes of this paper. Following recalculation of iron contents, all analyses were normalised to 100 wt. percent free of H_2O and CO_2.

Silica content

Analysed lavas from Java and Bali display almost continuous variation in K_2O/SiO_2 ratio (Whitford & Nicholls, this volume) and in the abundances of many trace elements. However, potassium-poor lavas of three adjacent volcanic centres in southwestern Java have trace element characteristics approaching those of the island arc tholeiitic association as defined by Jakeš & Gill (1970), and have been referred to this association by Whitford (1975). Lavas from two of these volcanoes (G. Guntur and G. Galunggung) have a silica mode of 51-52 percent, while lavas of the third volcano (G. Tjikurai) group around 56-58 percent SiO_2 (Fig. 1).

More potassic lavas, referred by Whitford (1975) to the calcalkaline association, have a silica mode of about 55 percent. The histogram for all analysed lavas shows a broad maximum, with a mode of about 55 percent SiO_2. Using the classification of Jakeš & White (1972), andesites (56-62 percent SiO_2) and basaltic andesites (52-56 percent SiO_2) are of approximately equal abundance, while basalts (less than 52 percent SiO_2) and particularly dacites (greater than 62 percent SiO_2) are much less abundant.

$Mg/Mg+Fe^{2+}$

The atomic ratio $Mg/Mg+Fe^{2+}$ (or 100 $Mg/Mg+Fe^{2+}$) has been used by Green & Ringwood (1967) as an indicator of magmatic differentiation, and to study compositional relations between coexisting liquids and crystals. A histogram of $Mg/Mg+Fe^{2+}$ values for lavas from Java and Bali is given in Figure 2.

The total range of $Mg/Mg+Fe^{2+}$ values covered is 0.64-0.33, with a mode for all analyses of about 0.48. Almost 50 percent of the relatively mafic lavas referred to the island arc tholeiitic association have $Mg/Mg+Fe^{2+}$ values greater than 0.50. This is true of only about 35 percent of lavas referred to the calcalkaline association.

A broad spread of points is generated when the parameters SiO_2 content and $Mg/Mg+Fe^{2+}$ are plotted against one another (Fig. 3). For example, lavas with 56 percent SiO_2 have $Mg/Mg+Fe^{2+}$ ranging from 0.64-0.40, while lavas with $Mg/Mg+Fe^{2+}$ of 0.50 have SiO_2 contents within the range 48-64 percent. Despite these wide ranges, the two parameters show an overall negative correlation.

* Σ FeO: total iron as FeO + analysed FeO + 0.9 Fe_2O_3.

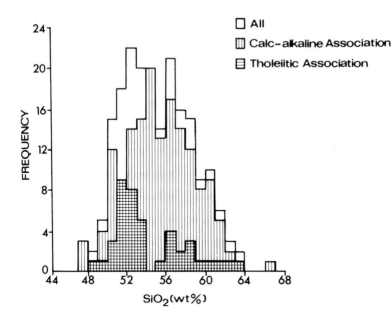

Fig. 1: Frequency diagram of silica contents (wt %) of analysed lavas from Java and Bali.

COMPARISON OF THE COMPOSITIONS OF LAVAS, AND LIQUIDS PRODUCED BY EXPERIMENTAL MELTING OF BASALT AND PERIDOTITIC COMPOSITIONS

The primary sources of magmas in island arcs are believed to be the basaltic crustal component of subducted lithosphere and the overlying wedge of peridotitic mantle (for recent reviews, see Boettcher, 1973; Wyllie, 1973; Ringwood, 1974). If equilibrium partial melting of these source materials occurs, then the compositions of the liquids produced will be constrained by equilibrium with amphibole, pyroxene, and garnet in amphibolite and eclogite representing the former oceanic crust, and with olivine, pyroxene, amphibole, or garnet in the peridotitic mantle. In each case, estimates of the compositions of liquids produced during experimental melting, or element distribution coefficients for equilibrium between one or more of the above crystalline phases and a range of liquids, are available, and may be used to allow comparison between natural and experimental compositions. Two cases — melting of eclogites and peridotitic compositions — are of major importance.

Eclogite in the former oceanic crust

Investigations of the nature of liquids formed by partial melting of quartz eclogite (Green & Ringwood, 1968; Green, 1972; Stern & Wyllie, 1973; Stern, 1974) have demonstrated that at pressures of about 30 kb the initial liquid to form is andesitic (approximately 60 percent SiO_2) under anhydrous conditions (T = ca. 1400°C) and rhyodacitic to rhyolitic (68-75 percent SiO_2) under water-saturated conditions (T = ca. 800°C). In each case, as the percentage of melting increases, liquid compositions become less silicic. Under hydrous conditions, an

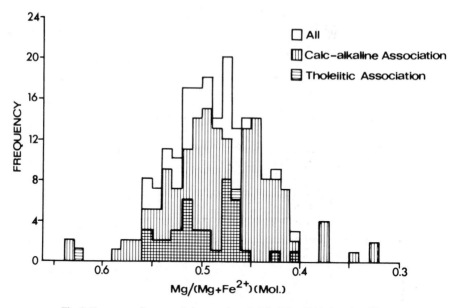

Fig. 2: Frequency diagram of the atomic ratio $Mg/Mg+Fe^{2+}$ of analysed lavas.

andesitic liquid forms at temperatures of 950-1000°C, and a basaltic andesite at 1000-1050°C. Green & Ringwood (1968) and Green (1972) considered that typical 'calcalkaline' compositions could be produced by eclogite melting, but Stern & Wyllie (1973) and Stern (1974) have stated that liquid compositions estimated for their partial melting experiments were anomalously rich in calcium.

Stern (1974) estimated the compositions of liquids formed by partial crystallisation of an olivine tholeiite with 5 percent H_2O at 30kb, at which pressure garnet and clinopyroxene were the dominant mineral phases present. At 900°C, the liquid composition (approximately 20 percent of liquid present) had 63 percent SiO_2 and a $Mg/Mg+Fe^{2+}$ value of about 0.2 (assuming $Fe/\Sigma FeO = 0.85$). At 1000°C, the SiO_2 content of the liquid (approximately 40 percent present) was much lower (an estimated 49-50 percent) and the $Mg/Mg+Fe^{2+}$ value was about 0.42. The starting basalt composition had 47.3 per cent SiO_2 and a $Mg/Mg+Fe^{2+}$ value of 0.62. Stern also calculated a trend for equilibrium crystallisation/partial melting for an average oceanic basalt, using mineral compositions derived from experiments on the above olivine tholeiite, and obtained liquid compositions with considerably higher silica contents. The compositions of both groups of liquids, estimated from the variation diagrams of Stern (1974, Fig. 3) are plotted in Figure 3. They define two distinct trends, A and B.

Peridotitic mantle

The compositions of liquids produced by partial melting of peridotitic compositions, particularly in the presence of water, are under debate. However, experimental data for the partitioning of Fe^{2+} and Mg between olivine and coexisting liquid (Green & Ringwood, 1967; Roeder & Emslie, 1970; Nicholls, 1974) indicate that liquids in equilibrium with olivine of supposed upper mantle materials (e.g. $Mg/Mg+Fe^{2+}$ = 0.84-0.94; Fujisawa, 1968) must be highly magnesian ($Mg/Mg+Fe^{2+}$ greater than, or equal to 0.61). The silica contents of these liquids depend strongly on total pressure and water content (or water fugacity) (Green, 1971, 1973; Nicholls, 1974; Mysen & Boettcher, in press) and may range from less than 40 percent to about 60 percent. Bulk analyses of glass and quench crystals coexisting with olivine ($Mg/Mg+Fe^{2+}$= 0.83-0.90) in charges run at water pressures of 5-15 kb and a temperature of 1000°C have 56-60 percent SiO_2 and $Mg/Mg+Fe^{2+}$ values of 0.68-0.77 (Nicholls, 1974, Table 10). At 1100°C and water pressures of 10 and 20 kb, liquid compositions calculated by Green (1973), assuming equilibrium with olivine $Mg/Mg+Fe^{2+}$= 0.89-0.90, have 56 and 48 percent SiO_2 and $Mg/Mg+Fe^{2+}$ values of 0.74 and 0.72 respectively. A possible curve of SiO_2 content versus $Mg/Mg+Fe^{2+}$ for liquids produced by partial melting of hydrous peridotite at about 10 kb is given in Figure 3 (trend C).

Melting of peridotitc compositions under water-undersaturated conditions gives rise to dominantly olivine-rich basaltic magmas (see e.g. Green & Ringwood, 1967; Green, 1971, Fig. 3). The value of

$$K_D = \left(\frac{Fe^{2+}}{Mg}\right)_{Olivine} \Bigg/ \left(\frac{Fe^{2+}}{Mg}\right)_{liquid} = 0.30$$

derived for olivine-rich basaltic compositions at 1 atm. (Roeder & Emslie, 1970) is smaller than that estimated by Nicholls (1974) for equilibrium involving water-saturated quartz-normative basaltic to andesitic liquids at pressures of 2-10 kb K_D = 0.4). Hence the $Mg/Mg+Fe^{2+}$ value of basaltic liquids formed by partial

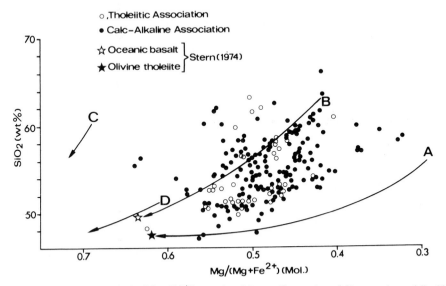

Fig. 3: Plot of SiO_2 versus $Mg/Mg+Fe^{2+}$ for analysed lavas. Curves A and B are estimated liquid composition paths for partial melting of olivine tholeiite with 5% H_2O, and an average oceanic basalt, at 30 kb total pressure (Stern, 1974). Curves C and D are estimated liquid composition paths for partial melting of peridotitic compositions under water-saturated and water-undersaturated conditions respectively, and total pressure of 10 kb (Nicholls, 1974; Green 1973). The arrows indicate the direction of increasing percentage of liquid.

melting of water-poor peridotite are probably lower than those for the water-saturated case (e.g. basaltic liquid with $Mg/Mg+Fe^{2+}$ of 0.61 in equilibrium with olivine, $Mg/Mg+Fe^{2+} = 0.84$). A possible SiO_2 versus $Mg/Mg+Fe^{2+}$ curve for partial melts of water-poor peridotite at pressures of about 10 kb is illustrated in Figure 3 (trend D).

Figure 3 illustrates clearly that all but three of the analysed lavas have $Mg/Mg+Fe^{2+}$ values too low (less than or equal to 0.59) for them to represent liquids originally in equilibrium with mantle olivines ($Mg/Mg+Fe^{2+}$ greater than or equal to 0.84). These lavas are also relatively poor in Ni (range = ca. 1-30 ppm) and Cr (= ca. 1-110 ppm). Two lavas from G. Tjerimai and one from G. Galunggung have $Mg/Mg+Fe^{2+}$ values of 0.62-0.64, compatible with equilibrium with olivine of the mantle, and relatively high Ni (25-100 ppm) and Cr (50-390 ppm). Silica contents are less diagnostic, since only about 10 percent of the lavas have SiO_2 greater than 60 percent, the highest value for experimental liquids demonstrably in equilibrium with olivine (Nicholls, 1974).

On the other hand, many of the lavas fall within the region of Figure 3 bounded by liquid compositions estimated for the partial melting of eclogite by Stern (1974). Most of the exceptions are relatively magnesian ($Mg/Mg+Fe^{2+}$ greater than or equal to 0.50), the most notable being the two most magnesian lavas from G. Tjerimai, which have about 56 percent SiO_2.

DISCUSSION

The relationships of Figure 3 appear to favour derivation of the bulk of lavas from the volcanoes of Java and Bali from magmas produced by melting of eclogite in subducted oceanic crust. However, other features of the geochemistry of these lavas argue against this conclusion, and against any mode of origin which involves garnet/liquid equilibria during partial melting or crystal fractionation processes. The most important feature is the relatively high content of heavier rare earth elements (HREE) in most analysed lavas (Fig. 4). As demonstrated by Gill (1974), liquids in equilibrium with eclogitic mineral assemblages are likely to be strongly depleted in HREE (Fig. 4) because of high values of HREE partition coefficients (e.g. Yb_{garnet}/Yb_{liquid} = 5-45; Harris & Nicholls, in preparation) for equilibrium between garnet and basaltic to andesitic liquids.

A second feature is the high proportion of relatively mafic compositions (47-56 percent SiO_2) amongst lavas with $Mg/Mg+Fe^{2+}$ values too low for direct derivation from peridotitic mantle (Fig. 3). The data of Stern (1974) show that production of similar compositions requires high degrees of melting (40-60 percent) of water-bearing eclogite. The requisite temperatures (1100-1200°C) are probably reached only when a subducted slab reaches minimum depths of 200-250 km (e.g. Toksöz, et al., 1971, Fig. 8). Many of the volcanoes under discussion are associated with Benioff zone depths of less than 200 km (Whitford & Nicholls, this volume, Fig. 1).

DERIVATION OF POSSIBLE PRIMARY MAGMA COMPOSITIONS

Most of the volcanoes under consideration for which a large number of analyses of lavas are available have produced relatively mafic compositions —

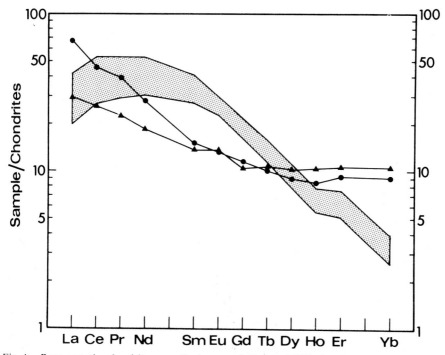

Fig. 4:   Representative chondrite-normalised rare earth element (REE) abundance patterns for lavas of Java and Bali, compared with abundances calculated by Gill (1974) for liquids produced by partial melting of ecologite. *Triangles* = Andesite (57% SiO2, 1.3% K2O), G. Semeru. *Dots* = Andesite (61% SiO2, 2.3% K2O), G. Sumbing. *Shaded area:* Range of REE abundances for andesitic liquids produced by 20-40% melting of eclogite of ocean floor basalt composition (Gill, 1974, Fig. 2).

basalt and basaltic andesite. Examples of these lavas, with $Mg/Mg+Fe^{2+}$ values of 0.50-0.64, are given in Table 1. Assuming that these lavas were derived ultimately from the partial melting of peridotitic mantle, rather than eclogite in subducted oceanic crust (above discussion), all but those of G. Galunggung and G. Tjerimai (Table 1, columns 1 and 4) are likely to have undergone appreciable crystal fractionation.

In the case of G. Papandajan, G. Slamet, G. Merapi, and G. Merbabu, trace element and Sr-isotope evidence indicates that some lavas may have been derived from magmas contaminated by crustal material (Whitford, 1975; Whitford & Nicholls, this volume). However, the contaminants (sediments or sub-arc crustal rocks) were presumably poor in Mg and Fe, and are unlikely to have caused appreciable reduction of $Mg/Mg+Fe^{2+}$ in mantle-derived basaltic magmas.

The simplest model for the derivation of the lavas of Table 1 which may be tested is based on the proposal that olivine was the major crystalline phase involved in fractionation. As discussed by Nicholls & Ringwood (1972, 1973), silica-saturated basaltic magmas produced by melting of hydrous peridotitic mantle beneath island arcs are likely to precipitate olivine before reaching the surface of the earth. Such a process appears to have taken place during the derivation of

Tabe 1. Major element chemistry of the most mafic analysed lavas of representative active volcanoes, Java[1].

	Gg[2]	Pp	Gu	Tj	Sm	Su	Sb	TP	Mb	Mp	Un
SiO2	49.73	57.2	50.65	56.16	49.8	51.5	52.23	51.9	50.45	50.91	48.42
TiO2	0.98	0.85	0.95	0.73	1.49	1.09	1.06	1.15	0.86	0.83	1.16
Al2O3	18.50	18.17	19.50	16.44	17.1	19.7	18.19	18.48	17.26	19.87	17.41
Fe2O3	1.57	1.31	1.46	1.30	1.74	1.52	1.50	1.62	1.72	1.45	1.69
FeO[3]	7.96	6.71	8.30	6.60	8.95	7.74	7.62	8.23	8.77	7.38	8.63
MnO	0.18	0.16	0.18	n.d.	0.21	0.19	n.d.	0.09	0.14	0.09	0.20
MgO	7.47	3.85	5.84	6.36	6.02	4.74	5.83	4.91	6.08	5.12	6.25
CaO	10.47	7.61	9.58	7.94	10.03	9.12	9.38	9.36	10.15	9.68	11.74
Na2O	2.82	3.07	3.06	2.89	3.24	3.29	2.97	3.11	2.78	3.12	2.88
K2O	0.33	1.13	0.48	1.59	1.34	1.09	1.25	1.12	1.81	1.56	1.62
Mg/Mg+Fe^{2+}[4]	0.63	0.51	0.56	0.63	0.55	0.52	0.58	0.52	0.56	0.55	0.56

[1] In order of increasing K 55 value (see Whitford & Nicholls, this volume).
[2] Gg — G. Galunggung; Pp — G. Papandajan; Gu — G. Guntur; Tj — G. Tjerimai; Sm — G. Slmaet; Su — G. Sundoro; Sb — G. Sumbing; TP — G. Tangkuban Prahu; Mb — G. Merbabu; Mp — G. Merapi; Un — G. Ungaran.
[3] Calculated on basis FeO = 0.85 Σ FeO.
[4] Calculated after above normalisation or iron.

tholeiitic associations in a number of island arcs, where olivine-poor basalts are associated with abundant basaltic andesites (e.g. in the Tonga-Kermadec arc — Ewart et al., 1973).

In the present paper, models based on the crystallisation and removal of olivine phenocrysts under conditions of both surface equilibrium (only the outermost shell of crystal in equilibrium with liquid) and bulk equilibrium conditions (total crystal in equilibrium with liquid — see e.g. Shaw, 1970) have been tested. Calculations simulating olivine fractionation under conditions of surface equilibrium were performed using a routine which takes the Fe^{2+}/Mg value for a given liquid composition, calculates the equilibrium olivine composition using a given Fe^{2+}/Mg partition coefficient, adds or subtracts a small amount of olivine, calculates a new Fe^{2+}/Mg value for the derivative liquid, and repeats these steps until a liquid in equilibrium with a desired olivine composition has been attained. In the present case, olivine was added to the lava compositions of Table 1 until liquids in equilibrium with three distinct olivine compositions ($Mg/Mg+Fe^{2+}$ = 0.84, 0.86, and 0.88) were obtained. The two values of the Fe^{2+}/Mg partition coefficient discussed above (K_D = 0.3 and 0.4) were used. The olivine increment used in addition calculations was progressively reduced until the total olivine added agreed to within about 0.5 wt percent with the results for subsequent reverse (subtraction) calculations.

A wide range of 'primary magma' compositions and percentages of olivine removed to produce the lavas of Table 1 was obtained. Assuming true fractional crystallisation (i.e. surface equilibrium) and K_D = 0.3, 'primary magmas' range from basalt to magnesian basaltic andesite compositions with about 7 percent MgO (in equilibrium with olivine with $Mg/Mg+Fe^{2+}$ = 0.84) to olivine-rich basalts with about 12 percent MgO (olivine $Mg/Mg+Fe^{2+}$= 0.88). If K_D = 0.4 is used, all compositions become more magnesian and more mafic, with as much as 17 percent MgO in compositions in equilibrium with olivine, $Mg/Mg + Fe^{2+} = 0.88$. The proportions of olivine to be removed to produce observed lavas range from zero (G. Galunggung and G. Tjerimai — equilibrium with olivine $Mg/Mg + Fe^{2+} = 0.84$), to 20 wt percent (G. Tangkuban Prahu — olivine $Mg/Mg + Fe^{2+} = 0.88$) if $K_D = 0.3$; the corresponding range for $K_D = 0.4$ is approximately 6-49 wt percent.

In calculations simulating olivine fractionation under conditions of bulk crystal/liquid equilibrium, olivine of the composition required for equilibrium with each lava composition of Table 1 was added until liquids in equilibrium with olivine with $Mg/Mg + Fe^{2+}$ of 0.84, 0.86, and 0.88 were produced. The proportions of olivine to be removed during the reverse bulk equilibrium fractionation process are similar to those for removal under surface equilibrium conditions for values of up to 5 wt percent. However, for higher proportions of olivine removed, the two sets of values are strongly divergent, and lava compositions produced by removal of 15-20 wt percent of olivine under surface equilibrium conditions require removal of twice that amount under bulk equilibrium conditions. This is because the composition of unzoned olivine removed under bulk equilibrium conditions is always less magnesian than the bulk composition of the continuously zoned crystals produced under surface equilibrium conditions.

Because the calculated 'primary magmas' are dominantly of olivine-rich basaltic composition, the value of K_D of 0.30 applicable to similar compositions (Roeder & Emslie, 1970) is probably to be preferred. Table 2 gives examples of the results obtained, including the range of 'primary magma' compositions in equilibrium with olivine of $Mg/Mg + Fe^{2+} = 0.84-0.88$, the proportions of olivine removed during crystallisation under surface equilibrium conditions to produce the lava compositions of Table 1, and the bulk (i.e. average) composition of the olivine. The magma compositions have the low TiO_2 and high Al_2O_3 of most island arc lavas. In some cases, CaO is lower than expected for olivine-rich basaltic compositions, and the production of the relevant lavas from a more realistic primary magma composition would probably involve fractionation of Ca-rich clinopyroxene as well as olivine. Most of the compositions have 48-50 percent SiO_2, and in all but their TiO_2 and Al_2O_3 contents resemble liquid compositions calculated by Green (1973) for melting of water-saturated pyrolite at 20 kb pressure (Table 2, composition A). The much more silicic compositions for G. Tjerimai and G. Papandajan resemble the liquid compositions calculated by Green for melting at 10 kb (Table 2, composition B). Because the maximum crustal thickness beneath Java is approximately 30 km (Ben Avraham & Emery, 1973), magma genesis at very shallow mantle depths is implied for these compositions.

These results demonstrate that under conditions of fractional (i.e. surface equilibrium) crystallisation the amount of olivine which must be removed from primary basalt to magnesian basaltic andesite magmas to produce the silica-saturated basalt and basaltic andesite lavas present in many island arcs is small. Removal of small amounts of olivine accompanied by Cr-spinel will have little effect on trace elements other than Ni and Cr, which are characteristically depleted in island arc lavas (e.g. Taylor, et al., 1969). In particular, little or no fractionation of rare earth elements will occur during olivine fractionation, and observed patterns may be related directly to abundances in the mantle source.

Table 2. Ranges of possible primary magma compositions for lavas of active volcanoes of Java, calculated assuming equilibrium with mantle olivine of $Mg/Mg+Fe^2$ = 0.84-0.88, and K D= 0.30 (see text).

	[1]Gg	Pp	Gu	[1]Tj	Sm	Su
SiO2	49.4 -49.1	55.7 -54.4	50.0 -49.1	56.0 -55.3	49.1 -48.2	50.5 -49.5
TiO2	1.0 - 0.9	0.8 - 0.7	0.9 - 0.8	0.7 - 0.7	1.4 - 1.3	1.0 - 0.9
Al2O3	18.0 -17.2	16.9 -15.5	18.5 -16.8	16.2 -15.6	16.0 -14.5	18.2 -16.6
Fe2O3	1.5 - 1.6	1.2 - 1.1	1.4 - 1.3	1.3 - 1.2	1.6 - 1.5	1.4 - 1.3
FeO	8.1 - 8.3	7.5 - 8.0	8.7 - 9.1	6.7 - 6.9	9.4 - 9.8	8.4 - 8.9
MnO	0.2 - 0.2	0.2 - 0.1	0.2 - 0.2		0.2 - 0.2	0.2 - 0.2
MgO	8.4 -10.3	6.7 - 9.9	7.7 -11.3	7.0 - 8.6	8.4 -12.1	7.5 -11.0
CaO	10.2 - 9.7	7.1 - 6.5	9.1 - 8.3	7.8 - 7.5	9.4 - 8.5	8.5 - 7.7
Na2O	2.8 - 2.6	2.9 - 2.6	2.9 - 2.6	2.9 - 2.7	3.0 - 2.7	3.1 - 2.8
K2O	0.3 - 0.3	1.1 - 1.0	0.5 - 0.4	1.6 - 1.5	1.3 - 1.1	1.0 - 0.9
$Mg/Mg+Fe^{2+}$	0.65- 0.69	0.61- 0.69	0.61- 0.69	0.65- 0.69	0.61- 0.69	0.61 - 0.69
Olivine[2] removed	2.6 - 7.8	7.8 -17.2	5.2 -15.9	1.5 - 5.8	6.7 -18.3	7.6 -18.2
$Mg/Mg+Fe^{2+}$[3]	0.85- 0.87	0.81- 0.84	0.82- 0.85	0.85- 0.87	0.82- 0.85	0.81 - 0.84

	Sb	TP	Mb	Mp	Un	A[4]	B[4]
SiO2	51.9 -50.9	50.9 -49.9	50.0 -49.0	50.5 -49.6	48.0 -47.3	56.1	48.5
TiO2	1.0 - 0.9	1.1 - 1.0	0.8 - 0.7	0.8 - 0.7	1.1 - 1.0	2.5	2.7
Al2O3	17.7 -16.2	17.0 -15.4	16.3 -14.8	19.0 -17.4	16.6 -15.0	12.4	12.2
Fe2O3	1.5 - 1.3	1.5 - 1.4	1.6 - 1.5	1.4 - 1.3	1.6 - 1.5	1.7	1.8
FeO	7.9 - 8.3	9.0 - 9.4	9.2 - 9.6	7.8 - 8.3	9.0 - 9.4	5.8	8.0
MnO		0.1 - 0.1	0.1 - 0.1	0.1 - 0.1	0.2 - 0.2	0.1	0.2
MgO	7.0 -10.3	8.0 -11.6	8.2 -11.9	7.0 -10.3	7.9 -11.6	9.3	11.3
CaO	9.1 - 8.3	8.6 - 7.8	9.6 - 8.7	9.2 - 8.5	11.2 -10.1	9.5	10.7
Na2O	2.9 - 2.6	2.9 - 2.6	2.6 - 2.4	3.0 - 2.7	2.8 - 2.5	2.0	2.1
K2O	1.2 - 1.1	1.0 - 0.9	1.7 - 1.5	1.5 - 1.4	1.6 - 1.4	0.5	0.5
$Mg/Mg+Fe^{2+}$	0.61- 0.69	0.61 - 0.69	0.61 - 0.69	0.61- 0.69	0.61 - 0.69	0.74	0.68
Olivine[2] removed	3.0 -12.6	8.7 -20.1	5.9 -17.3	5.0 -14.6	4.7 -15.8		
$Mg/Mg+Fe^{2+}$	0.83- 0.86	0.81- 0.84	0.82- 0.85	0.82- 0.85	0.83- 0.85		

[1] Corresponding most mafic lava composition (Table 1) has $Mg/Mg+Fe^{2+}$ greater than that required for equilibrium with olivine, $Mg/Mg+Fe^{2+}$ = 0.84. Calculated compositions given for equilibrium with olivine, $Mg/Mg+Fe^{2+}$ = 0.86 and 0.88 only.

[2] Weight percent olivine to be removed during fractional crystallisation (i.e. surface equilibrium conditions) in order to produce the corresponding most mafic lava composition.

[3] $Mg/Mg+Fe^{2+}$ value of bulk (average) olivine removed.

[4] A and B: Calculated compositions of liquids produced by 28 and 27% melting of pyrolite at 1100°C, P_{H2O} = 10 and 20 kb respectively (Green, 1973).

The results also imply that peridotitic mantle is the major source of magmas in the western Sunda arc. However, spatial variation in K, Rb, Sr, light REE etc. (Whitford, 1975; Whitford & Nicholls, this volume) appears to require an independent source of these elements within the mantle. This source is probably the crustal component of subducted oceanic crust, and movement of the above elements from it to the overlying mantle takes place via hydrous fluids and/or silicic melts (Ringwood, 1974; Nicholls, 1974).

PRODUCTION OF MORE SILICIC LAVA COMPOSITIONS

Representative variation diagrams (SiO_2 and CaO as a function of $Mg/Mg+Fe^{2+}$) for the products of individual volcanoes are given in Figure 5. The four oxides involved are those which vary most widely, and which are most diagnostic in providing evidence for the nature of the crystalline phases controlling differentiation.

In general, SiO_2 is negatively correlated, and CaO positively correlated with $Mg/Mg+Fe^{2+}$. The slopes of the variation curves tend to be steeper for volcanoes belonging to the calcalkaline association (G. Ungaran, Fig. 5D) than for those of the island arc tholeiitic association (G. Galunggung, Fig. 5A), and also tend to increase towards the more silicic products of each volcano. In some cases (e.g. G. Sumbing, Fig. 5C) apparent reversals of the general tendency of $Mg/Mg+Fe^{2+}$ to decrease with increasing SiO_2 occur, in compositions with 58-62 percent SiO_2. In several cases the reversal is associated with pronounced changes in other chemical parameters (toward more potassic compositions in the case of G. Guntur — see Whitford & Nicholls, this volume) and/or the mineralogy of the lavas (from pyroxene- to amphibole-bearing assemblages in both G. Guntur and G. Sumbing). Where pronounced discontinuities occur in the trends of components such as K_2O it is probable that no simple genetic relationship applies throughout the series.

On the assumption that continuous major-element variation indicates a genetic relationship between the lavas of each volcano, an attempt has been made to match the observed trends by removal of one or more of the phases olivine, clinopyroxene, amphibole, plagioclase, and magnetite — those phases which are commonly present as phenocrysts in the lavas — from the lava composition of highest $Mg/Mg+Fe^{2+}$ in each volcano (Table 1), and in some cases, a second composition of lower $Mg/Mg+Fe^{2+}$. The mineral compositions used were those of olivines ($Mg/Mg+Fe^{2+} = 0.80-0.85$), clinopyroxenes (0.82-0.87), and amphiboles (0.66-0.72, on the basis of $FeO/\Sigma FeO = 0.75$) from an experimentally crystallised tholeiitic basalt (Holloway & Burnham, 1972). The olivine compositions are similar to those predicted to be in equilibrium with liquids with $Mg/Mg+Fe^{2+}$ values of about 0.65-0.55 using an Fe^{2+}/Mg partition coefficient, K_D, of 0.30. The other phases used were calcic plagioclase (An_{90-75}), orthopyroxene ($Mg/Mg+Fe^{2+} = 0.81-0.86$), and titanomagnetite with 50 mol. percent ulvöspinel. In each case, a fixed proportion (2, 5 or 10 wt percent) of each phase was removed in turn from the lava compositions under consideration, and the resulting changes in composition were expressed as vectors on the variation diagrams. Because the major element compositions of the natural phenocrysts have been determined by microprobe analysis for only a few of the lavas considered in this paper, a more rigorous approach to the calculation of fractionation trends, such as the least-squares mixing calculations used by Lowder & Carmichael (1970) in their study of the lavas of the Talasea Peninsula, is not possible at present.

In the case of volcanoes whose lavas plot as curves with gentle slopes on the SiO_2 and CaO versus $Mg/Mg+Fe^{2+}$ diagrams, e.g. G. Galunggung (Fig. 5A), the observed trends could have resulted from fractionation of clinopyroxene (probably with small amounts of accompanying olivine) or amphibole. Modal mineralogy of G. Galunggung lavas supports the former alternative, or possibly removal of

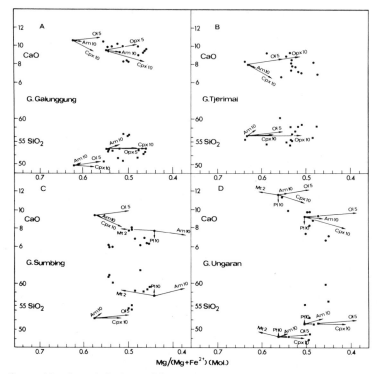

Fig. 5: Compositional trends for lavas of four representative volcanic centres in Java, compared with trends produced by removal of fixed proportions of phenocryst phases from one or more of these lavas. Vectors representing the changes in composition produced are labelled to show the indentity of the phase removed. Ol = olivine. Cpx = clinopyroxene. Opx = orthopyroxene. Am = amphibole. Mt = titanomagnetite. Pl = plagioclase. The proportion removed is given in wt. percent (2, 5, 10).

orthopyroxene and clinopyroxene. Where slopes of variation trends are steeper, e.g. G. Ungaran (Fig. 5D), removal of clinopyroxene or amphibole does not satisfactorily explain the rapid decrease in CaO. It is necessary to introduce a phase or phases which either reduce CaO without reducing $Mg/Mg+Fe^{2+}$ (plagioclase) or which decrease the rate of $Mg/Mg+Fe^{2+}$ change without seriously offsetting CaO decrease (1-2 percent of titanomagnetite can produce the desired effect). In cases where apparent reversals of the $Mg/Mg+Fe^{2+}$ trend occur, e.g. G. Sumbing (Fig. 5C), removal of small amounts of magnetite, along with clinopyroxene or amphiboles, should satisfactorily explain the trends.

Quantitative evaluation of the proposed fractionation schemes in terms of trace element abundances is difficult, as partition coefficients are likely to vary as a function of pressure, temperature, liquid composition, and in some cases oxidation state, and the influence of these variables is poorly known. Qualitatively, removal of plagioclase as proposed above could lead to derivative liquids with negative europium (Eu) anomalies. Lavas of volcanoes in which plagioclase fractionation appears necessary on grounds of major element variation (G. Sumbing, G. Ungaran — Fig. 5) do not show negative Eu anomalies (see Fig. 4). This may simply mean that the quantity of plagioclase removed has been insufficient to produce detectable anomalies. Removal of more than a few percent of titanomagnetite can cause serious depletion of vanadium (V) in derived liquids (Taylor et al., 1969). Andesites and dacites of Java are commonly depleted in V with respect to associated more mafic compositions, a relationship which is in qualitative accord with the idea of fractionation of small amounts of magnetite suggested above.

SUMMARY AND CONCLUSIONS

(a) The dominant types amongst analysed young lavas in the islands of Java and Bali are basaltic andesite and andesite, with a silica mode of about 55 percent. Lavas with around 60 percent SiO_2, considered to be dominant in many island arcs, are relatively uncommon.

(b) Almost all of the analysed lavas have $Mg/Mg+Fe^{2+}$ values too low (less than 0.60) for them to represent unmodified products of the partial melting of peridotitic mantle. However, relatively high contents of heavy REE throughout the spectrum of compositions argue against equilibrium with a garnet-bearing (eclogitic) residue in subducted oceanic crust (and also against high-pressure fractionation of garnet from mafic primary magmas).

(c) If the major primary magmas were basaltic, and were produced by melting of peridotitic mantle enriched to some extent in water, potassium and related trace elements, the removal of small amounts (less than 15 wt percent) of liquidus olivine and Cr-spinel could produce many of the observed range of basalts and basaltic andesites, without affecting trace elements other than Ni and Cr.

(d) Further differentiation to produce andesitic to dacitic compositions probably involves phases — olivine, clinopyroxene, orthopyroxene, amphibole, plagioclase, magnetite — which crystallise at relatively low pressure (less than 10 kb).

ACKNOWLEDGEMENTS

The authors thank M. T. Zen (Bandung Institute of Technology), K. Kusumadinata (Geological Survey of Indonesia), and M. Koesmono (Padjadjaran State University, Bandung) for their assistance with fieldwork, and for the donation of samples. P. H. Beasley, E. H. Pedersen, and N. G. Ware provided technical assistance in petrographic and analytical studies.

REFERENCES

Ben Avraham, Z., & Emery, K.O., 1973: Structural framework of the Sunda Shelf. *Bull. Am. Assoc. Petrol. Geol., 57*, pp. 2323-2366.

Boettcher, A.L., 1973: Volcanism and orogenic belts — the origin of andesites. *Tectonophysics, 17*, pp. 223-240.

Ewart, A., Bryan, W.B. & Gill, J.B., 1973: Mineralogy and geochemistry of the younger volcanic islands of Tonga, S.W. Pacific. *J. Petrology, 14*, pp. 429-465.

Fudali, R.F., 1965: Oxygen fugacities of basaltic and andesitic magmas. *Geochim. cosmochim. Acta. 29*, pp. 1063-1075.

Fujisawa, H., 1968. Temperature and discontinuities in the transition layer within the earth's mantle: Geophysical application of the olivine-spinel transition in the· Mg_2SiO_4 — Fe_2SiO_4 system. *J. geophys. Res., 73*, pp. 3281-3294.

Gill, J.B., 1974: Role of underthrust oceanic crust in the genesis of a Fijian calc-alkaline suite. *Contr. Miner. Petrol. 43*, pp. 29-45.

Green, D.H., 1971: Composition of basaltic magmas as indicators of conditions of origin: application to oceanic volcanism. *Phil. Trans. Roy. Soc. Lond. A, 268,* pp. 707-724.

Green, D.H., 1973: Experimental melting studies on a model upper mantle composition at high pressure under water-saturated and water-undersaturated conditions. *Earth Planet. Sci. Lett., 19,* pp. 37-53.

Green, D.H. & Ringwood, A.E., 1967: The genesis of basaltic magmas. *Contr. Miner. Petrol. 15,* pp. 103-190.

Green, T.H., 1972: Crystallization of calc-alkaline andesite under controlled high-pressure hydrous conditions. *Contr. Miner. Petrol. 34,* pp. 150-166.

Green, T.H. & Ringwood A.E., 1968. Genesis of the calc-alkaline igneous rock suite. *Contr. Miner Petrol. 18,* pp. 105-162.

Harris, K.L., & Nicholls, I.A. (in preparation). Experimental study of the partitioning of selected rare earth elements between garnet, clinopyroxene, amphibole, and basaltic and andesitic melts.

Holloway, J.R. & Burnham, C.W., 1972: Melting relations of basalt with equilibrium water pressure less than total pressure. *J. Petrology, 13,* pp. 1-29.

Jakeš, P. & Gill, J.B., 1970: Rare earth elements and the island arc tholeiitic series. *Earth Planet. Sci. Lett., 9,* pp. 17-28.

Jakeš,,P., & White, A.J.R., 1972: Major and trace element abundances in volcanic rocks of orogenic areas. *Bull. Geol. Soc. Am., 83,* pp. 29-40.

Lowder, G.G. & Carmichael, I.S.E., 1970: The volcanoes and caldera of Talasea, New Britain: Geology and petrology. *Geol. Soc. Am. Bull, 81,* pp. 17-38.

Mysen, B.O. & Boettcher, A.L., 1975: Melting of a hydrous mantle: II. Geochemistry of crystals and liquids formed by anatexis of mantle peridotite at high pressures and high temperatures as a function of controlled activities of water, hydrogen and carbon dioxide. *J. Petrology* (in press).

Neumann van Padang, M., 1951. *Catalogue of the active volcanoes of the world. Part 1, Indonesia.* Int. Volcan. Ass., Italy.

Nicholls, I.A., 1974: Liquids in equilibrium with peridotitic mineral assemblages at high water pressures. *Contr. Miner. Petrol. 45,* pp. 289-316.

Nicholls, I.A & Ringwood, A.E., 1972: Production of silica-saturated tholeiitic magmas in island arcs. *Earth Planet. Sci. Lett. 17,* pp. 243-246.

Nicholls, I.A. & Ringwood, A.E., 1973: Effect of water on olivine stability in tholeiites and the production of silica-saturated magmas in the island arc environment. *J. Geol., 81,* pp. 285-300.

Ringwood, A.E., 1974: The petrological evolution of island arc systems. *J. Geol. Soc. Lond., 130,* pp. 183-204.

Roeder, P.L., & Emslie, R.F., 1970: Olivine-liquid equilibrium. *Contr. Miner. Petrol., 29,* pp. 275-289.

Shaw, D.M., 1970: Trace element fractionation during anatexis. *Geochim. cosmochim. Acta., 34,* pp. 237-242.

Stern, C.R., 1974: Melting products of olivine tholeiite basalt in subduction zones. *Geology,* May 1974, pp. 227-230.

Stern, C.R. & Wyllie, P.J., 1973: Melting relations of basalt-andesite-rhyolite-H2O and a pelagic red clay at 30 kb. *Contr. Miner. Petrol., 42,* pp. 313-323.

Taylor, S.R., Kaye, M., White, A.J.R., Duncan, A.R., & Ewart, A., 1969: Genetic significance of Co, Cr, Ni, Sc and V in andesites. *Geochim. cosmochim. Acta., 33,* pp. 275-286.

Toksöz, M.N., Minear, J.W. & Julian, B.R., 1971: Temperature field and geophysical effects of a downgoing slab. *J. geophys. Res., 76,* pp. 1113-1138.

Whitford, D.J., 1975: Strontium isotopic studies of the volcanic rocks of the Sunda arc, Indonesia, and their petrogenetic implications. *Geochim. cosmochim. Acta. 39,* pp. 1287-1302.

Wyllie, P.J., 1973: Experimental petrology and global tectonics — a preview. *Tectonophysics, 17,* pp. 189-209.

SEISMIC SURVEILLANCE OF VOLCANOES IN PAPUA NEW GUINEA

N.O. MYERS

Geological Survey of Papua New Guinea, Volcanological Observatory,
P.O. Box 386, Rabaul, Papua New Guinea.

ABSTRACT

Several types of seismic instruments are presently in use for volcanological surveillance in Papua New Guinea. At Rabaul, on New Britain, a seismic network of eight telemetry stations has been established in and around the caldera, which permits detection and location of small local seismic events occurring within the caldera. A newly designed radio-telemetred seismograph comprising a remote transmitter and a base station with visual recording, timing, and power equipment has become the standard seismograph system for volcanic surveillance at new observatories, such as those at Mount Lamington and Karkar volcanoes. Remote outstations of the same design are in use at the older observatories at Rabaul and Esa'Ala. These older observatories, and that at Manam Island, still employ standard observatory seismographs with photographic recording. A light-weight portable seismograph based on a visual recorder and a primary cell battery supply has been progressively developed since 1971. Its design is based on the special needs of field seismic studies at active volcanoes. A small, simple tape-recording seismograph presently under development has been used during an eruption of Langila to collect sample records of volcanic tremor bursts for computer frequency analysis. An inexpensive seismic event counter has been developed for registering earthquakes. Its sensitivity is set at about the level of the smallest humanly detectable earthquake. This instrument is intended for installation at volcanoes which are not monitored by permanent seismograph stations.

INTRODUCTION

Regular monitoring of volcanoes in Papua New Guinea started as a result of eruptions of Vulcan (Raluan) and Tavurvur (Matupi) within the caldera at Rabaul in 1937 (Fisher, 1939a) which caused the deaths of several hundred people. Routine observations commenced in Rabaul in late 1937, and a permanent observatory with a seismograph was established in 1940 (Fisher, 1954). Ground investigations at several other volcanoes in Papua New Guinea were also carried out at that time (Fisher, 1939b). The observatory was destroyed during the Second World War, although observations were continued for some time during the Japanese occupation (T. Kizawa, pers. comm.; see also Fisher, this volume).

With the appointment of G.A.M. Taylor as volcanologist, volcanic studies in Papua New Guinea resumed and observations recommenced at Rabaul in 1950. By the end of 1953 the Volcanological Observatory at Rabaul had been re-erected on the pre-war site, and its equipment included a standard observatory Benioff seismograph with a photographic recording system. G.A.M. Taylor also planned two new permanent volcano observatories which were established on Manam Island in 1964 and at Esa'Ala in 1966 (Fig. 1). These installations depended on the observatory technology of that time, and required substantial buildings with underground vaults on solid rock, a full-time electricity supply, and easy access by station personnel. The actual location of a site in relation to the volcano under surveillance necessarily became a compromise. Handicaps lay in the continued use of photographic recording at these observatories, and in the portable Willmore seismograph which was used for field studies.

Seismographic equipment specifically designed for volcano surveillance came into operation in the late 1960s with the installation of the Rabaul harbour network

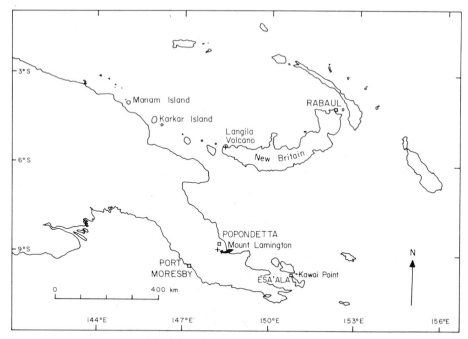

Fig. 1: Locality map of Papua New Guinea.

which had been conceived by G.A.M. Taylor. This used telemetry techniques and a direct-writing recording system. Other new instruments were designed and put into service in the 1970s, including a telemetry system which was employed at new observatories established at Mount Lamington in 1971 and Karkar Island in 1975 (Fig. 1).

This paper aims to give a brief outline of the techniques and equipment designs used in seismic surveillance of volcanoes in Papua New Guinea at the present day. Equipment such as the TM-1 telemetry system, the portable seismograph, the seismic event counter, and the tape seismograph, have been designed and manufactured by the Volcanological Section of the Geological Survey of Papua New Guinea, headquartered in Rabaul, which is responsible for volcano surveillance in the country.

RABAUL HARBOUR NETWORK

Equipment for the original five stations of the network at Rabaul (Fig. 2) was designed by Professor G. Newstead and his associates of the University of Tasmania, who later directed the manufacture and installation of the various units.

Since it was the intention to establish each outstation as a permanent installation, equipment design was made on the same basis. Stations WAN, SUL, RAL, and TAV are connected to the central observatory by underground cable whereas VUL is connected by VHF radio (Fig. 2). Equipment at each outstation includes a Benioff portable seismometer, a seismic amplifier, a sub-carrier modulator, and a remotely controlled stepped attenuator and calibration system.

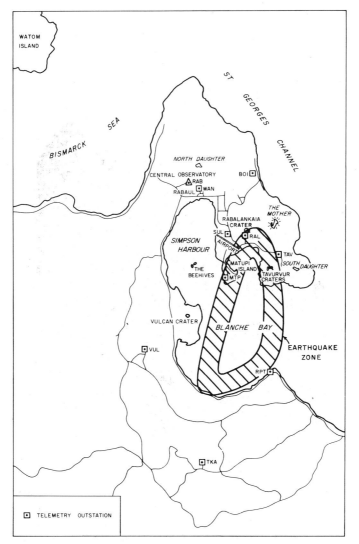

Fig. 2: Rabaul area showing central observatory (RAB), telemetry outstations, roads, and the approximate extent of earthquakes in the harbour area.

Power for each outstation except VUL is provided via cables by regulated supplies at the central observatory; power to VUL was originally supplied by a diesel generating set near the site. Each outstation signal is recorded on a separate Helicorder recorder at the central observatory.

Establishment of the network allowed small earthquakes occurring within the caldera to be located for the first time. However, the distribution of these events was such that many of them lay too far to the south of the network for reliable locations (R.J.S. Cooke, pers. comm., see Fig. 2) as the stations were grouped around the northern perimeter of the caldera.

Fig. 3: Radio-telemetry outstation (TKA) of the Rabaul harbour network, showing seismometer housing in the foreground.

Because of the need for more stations in Rabaul and at other volcanoes in Papua New Guinea, it was decided to undertake development of a new radio telemetry system (see below). This equipment comprised a single-channel sub-carrier system using 27 MHz amplitude-modulated radio equipment for the bearer link. The first new unit was installed in 1970 at TKA (Fig. 2) 16 km south of Rabaul. During 1972 an improved system using phase-modulated VHF radio was designed (see below), and the TKA outstation was re-equipped with the improved equipment (Fig. 3). New outstations at MTP and RPT (Fig. 2) in the Rabaul caldera were installed using the older equipment in 1972, and the SUL outstation was closed. VUL outstation was re-equipped in 1973 with the new equipment. An eighth outstation, BOI (Fig. 2), was opened in 1973, using a line-operated version of the new equipment.

TYPE TM-1 TELEMETRY SYSTEM

The telemetry system used in the original stations of the Rabaul harbour network has a good record of service, but its design makes it unsuitable for installation in many localities where suitable mains power is unavailable; in addition, the scale of costs for stations of the original type was too great for a developing country which hopes to expand operations to cover many unmonitored volcanoes.

Development of a new telemetered seismograph was undertaken for use with either line or radio bearer equipment (see Appendix for technical details). General specifications for a new system required that the seismometer, telemetry transmitter, and associated equipment, be capable of operating remotely for long periods without protective housing, up to 30 km from a recording station located where mains electricity is available. The equipment must also be transportable and easy to install. As it was intended to use direct-writing visual recorders, a wide-band system response, substantially flat to seismic signals, was considered unsuitable because of the poor signal-to-noise ratio caused by microseismic activity in the low-frequency part of the spectrum. It was believed that a bandpass response of 2 Hz to 10 Hz would yield more information because improved signal-to-noise ratio would permit operation at a higher magnification.

The first units were installed at Tanaka (TKA), south of Rabaul, in 1970 (see above), and at Mount Lamington volcano (Fig. 1) in 1971 where the radio receiver and recorder were located at Popondetta about 20 km to the north. The original equipment comprised a single-channel sub-carrier system using HF (27 MHz) amplitude-modulated radio as the bearer link. During 1972, improved equipment using phase-modulated VHF radio was constructed and used to re-equip the two stations.

This equipment was developed further to provide two information channels on a single bearer, to meet the anticipated need for greater dynamic range in the event of high levels of activity during an eruption. The second channel was to carry a strong-motion (i.e. low-sensitivity) seismic signal, which could be selected at the recording station when required. This second channel would also be available in other potential uses for an independent seismic signal, such as that obtained from a second seismometer component or from an incoming telemetry signal which was to be relayed from another location.

Four twin-channel systems have so far been installed, all in the sensitive/strong-motion mode: at Tanaka (TKA) as part of the Rabaul network, at Kawai Point 12 km from Esa'Ala (ESB), and at Mount Lamington (LMG) and Karkar Island (Fig. 1) with satisfactory results.

The sub-carrier equipment described is of simple design, inexpensive, and easily constructed. It is very reliable in operation, and is in fact generally more reliable than associated radio equipment. Where radio is employed, batteries of air-cell type sufficient for one year's operation of the remote equipment are normally installed. The equipment is also versatile, being capable of operation in sensitive/strong-motion mode, as a two-seismometer system, and as a radio relay, and can be adapted to land-line telemetry.

PORTABLE SEISMOGRAPH

A small field seismograph developed in 1970-71 has been progressively improved on the basis of field experience. It is based on a direct-writing recorder, and a primary cell battery supply. The recorder is a Sprengnether R6034 smoked-paper unit with chart speeds of 25, 50, and 100 mm/minute selectable with change gears. The preamplifier is a type developed at Rabaul Observatory. Attenuators provide signal attenuation in 6 dB steps from 0 dB to 54 dB. Other facilities provided are: a relative calibration device to permit magnification to be monitored as the battery supply ages, and a time-signal amplifier for chronographic recording of broadcast radio time-signals. Except for the battery, all units including a Bulova type TE-11 chronometer and a radio time-signal receiver are mounted in one case. A Willmore Mk1 seismometer is normally used with this seismograph.

PORTABLE TAPE-RECORDING SEISMOGRAPH

A small and simple tape-recording seismograph, which is still under development, was designed specifically to record samples of volcanic tremor bursts during an eruption of Langila volcano (Fig. 1) for computer frequency analysis. The instrument is intended for short-term recordings and is not suitable for continuous seismic surveillance. The original equipment consisted of a monoaural portable cassette tape recorder and a Willmore Mk1 seismometer whose output signal is frequency-modulated with a centre frequency of 800 Hz, and recorded on the tape recorder. Demodulated replay of tapes on a Helicorder with a chart speed of 180 mm/minute produced records of higher quality than those that could be obtained on the visual recording portable seismograph described above. The effects of 'wow' and 'flutter' on these records was surprisingly small and was not observable. After continued use, however, tape-speed variations became quite apparent, so a form of electronic compensation was added, with fair success.

The compensation signal is a recorded steady tone of 6700 Hz which on playback is demodulated by an FM discriminator, whose output is correctly phased and mixed with the demodulated seismic signal, which then cancels the greater proportion of the wow and flutter component of the seismic signal.

SEISMIC EVENT COUNTER

An inexpensive event counter has been developed for volcanic surveillance. It is intended to count 'felt' earthquakes at places where there is no seismograph. The trigger sensitivity is set at about the level of the smallest humanly detectable earthquake; thus, checks may be made of 'felt' events without the loss of data which inevitably occurs with human observers because of their different states of movement, alertness, wakefulness, etc. The events registered on the counter are classified as probably volcanic or probably tectonic on the basis of recordings at the closest seismograph stations.

The sensor is an omni-directional pendulum with an adjustable period from 0.4 to 0.8 seconds. The indicator is a five-digit electro-mechanical counter, modified for operation on a 12-volt supply. The electronics are minimal, comprising a timer and relay driver stage. Once triggered, the timer is inhibited until resetting after an interval of approximately 50 seconds, so that only one unit per earthquake will be counted in all but exceptional cases.

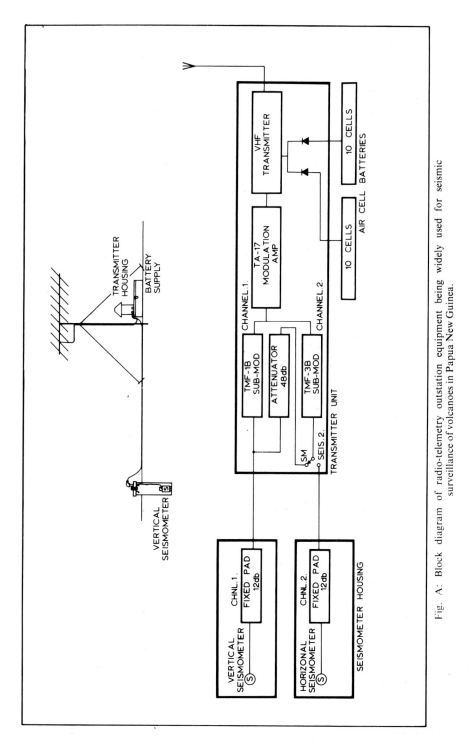

Fig. A: Block diagram of radio-telemetry outstation equipment being widely used for seismic surveillance of volcanoes in Papua New Guinea.

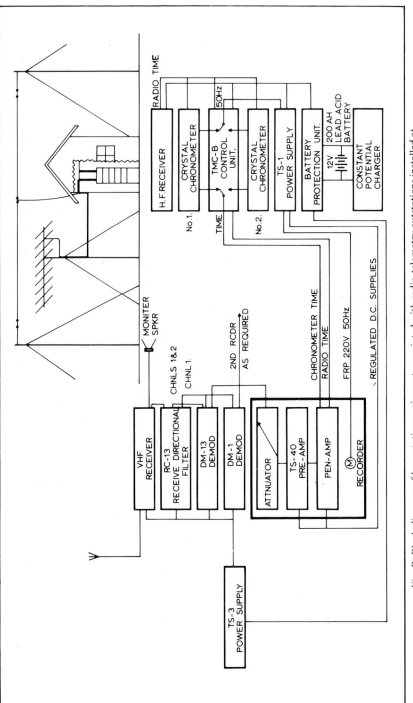

Fig. B: Block diagram of base station equipment associated with radio-telemetry outstations installed at Mount Lamington, Kawai Point, and Karkar Island.

ACKNOWLEDGEMENTS

The author thanks M. Tiria for drafting some of the illustrations, R.J.S. Cooke for reading the manuscript, and E. Kent for typing the manuscript. The Chief Government Geologist of the Geological Survey of Papua New Guinea authorised publication of this paper.

REFERENCES

Fisher, N.H., 1939a: Geology and volcanology of Blanche Bay etc. *Terr. New Guinea Geol. Bull. 1.*
Fisher, N.H., 1939b: Volcanoes of the Territory of New Guinea. *Terr. New Guinea Geol. Bull. 2.*
Fisher, N.H., 1954: Report of the sub-committee on volcanology 1951. *Bull. Volc. 15,* pp 71-79.

Appendix. *Technical description of the TM-1 seismograph system*

The following description applies to the sensitive/strong-motion mode of operation of the Type TM-1 twin-channel telemetry system. The two channels of seismic information are derived from the output of a vertical seismometer (Willmore Mk2 or Geotech S13) by splitting the signal, and applying 48 dB additional attenuation to the strong-motion channel (Fig. A). The two seismic signals are amplified and used to frequency-modulate voltage-controlled oscillators to produce sub-carrier signals centered on 290 Hz for channel 1 (sensitive) and 1960 Hz for channel 2 (strong motion). The sub-carriers are then mixed for transmission on a single bearer.

At the recording station the mixed sub-carrier signals are separated by means of directional filters (Fig. B). A switch enables selection of either channel for demodulation and recording. Provision is also included for use of a second demodulator and recorder when two seismometers are employed. A preamplifier and a stepped attenuator are interposed between the demodulator and recorder. The recorder is a Sprengnether R6038 smoked-paper unit which can be operated at chart speeds of 60, 120, and 240 mm/minute for recording periods of 24, 12 and 6 hours.

The sensitivity and range of this system is adequate for most purposes. With all attenuation removed (that is, the seismometer attenuator fixed pad reduced to 0 dB; Fig. A), and the recorder attenuator and amplifier gain settings for maximum amplification (see Fig. B), a seismic signal of about 4 μV peak-to-peak amplitude at 2.5 Hz at the seismometer output, will write a record of 15 mm peak-to-peak amplitude. As the maximum signal which can be applied to the sub-carrier modulator is 3 mV peak-to-peak, the dynamic range is approximately 58 dB. Therefore when used in the strong-motion mode (additional 48 dB attenuation) a total range of 106 dB is available.

When the system is adapted to land-line telemetry the equipment is modified by replacing the modulation amplifier (see Fig. A), with a low-power line amplifier. In this installation all units except the battery are mounted within the seismometer housing to make a very compact unit. Power requirements are small: approximately 0.4 W, which permits the use of a 12-volt battery comprising medium-size Leclanche cells arranged in series-parallel, sufficient for 6 months' supply. The power consumption of a remote outstation using radio is approximately 2.6 W (0.22 A at 12-volt).

LATE CAINOZOIC VOLCANISM AND PLATE TECTONICS AT THE SOUTHERN MARGIN OF THE BISMARCK SEA, PAPUA NEW GUINEA

R. W. JOHNSON

Bureau of Mineral Resources, P.O. Box 378, Canberra City, A.C.T. 2601

ABSTRACT

Two Late Cainozoic volcanic arcs can be recognised at the southern margin of the Bismarck Sea, Papua New Guinea. Both arcs provide striking examples of the geodynamic complexity to be expected in regions characterised by small plates whose instantaneous poles of rotation are nearby. A western arc is associated with the boundary between the South Bismarck and Indo-Australian plates. The chemical compositions of its volcanic rocks change along the arc — i.e., in a direction parallel to the strike of a postulated subducted lithospheric slab. These changes can be explained by identifying Late Cainozoic poles of rotation in the northwestern part of mainland Papua New Guinea, and by postulating eastwardly increasing rates of plate convergence. An eastern volcanic arc is associated with the boundary between the South Bismarck and Solomon Sea plates. The volcanoes are arranged in an unusual zig-zag pattern, and the compositions of the volcanic rocks change with increasing depths to the northward dipping New Britain Benioff zone — i.e., in a direction at right-angles to the strike of the Benioff zone and to the axis of the New Britain submarine trench. The existence of a thrust slice in the northwestern corner of the Solomon Sea is postulated to account for the distribution pattern of the eastern-arc volcanoes.

INTRODUCTION

Petrological studies of island-arc volcanoes have been strongly influenced by the theory of plate tectonics which proposes that lithospheric plates are subducted beneath the arcs (e.g. Wyllie, 1973). Recent studies favour the concept that island-arc magmas are formed above the downgoing slabs by partial melting of upper mantle peridotite under hydrous conditions. Partial melting of the peridotite is thought to take place through access of volatiles produced by dehydration of the crustal portions of the slabs, or of volatile-rich siliceous melts generated by partial fusion of the slabs (e.g. McBirney, 1969; Green & Ringwood, 1972; Wyllie, 1973; Stern & Wyllie, 1973; Nicholls & Ringwood, 1973; Ringwood, 1974). An understanding of the occurrence, periodicity, and chemistry of volcanism in any island-arc would therefore appear to be critically dependent upon the formulation of a geodynamic plate model.

This paper proposes a plate model for the 1000 km-long chain of volcanoes that borders the southern margin of the Bismarck Sea (Fig. 1). It summarises conclusions reached in another report, which includes a more complete presentation and discussion of petrological data (Johnson, in prep.).

On a global scale, Papua New Guinea straddles part of the boundary between the Indo-Australian and Pacific places plates (Le Pichon, 1970; Denham, 1973). In detail, however, the configurations of plate boundaries are more complicated, and three independent studies suggested that the Late Cainozoic geodynamics of the region has been greatly influenced by at least two minor plates, trapped between the two larger ones (Johnson & Molnar, 1972; Curtis, 1973; Krause, 1973).

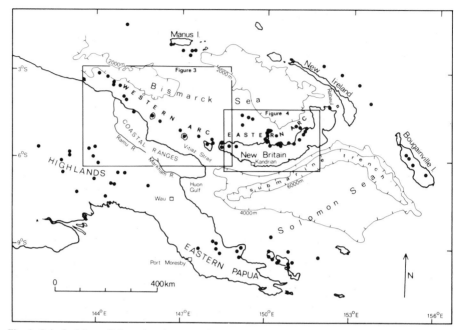

Fig. 1: Principal Late Cainozoic volcanoes of Papua New Guinea (solid dots), including those of the 'western' and 'eastern' arcs shown in Figs. 3 and 4, respectively. Squares are settlements (the site of Rabaul township is a volcano).

VOLCANO DISTRIBUTION

The Late Cainozoic volcanoes of the south Bismarck Sea region are associated with the southern margin of the well-defined minor plate that underlies the southern part of the Bismarck Sea. Following the terminology of Johnson & Molnar (1972), this plate will be called 'South Bismarck' (Fig. 2). Its southern margin is adjacent to the Indo-Australian plate in the west, and to the 'Solomon Sea' plate in the east. The south Bismarck Sea volcanoes are therefore associated with two plate boundaries, and although it is convenient to refer to all the volcanoes as the 'Bismarck volcanic arc' (e.g. Johnson et al., 1973; Cooke et al., this volume), it is proposed here that *two* volcanic arcs should be recognised — each one corresponding to a different plate boundary and to different geodynamic events during the Late Cainozoic. On the basis of chemical composition (see later) and distribution, the two volcanic arcs are:

1. A *western* arc, extending from the Schouten Islands, in the west, to the Cape Gloucester area of west New Britain, in the east (Fig. 3). These volcanoes are considered to be associated with a contact zone between the South Bismarck and Indo-Australian plates, whose nature is discussed below.

2. An *eastern* arc, which includes the volcanoes of the Witu Islands, Willaumez Peninsula, and north coast of central New Britain (Fig. 4). Discussion below proposes that these volcanoes are associated with the South Bismarck/Solomon Sea plate boundary, although this apparently simple correlation is shown to be complicated by a possible thrust slice in the northwestern corner of the Solomon Sea.

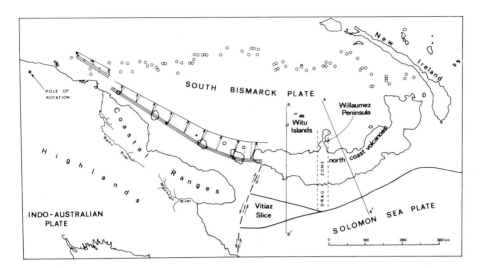

Fig. 2: Tectonic features at the southern margin of the Bismarck Sea. Open circles represent epicentres of earthquakes that define the northern margin of the South Bismarck plate (after Denham, 1973). This northern margin is thought to continue southeastwards and reach the Solomon Sea plate at a triple junction south of New Ireland (see: Johnson & Molnar, 1972; Curtis, 1973: Krause, 1973). The solid lines marking the margins of the Solomon Sea plate and Vitiaz slice are the axes of submarine trenches south of New Britain (cf. Fig. 1). The western ends of these trenches are visualised as terminating at a transcurrent plate boundary (curved dashed line) the northern end of which is thought to mark the eastern end of the near-vertical slab beneath the western arc, and which may have been responsible for the apparent offset between the coastal ranges and New Britain in Tertiary times. This postulated plate boundary is drawn as the northward extension of the zone of minor seismicity found in eastern Papua (e.g. Milsom, 1970), but is located with some reservation as there is no direct evidence for its existence on the floor of the northwestern Solomon Sea (in fact, the boundary may be a *zone* of deformation, and its position could have changed throughout the Late Cainozoic). The three closely spaced lines north of the coastal ranges are hypothetical dimensionless strike lines for the upper surface of the postulated northward-dipping slab beneath the western arc. The set of arrrows illustrates the effect of plate convergence about a single position for a series of instantaneous poles of rotation (other positions in the northwest part of mainland Papua New Guinea have similar effects) — i.e.. that those points at the upper surface of the downgoing slab farther from the pole move greater distances for any period of rotation about the pole; in other words, rates of plate convergence increase away from the pole. The straight lines accompanying each arrow are normal to the strike lines. Note that because of the differences in orientation of the strike lines on either side of Long Island, the transcurrent components of movement for this particular pole position change from right-lateral west of the island to left-lateral east of it. A-A' and B-B' show locations of cross-sections in Figs 7A and B, respectively. See text for explanation of 'drag zone'.

The volcanoes of the Rabaul area (Fig. 1) may also be associated with the South Bismarck/Solomon Sea plate boundary, but they are considered to be distinct from the eastern arc and are therefore not considered here (see Heming, 1974). Neither does this paper deal with the volcanic centres of Andewa and Schrader in west New Britain (Fig. 4), and with Uvo in the Adelbert Range (Fig. 3), as they appear to be older (at least in part) than the other volcanoes, and are not sufficiently well surveyed. It is emphasised that a study of the petrology and ages of rocks from Andewa, Schrader, and Uvo could necessitate some revision to the geodynamic interpretations given below.

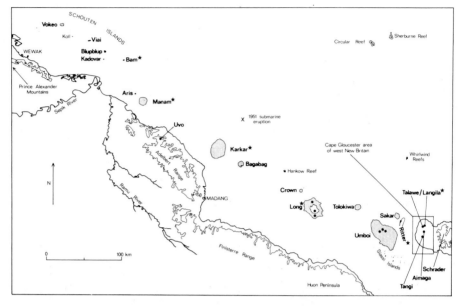

Fig. 3: Principal volcanoes of the western arc. Islands are shown stippled or solid, and volcanic islands are indicated by bold lettering. Filled circles are prominent volcanic centres on mainland Papua New Guinea and New Britain, and on Long and Umboi Islands. Stars indicate volcanoes which have erupted during the last one-hundred years. Langila is a cluster of satellite craters on the eastern flank of Talawe. The outlines of Schrader, and of the coastal ranges of mainland Papua New Guinea, are shown by an approximate 330 m contour.

GEOLOGY OF THE WESTERN ARC

Description

The volcanoes of the western arc form an extremely narrow zone (Fig. 3). The volcanoes are close together at the eastern end of the arc, but farther west they are more widely separated: the largest gap is 120 km — between Manam and Karkar Islands. In addition, a comparison of the volumes of the volcanoes shows that the Schouten Islands, in the extreme west, are smaller than most other western arc volcanoes. These relationships suggest therefore that, in broad terms, the volume of volcanic rocks decreases from east to west along the western arc.

Three other features shown in Figure 3 are also noteworthy: (1) the volcanoes of the Schouten Islands form a chain which is offset about 25 km northeastwards from the extension of the line that can be drawn between Aris, Manam, Karkar, and Bagabag Islands; (2) Long Island, Crown Island and Hankow Reef, and Tolokiwa Island and the long axis of Umboi Island constitute two parallel lines which intersect the general trend of the western arc at about 30°; and (3) in the Cape Gloucester area of west New Britain, Tangi, Aimaga, and Talawe volcanoes form a north-south line. As shown in Figure 3, six subaerial volcanoes and one submarine centre have had eruptions recorded by literate observers during the last one-hundred years.

About 500 rocks from the western arc volcanoes have been studied in thin section by the writer. In addition, 143 major-element chemical analyses have been obtained, including those published by Morgan (1966). These data indicate that the most common rock types in the western arc are hypersthene-normative basalts and low-silica andesites (containing less than 57 wt percent SiO_2). High-silica andesites are also present, but dacites (more than 62 percent SiO_2) are rare and rhyolites (more than 70 percent SiO_2) have not been found. Basalts appear to be absent from the Schouten Islands.

The chemical analyses show that pronouced changes in chemistry take place along the western arc. For example, in Figure 5, alkali contents are shown to increase generally from zone A, in the extreme west, to zone B, comparing rocks of the same silica content; they increase abruptly to maximum values in the rocks of zone C, but *decrease* in those of zone D, most of which nevertheless contain higher total-alkali contents than most zone rocks. In other words, the lowest total-alkali contents are shown by rocks from the Scouten Islands (comparing rocks with the same silica content), and the highest values are those of rocks from Long and Tolokiwa Islands. Similar maxima and minima are shown when the following oxides (and oxide ratio) are plotted against silica — K_2O, Na_2O, K_2O/Na_2O, TiO_2, P_2O_5.

The compositions of these rocks are the same as those of many hypersthene-normative volcanic rock suites associated with active island-arcs in many parts of the world. By analogy, and because of the high frequency of associated intermediate-focus earthquakes, it can be inferred that the western arc has developed at a convergent plate boundary where subduction has taken place. The supporting evidence, however, is incomplete as a Benioff-zone-like feature is associated with only part of the arc, and because a submarine trench appears to be absent.

Studies of intermediate-depth earthquakes recorded during the last 10 to 20 years show: a pronounced concentration of epicentres in the strait between Long Island and the north coast of mainland Papua New Guinea; a decrease in the density of these epicentres to the east and west of Long Island; and an apparent absence of intermediate earthquakes beneath the volcanoes west of Karkar Island (e.g. Denham, 1969; Johnson et al., 1971; Curtis, 1973). The intermediate-depth earthquake foci east of Karkar Island define a steeply inclined northward-dipping seismic zone, which extends from depths of about 120 km down to more than 200 km and whose dip increases down-dip to more than 80° beneath the arc. Focal mechanism solutions for a few of these earthquakes give steep, dip-slip underthrust senses of motion (Ripper, 1975).

Many shallow- and intermediate-depth earthquakes take place beneath the coastal ranges of mainland Papua New Guinea to the south, but these do not appear to define a Benioff zone (Johnson et al., 1971). Rare intermediate-focus earthquakes beneath the Highlands-Wau region (Fig. 1) suggest the presence of lithospheric material at these depths, but records are required for a much longer time span before these earthquakes become sufficiently numerous to clearly define any possible inclined seismic zone (see also Fig. 6 and discussion below). There is a pronounced deficiency of intermediate-depth events beneath the Adelbert Range, opposite the gap in the volcanic arc between Manam and Karkar Islands (R. J. S. Cooke, pers. comm., 1974).

In contrast to the eastern arc, which is associated with the New Britain submarine trench (see below), the available bathymetry suggests that no trench-like feature is assocated with the western arc. Unpublished seismic reflection profiles obtained by the Bureau of Mineral Resources in 1970 revealed the presence of extremely thick sedimentary accumulations between the western arc and the north coast of the mainland and north of the western arc, but it is doubtful if either of these can be regarded as an infilled submarine trench (see also Connelly, 1974).

Discussion

Dewey & Bird (1970) described the mainland of Papua New Guinea south of the western arc as a probable example of a region where a continent had collided with an island-arc. They suggested that this collision caused a subduction zone to change its polarity from an original northward dip beneath the island-arc, to a southward one — i.e., the Benioff zone 'flipped' (cf. McKenzie, 1969). Johnson & Molnar (1972), Karig (1972), and Hamilton (1972) all supported this interpretation. The geology of the region lends support to the interpretation that the Adelbert and Finisterre Ranges and Huon Peninsula (i.e. the coastal ranges, Figs 1, 2) represent a welt of island-arc rocks on the northern edge of the continental mass beneath the Highlands of Papua New Guinea (Thompson & Fisher, 1957; BMR, 1972; Bain, 1973; Robinson et al., 1974). However, there is little support for the concept of a reversal of arc polarity.

Firstly, as discussed above, it has yet to be clearly established that mainland Papua New Guinea is underlain by a single, southward-dipping Benioff zone at the present day. In addition, it is unlikely that a southward-dipping seismic zone existed beneath the coastal ranges and Ramu-Markham Valleys (Figs 1, 2) in earlier parts of the Quaternary. Worldwide observations of many island-arcs and continental margins indicate that volcanoes overlie those parts of inclined Benioff zones which are deeper than 70-100 km (e.g. Dickinson, 1970). If a southward-dipping Benioff zone existed beneath the northern part of mainland Papua New Guinea, Quaternary volcanoes would be expected in, and immediately south of, the Ramu and Markham Valleys, and not in a narrow zone north of the coastal ranges. (The Quaternary volcanoes of the Highlands region, Fig. 1, do not appear to be directly related to a downgoing slab — see Mackenzie, this volume).

The claim of Johnson & Molnar (1972), Curtis (1973), and Krause (1973) that the Ramu and Markham Valleys represent the present-day boundary between the Indo-Australian and South Bismarck plates is also open to question. Firstly, there is no concentration of earthquakes beneath the valleys. Rather, the epicentres are spread more or less evenly over a wide zone covering the coastal ranges, the Ramu and Markham Valleys, and the northern foothills of the Highlands region (see, for example, Denham, 1969). Secondly, extensive Holocene faulting has not been reported from the Ramu and Markham Valleys, and there is no evidence that Holocene faults are more common in these valleys than they are in other parts of the wide zone of present-day earthquake acitivity (BMR, 1972). The Ramu and Markham Valleys may represent the site of a *former* plate boundary which was destroyed by a continent/island-arc collision (see below), but the seismic and geological evidence does not demonstrate that the valleys necessarily represent a *present-day* plate boundary.

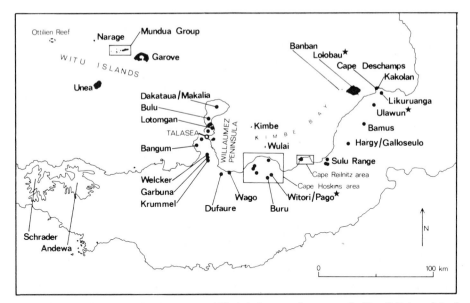

Fig. 4: Principal volcanoes of the eastern arc. Filled circles are volcanoes on the New Britain mainland. Island volcanoes are shown solid. Dakataua, Witori, and Hargy are volcanoes containing calderas within which the volcanoes Makalia, Pago, and Galloseulo, respectively, have been built. 330 m-contour shows the outline of Schrader and Andewa volcanic complexes. Stars indicate volcanoes which have erupted during the last one-hundred years (Makalia and Bamus have probably also erupted during this period).

GEOLOGY OF THE EASTERN ARC

Description

Unlike the western arc, the eastern arc of the New Britain region appears, on first inspection, to be part of the more familiar kind of island-arc. The arcuate mountainous axis of New Britain island is a geanticlinal ridge composed mainly of Tertiary volcanic and intrusive rocks and limestones similar in type and age to those of the mainland coastal ranges (Ryburn et al., in prep.; BMR, 1972). A submarine trench showing a maximum depth of more than 8000 m parallels the island off the south coast (Fig. 1). A Benioff zone dips northwards at about 70° beneath New Britain, and shows earthquake foci up to 565 km deep (Denham, 1969; Johnson et al., 1971; Curtis, 1973); the Quaternary volcanoes have been built up on the northern concave side of the island over the deeper parts of the Benioff zone (Fig. 4). The thickness of crust beneath New Britain ranges from about 25 km in the north to about 45 km along the south central coast (Finlayson & Cull, 1973; Wiebenga, 1973).

Despite these familiar island-arc characteristics, two features of the geology are unusual. Firstly, the distribution of the volcanoes is anomalous. Instead of forming continuous chains parallel to the entire lengths of the axes of New Britain and the submarine trench, the volcanoes show a pronounced zig-zag distribution pattern (Fig. 4). The zig-zag consists of (1) the east-west zone of the Witu Islands (excluding Unea Island), (2) the north-south chain of volcanoes on Willaumez Peninsula, and (3) the zone of prominent volcanoes along the north central coast of New Britain, between Willaumez Peninsula and Likuruanga (Fig. 4).

Secondly, the bathymetry of the northwestern corner of the Solomon Sea shows that west of 150°E the New Britain trench shallows and splits into two branches (e.g. Krause et al., 1970; Mammerickx et al., 1971). A northern branch extends into Vitiaz Strait, and a southern one trends towards Huon Gulf. An unpublished BMR seismic reflection profile run south from near Kandrian (Fig. 1), shows that the southern branch is deeper than the northern one, and contains a thicker sedimentary sequence.

These two anomalous features show some noteworthy correlations (Fig. 2). Firstly, the Witu Islands zone lies north of the double-trench feature west of 150°E. Secondly, the trend of Willaumez Peninsula passes southwards through the point where the New Britain trench splits into two. Lastly, the north central coast volcanoes lie north of the deeper, single-axis part of the New Britain trench.

About 700 volcanic rocks from the eastern arc have been examined in thin section by the writer. Including the chemical analyses published by Lowder & Carmichael (1970), Johnson et al. (1972), and Blake & Ewart (1974), a total of 205 eastern arc rocks have been analysed for major elements. As in the western arc, low-silica andesite is common in the eastern arc. However, unlike the western arc rocks basalts are much less common, high-silica andesites and dacites are more common, and rhyolites are present, although rare. Except for these differences in the relative abundance of rock-types, the rocks of both the eastern and western arcs are chemically similar; all of them are hypersthene-normative.

Whereas the western arc shows systematic changes in rock compositions *along* the volcanic chain, the eastern arc volcanoes show changes *across* the arc — that is, in a direction at right-angles to the axis of the submarine trench, and concomitant with changing depths to the underlying Benioff zone. In Figure 5 it can be seen that in rocks with the same silica content, total-alkali contents increase progressively northwards between zone E and zone H through the intermediate zones F and G (note also that rocks from the northern part of Willaumez Peninsula differ in composition from those in the southern part). This progressive increase in alkali contents is the same as those identified in many island-arcs as distance from the submarine trench and depth to the Benioff zone increase (e.g. Rittmann, 1953; Kuno, 1959). Johnson (in prep.) shows that K_2O and Na_2O (independently), and TiO_2 and P_2O_5, also generally change northwards in the eastern arc.

Discussion

The New Britain island-arc is widely accepted as a region where the Solomon Sea and South Bismarck plates converge, the former being consumed along the New Britain trench (Isacks & Molnar, 1971; Johnson & Molnar, 1972; Curtis, 1973; Krause, 1973)*. However, if plate convergence is taking place in this region, and the Solomon Sea plate is disappearing beneath New Britain, the branching of the submarine trench west of 150°E indicates that subduction cannot be regarded as a simple case of one plate disappearing beneath another along a single line. In addition, the correlation proposed above between the submarine features south of New Britain island and volcano distribution north of it suggests that the 'double trench' feature west of 150°E has exerted a pronounced effect on magma genesis beneath New Britain.

*Wiebenga (1973) presented an alternative interpretation that the New Britain trench is a tensional feature.

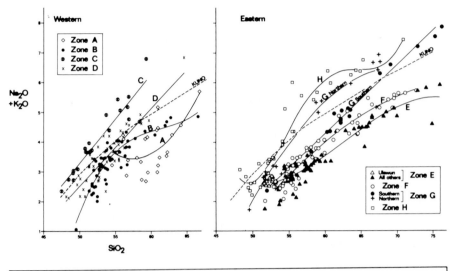

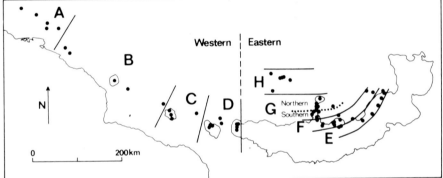

Fig. 5: Weight percent Na2O + K2O *versus* SiO2 diagrams (upper) for rocks from western and eastern arcs, using the division of volcanoes into zones A to H shown in the lower diagram. The dashed line labelled 'KUNO' in the upper diagrams separates the 'high-alumina basalt' (upper) and 'tholeiitic' compositional fields of Kuno (1959). Solid lines A to H are regression lines for points representing rocks from each of the zones A to H, respectively. The regression lines were obtained using a program which calculates a progressive series of regression lines, increasing in degree of the equation (linear, quadratic, cubic, etc.), until a line of best fit is obtained (see Johnson, in prep.); a measure of the closeness of fit of points to the corresponding regression line is given by the following values which are nearer unity where the closeness of fit is greater: zone A, 0.40; zone B, 0.79; zone C, 0.88; zone D, 0.81; zone E, 0.89; zone F, 0.93; zone G (Southern), 0.99; zone G (Northern), 0.95; zone H, 0.98. The rocks of Ulawun, a zone E volcano, are chemically more similar to the rocks of zone F than to those of zone E.

SPECULATIONS ON PLATE KINEMATICS

Rates of slab descent

In this section, some speculative interpretations are presented on the development of the western and eastern arcs, based on the initial conclusion that each one is underlain by a subducted lithospheric slab. It is envisaged that these slabs are colder than the mantle material through which they have descended, and that they are capable of extracting heat from the surrounding mantle (e.g. Oxburgh & Turcotte, 1970; Minear & Toksöz, 1970a,b; Griggs, 1972; Le Pichon et al.,

1973). Furthermore, it is accepted that, all other factors being more or less equal, rates of descent will determine the effectiveness of the slabs as 'heat sinks' — that slowly descending slabs will heat up more readily than quickly descending ones with the same composition and thickness. Thus, rates of slab descent will govern the thermal regimes (the shapes of isotherms) of the slabs and of those parts of the upper mantle above the slabs where magmas are believed to form. In addition, if intermediate- and deep-focus earthquakes take place because of mechanical failure in the cold interiors of the downgoing slabs, deeper earthquakes will take place more readily in those parts of the slabs which have remained colder to greater depths — that is, in slabs which have been subducted more rapidly.

Development of the western arc

The events which are thought to have led to the formation of the western arc are summarised in Figure 6. The most important points to note are: (1) that a Miocene continent/island-arc collision destroyed the plate boundary which had existed south of the Adelbert-Huon island-arc; (2) that post-Miocene plate convergence and foreshortening caused the uplift of the Highlands and the associated welt of island-arc rocks (the coastal ranges); and (3) caused the steepening of the subducted slab which continued to descend beneath the South Bismark plate, resulting in the Quaternary volcanism of the western arc. It is also important to note that no single line at the Earth's surface is identified as the present-day plate boundary in this region; rather, the contact between the Indo-Australian and South Bismarck plates is visualised as a broad zone of foreshortening characterised by earthquakes which show compressional focal mechanism solutions, but which beneath mainland Papua New Guinea are not concentrated in a single, well-defined inclined seismic zone.

The 3-dimensional form of the subducted slab beneath the entire western arc cannot be deduced directly because of the apparent absence of intermediate- and deep-focus earthquakes west of Karkar Island. However, the distribution pattern of the volcanoes may be taken as circumstantial evidence for establishing the general shape of the slab throughout the Late Cainozoic.

The three closely-spaced parallel lines in Figure 2 are hypothetical strike-lines for the upper surface of the subducted slab. They are drawn parallel to the general trend of the western arc, and are dimensionless, although they might correspond with depths between about 150 and 250 km. Two features are noteworthy. Firstly, the lines are straight between the Schouten Islands and Long Island, but east of Long they curve towards the east (Fig. 2). Secondly, the offset of the Schouten Islands line suggests that a comparable displacement may exist in the underlying slab.

Using this plate configuration, and accepting the proposal of Krause (1973) that the instantaneous poles of rotation for the Indo-Australian and South Bismarck plates could be in the northwestern part of mainland Papua New Guinea, it can be shown that for any finite period of time the magnitudes and directions of the relative velocity vectors change progressively along the contact of the two plates at depth. For example, using the pole position in Figure 2, rates of plate convergence increase regularly between the Schouten Islands and the Cape Gloucester area because of the increasing distance from the pole. The direction of

Fig. 6: Simplified scheme for the development of the western arc. Sketches 1 and 2 are based largely on the geological history given by Robinson et al. (1974). The sketches are highly diagrammatic and for the sake of clarity they ignore the complex geological events that took place at the northern edge of, and within, the continental mass (heavy stipple) now represented by the Highlands-Wau region (cf. Fig. 1). These events include, for example, the emplacement of the Marum Basic Belt, extensive magmatism in Middle Miocene times, and widespread volcanism in the Quaternary (see, for example: Page & McDougall, 1970; BMR, 1972; Mackenzie & Chappell, 1972; Bain, 1973; Mackenzie, this volume). At least· some of the events are possibly related to underthrusting at the northern edge of the continent, as shown in sketches 2 and 3 by the dotted line (cf. Fig. 6 of Johnson & Molnar, 1972; present-day earthquake activity gives little indication of the existence of underthrust lithosphere beneath the Highlands-Wau region, except for a few intermediate-depth earthquakes in the Wau area).

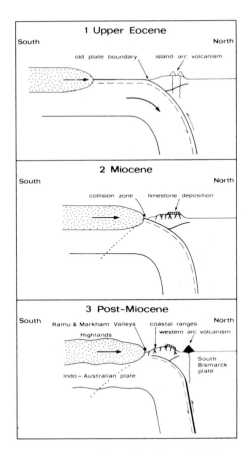

1. During the Upper Eocene and Oligocene the leading edge of the northward-moving Australian continent, carried on the Indo-Australian plate, moved towards a zone of northward subduction where oceanic crust was consumed beneath an Adelbert-Huon island-arc (light stipple). The 'Finisterre Volcanics' (Robinson et al., 1974) are believed to represent island-arc magmas related to the downgoing portion of the slab.

2. By Lower Miocene times it is envisaged that the leading edge of the continent had collided with the island-arc, and that subduction and island-arc volcanism had effectively ceased. Plate convergence continued, however, and movements were taken up on faults in the collision zone (represented by the present-day Ramu and Markham Valleys), on block-faults in the island-arc, and on faults at the northern edge of the continental mass. Limestones were deposited on fault blocks in the northern part of the island-arc in the Lower and Middle Miocene and, to a lesser extent, in the Upper Miocene.

3. Convergence continued throughout post-Miocene times, resulting in the uplift, warping, and foreshortening of the continental mass and its new welt of island-arc rocks (the coastal ranges). The subducted slab beneath the old island-arc was steepened, and movement of the Indo-Australian plate beneath the South Bismarck plate continued, giving rise to Quarternary volcanism in the western arc (see text).

plate convergence will not, however, be at right-angles to the strike lines given in Figure 2, but will contain transcurrent components of motion. West of Long Island this transcurrent component will be right-lateral, but east of Long Island it will change to left-lateral, as distance from the pole of rotation increases, because of the change in orientation of the strike lines. Similar changes in the rate of plate convergence (and the sense of transcurrent motion) can be proposed using many other pole positions in the northwestern part of mainland Papua New Guinea.

This interpretation is consistent with the known distribution of earthquakes beneath the western arc. Intermediate-focus earthquakes are common east of

Karkar Island, perhaps because rates of plate convergence are greater there and cold lithosphere has been subducted to greater depths than in the west. In addition, because the concentration of earthquakes south of Long Island coincides with the change in orientation of the strike lines, it is possible these earthquakes are due to high stresses set up where the slab is strongly bent. Rates of plate convergence are lowest in the west, where subduction may be so slow that the downgoing slab is no longer sufficiently cold to produce intermediate-focus earthquakes when it reaches these depths.

If rate of subduction is a primary influence on the thermal regimes of the slab and the overlying upper mantle (see above), the conclusion that rates of plate convergence change progressively along the western arc has important consequences for theories of magma genesis. Because of the changes in convergence rates, the patterns of isotherms may be different in all sections drawn perpendicular to the western arc. Slab-derived water and melts may therefore rise from the downgoing slab at different depths in each section, and may lead to partial melting at different depths in the overlying mantle. Magmas beneath each part of the arc may therefore be generated under unique sets of conditions of pressure, temperature, contents of volatiles, composition and volume of the slab melts (which may mix with the mantle-derived magmas), etc. The possible interactions between these variables are, of course, so complex that it is impossible to elucidate them satisfactorily at the present time. Indeed, additional factors may contribute to even greater complexities — e.g., compositional heterogeneities in the slab and upper mantle, lateral changes in thickness of the slab, lateral differences in thermal properties of the slab and overlying mantle, and changes in the positions of the instantaneous poles of rotation throughout the Late Cainozoic. However, it is proposed that in broad terms the changes in rock compositions along the western arc are functions of differences in the thermal regimes beneath each part of the arc, and that these, in turn, depend upon differences in rates of plate convergence and subduction along the western arc.

Finally, it is noted that these interpretations are consistent with the apparent increase in the volume of subaerial volcanic rocks from west to east along the western arc. In the east, the volumes of slab-derived water and melts may be greater than those produced from the slab in the west because, during any finite period of time, a greater supply of subducted, water-rich crustal rocks will be made available for dehydration or fusion. In the east, these larger volumes are likely to induce more extensive partial melting of mantle peridotite compared with that in the west, and possibly therefore lead to the eruption of a proportionally greater volume of magma.

Development of the eastern arc

A cross-section through the eastern arc east of 150°E shows all the familiar features of an island-arc (Fig. 7A). The volcanoes (zones E and F of Fig. 5) overlie the shallow part of the New Britain Benioff zone, and their rocks are low in alkalis (Fig. 5), corresponding to rocks called 'tholeiitic' by others (e.g. Tilley, 1950; Kuno, 1959; Jakeš & Gill, 1970). These rocks may have originated in the manner proposed by Nicholls & Ringwood (1973) — i.e. sufficient volumes of water were released from the downgoing slab by subsolidus breakdown of amphibole, and rose to parts of the overlying upper mantle where temperatures were greater than those

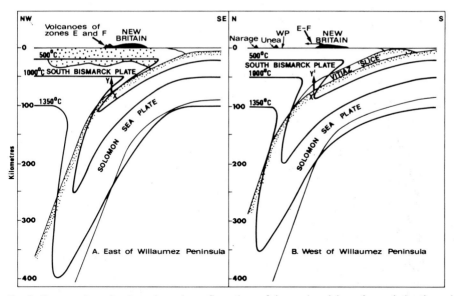

Fig. 7: Cross-sections showing schematic configurations of downgoing slabs and speculative thermal regimes east (A) and west (B) of Willaumez Peninsula. Locations of cross-sections are shown in Fig. 2. Light stippling shows hydrated crustal rocks in the upper parts of the slabs. Coarse stippling indicates thickness of crust in cross-section A, which coincides with the crustal profile K-L given by Finlayson & Cull (1973, Fig. 9b; crustal thicknesses beneath west New Britain are undetermined). WP and E-F represent the northern end of Willaumez Peninsula, and zones E and F (Fig. 5), respectively, projected from the east onto cross-section B. X-Y and Z'-Y' are equivalent ascent paths of slab-derived water (see text for details). The flat parts of the isotherms are taken from the 'undisturbed' mantle geotherm of Griggs (1972).

of the appropriate peridotite-water solidus (see Green, 1972). For example, in Figure 7A, water may be released from the slab at X and rise, and if sufficient water reaches Y at 50 km depth it may cause melting there, because Y is above the temperature of the water-saturated peridotite-water solidus at 15 kilobars (Green, 1972). Subsequent fractionation of these primary magmas may then lead to the spectrum of compositions represented by the rocks of zones E and F.

In contrast, the cross-section west of $150^\circ E$ shows features markedly different from those in the eastern section (Fig. 7B): two trenches appear to be present; volcanoes equivalent to those of zones E and F are absent; and the Witu Islands represent the only known eastern-arc volcanoes which overlie the deepest part of the New Britain Benioff zone. One possible interpretation of these features follows, but it is largely speculative, and should be regarded as tentative.

It is proposed that the branching of the submarine trench indicates that underthrusting may have taken place along two boundaries throughout the Late Cainozoic, and that the area of sea-floor between the two branches is the upper end of a thrust slice that dips northwards beneath New Britain. This 'Vitiaz' slice (Fig. 2) is envisaged as a sliver of crust and upper mantle — perhaps only 20 km thick — underlain by the crust of the subducted Solomon Sea plate (Fig. 7B). Down-dip, the Vitiaz slice is visualised as thinning out, extending no further than about 70 or 80 km north of the north coast of New Britain, and no deeper than about 100 km. The slice is presumed to have underthrust an unknown length of sea-floor (Fig. 7B).

In broad terms, it may be suggested that at depths less than 100 km the effect of the Vitiaz slice has been to complement the effect of the downgoing Solomon Sea plate as a heat-sink; the combined effect of both the slice and the plate may have been to absorb more heat from the surrounding mantle than would have been absorbed by the Solomon Sea plate alone. In addition, if it is true that the present-day pole of rotation for the South Bismarck and Solomon Sea plates is west of New Britain (Krause, 1973), it follows that rates of subduction west of Willaumez Peninsula are lower than those east of it, and therefore at depths greater than about 100 km more heat may have been extracted by the slab in the east than in the west.

The speculative isotherms shown in Figure 7 reflect these proposed differences in the capacities for heat absorption by the slabs on either side of Willaumez Peninsula. The isotherms in the eastern section extend deeper within the slab than those in the west, and parts of the mantle above the slab in the west at depths less than about 100 km are cooler than equivalent parts in the east. Thus, in Figure 7B, water derived from X' (equivalent to X in Fig. 7A) will, on rising to Y' (equivalent Y) at a depth of 50 km (15 kb) reach mantle whose temperature is lower than that of the water-saturated peridotite-water solidus (Green, 1972). No amount of water will cause melting, and therefore no volcanism will be expected along this part of the north coast of New Britain (Fig. 7B). To the north, however, water or water-rich melts are presumably derived from the slab at a few hundred kilometres depth. Experiments are, at present, unable to predict the composition of these melts and their effect on the overlying mantle, but mantle melting perhaps takes place (to give rise to the volcanism of the Witu Islands) because the temperatures of the slab melts generated from the Solomon Sea plate west of Willaumez Peninsula are higher than those of the melts which might have been generated from the plate east of the Peninsula (compare the position of the $1350^{\circ}C$ isotherm in Fig. 7).

The Vitiaz slice may also provide a possible explanation for the existence of Willaumez Peninsula, whose southward extension intersects the point where the New Britain trench divides (Fig. 2). If underthrusting has taken place along both branches of the trench west of $150^{\circ}E$, rates of underthrusting at the Vitiaz/South Bismarck boundary will have been much less than rates at the Solomon Sea/South Bismarck plate boundary, because the combined rates of convergence for the two boundaries west of $150^{\circ}E$ are thought to less than the rate east of $150^{\circ}E$ (see above). Thus, the region beneath Willaumez Peninsula, and beneath the South Bismarck plate, can be visualised as one in which the downgoing Solomon Sea plate east of $150^{\circ}E$ may be dragged past the underthrust Vitiaz slice. If so, releases in total-pressure caused by shearing could have taken place in the 'drag-zone' (Fig. 2) beneath Willaumez Peninsula, producing melting of subducted oceanic crust in a comparatively narrow, north-south zone. Perhaps, too, shear-strain heating (cf. Minear & Toksöz, 1970a,b) was enhanced in this zone. Rise of the slab melts into the overlying upper mantle might have caused partial melting of peridotite, and the formation of primary magmas, which rose, fractionated, and produced the volcanoes of Willaumez Pensinsula. In this interpretation, therefore, Willaumez Peninsula does not represent a fault in the South Bismarck plate, but rather is the surface expression of differential movements *beneath* the South Bismarck plate.

CONCLUDING REMARKS

The volcanoes of the south Bismarck Sea are considered to be related to convergent plate boundaries which, throughout the Late Cainozoic, have been characterised by unusually complex relative movements. This paper has attempted to provide an explanation for the distribution of the volcanoes using orthodox principles of plate theory, but the interpretations should be regarded as tentative, as considerably more geophysical and geological data should be collected before any plate model for this region can be regarded as secure.

ACKNOWLEDGEMENTS

The writer gratefully acknowledges discussions with colleagues in the Geological Survey of Papua New Guinea and the Bureau of Mineral Resources. Incisive criticisms by J. S. Milsom of a draft manuscript of this paper have helped greatly in clarifying some of the geodynamic interpretations. A review by R. J. Arculus is also gratefully acknowledged. This paper is published with the permission of the Director of the Bureau of Mineral Resources, Canberra.

REFERENCES

Bain, J. H. C., 1973: A summary of the main structural elements of Papua New Guinea. *in* Coleman, P. J. (Ed.), *The Western Pacific: island arcs, marginal seas, geochemistry.* Univ. W. Aust. Press, pp. 147-161.

Blake, D. H. & Ewart, A., 1974: Petrography and geochemistry of the Cape Hoskins volcanoes, New Britain, Papua New Guinea. *J. geol. Soc. Aust., 21,* pp. 319-332.

BMR, 1972: 1:1 million-scale geological map of Papua New Guinea. *Bur. Miner. Resour. Aust.*

Connelly, J. B., 1974: A structural interpretation of magnetometer and seismic profiler records in the Bismarck Sea, Melanesian Archipelago. *J. geol. Soc. Aust., 21,* pp. 459-469.

Curtis, J. W., 1973: Plate tectonics and the Papua New Guinea-Solomon Islands region. *J. geol. Soc. Aust., 20,* pp. 21-26.

Denham, D., 1969: Distribution of earthquakes in the New Guinea-Solomon Islands region. *J. geophys. Res., 74,* pp. 4290-4299.

Denham, D., 1973: Seismicity, focal mechanisms and the boundaries of the Indian-Australian plate. *in* Coleman, P. J. (Ed.), *The Western Pacific: island arcs, marginal seas, geochemistry.* Univ. W. Aust. Press, pp. 35-53.

Dewey, J. F., & Bird, J. M., 1970: Mountain belts and the new global tectonics. *J. Geophys. Res., 75,* pp. 2625-2647.

Dickinson, W. R., 1970: Relations of andesites, granites, and derivative sandstones to arc-trench tectonics. *Rev. Geophys. Space Phys., 8,* pp. 813-860.

Finlayson, D. M. & Cull, J. P., 1973: Strucutural profiles in the New Britain-New Ireland region. *J. geol. Soc. Aust., 20,* pp. 37-48.

Green, D. H., 1972: Magmatic activity as the major process in the chemical evolution of the Earth's crust and mantle. *Tectonophysics, 13,* pp. 47-71.

Green, T. H. & Ringwood, A. E., 1972: Crystallisation of garnet-bearing rhyodacite under high pressure hydrous conditions. *J. geol. Soc. Aust., 19,* pp. 203-212.

Griggs, D. T., 1972: The sinking lithosphere and the focal mechanism of deep earthquakes. *in* Robertson, E. C. (Ed.), *The nature of the solid earth.* McGraw-Hill, New York, pp. 361-384.

Hamilton, W., 1972: Tectonics of the Indonesian region. *U.S. geol. Surv. Project Report. Indonesian Investigations.*

Heming, R. F., 1974: Geology and petrology of Rabaul caldera, Papua New Guinea. *Bull. geol. Soc. Amer., 85,* pp. 1253-1264.

Isacks, B. & Molnar, P., 1971: Distribution of stresses in the descending lithosphere from a global survey of focal-mechanism solutions of mantle earthquakes. *Rev. Geophys. Space Phys., 9,* pp. 103-174.

Jakeš, P. & Gill, J. B., 1970: Rare earth elements and the island arc tholeiitic series. *Earth Planet Sci. Lett., 9,* pp. 17-33.

Johnson. R. W., in prep.: Late Cainozoic volcanoes at the southern margin of the Bismark Sea, Papua New Guinea. Part 1: Distribution and major element chemistry.

Johnson, R. W., Davies, R. A. & White, A. J. R., 1972: Ulawun volcano, New Britain. *Bur. Miner. Resour. Aust. Bull.* 142, 42p.

Johnson, R. W., Mackenzie, D. H. & Smith, I. E., 1971: Seismicity and late Cenozoic volcanism in parts of Papua-New Guinea. *Tectonophysics, 12,* pp. 15-22.

Johnson, R. W., Mackenzie, D. E., Smith, I. E. & Taylor, G. A. M., 1973: Distribution and petrology of late Cenozoic volcanoes in Papua New Guinea. *in* Coleman, P.J. (Ed.), *The Western Pacific: island arcs, marginal seas, geochemistry.* Univ. W. Aust. Press, pp. 523-533.

Johnson, T. & Molnar, P., 1972: Focal mechanisms and plate tectonics of the southwest Pacific. *J. geohpys. Res., 77,* pp. 5000-5032.

Karig, D. E., 1972: Remnant arcs. *Bull. geol. Soc. Amer., 83,* pp. 1057-1068.

Krause, D. C., 1973: Crustal plates of the Bismarck and Solomon Seas. *in* Fraser, R. (Ed.), *Oceanography of the South Pacific 1972.* N.Z. Nat. Comm. UNESCO, Wellington, 1973, pp. 217-280.

Krause, D. C., White, W. C., Piper, D. J. W. & Heezen, B. C., 1970: Turbidity currents and cable breaks in the western New Britain trench. *Bull. geol. Soc. Amer., 81,* pp. 2153-2160.

Kuno, H., 1975: Origin of Cenozoic petrographic provinces of Japan and surrounding area. *Bull. Volcanol., 20,* pp. 37-76.

Le Pichon, X., 1968: Sea-floor spreading and continental drift. *J. geophys. Res., 73,* pp. 3661-3697.

Le Pichon, X., 1970: Correction to paper by Xavier Le Pichon 'Sea-floor spreading and continental drift'. *J. geophys. Res., 75,* pp. 2793.

Le Pichon, X., Francheteau, J. & Bonnin, J., 1973: *Plate tectonics.* Developments in geotectonics, 6, Amsterdam, Elsevier.

Lowder, G. G. & Carmichael, I. S. E., 1970: The volcanoes and caldera of Talasea, New Britain: geology and petrology. *Bull. geol. Soc. Amer., 81,* pp. 17-38.

Mackenzie, D. E. & Chappell, B. W., 1972: Shoshonitic and calc-alkaline lavas from the Highlands of Papua New Guinea. *Contr. Mineral. Petrol., 35,* pp. 50-62.

Mammerickx, J., Chase, T. E., Smith, S. M. & Taylor, I. L., 1971: Bathymetry of the South Pacific, chart No. 11. *Scripps Instn Oceanogr. and Inst. Mar. Resour.*

McBirney, A. R., 1969; Compositional variations in Cenozoic calc-alkaline suites of Central America. *State Oreg. Dept Geol. Mineral Ind. Bull., 65,* pp. 185-189.

McKenzie, D. P., 1969: Speculations on the consequences and causes of plate motions. *Geophys. J.R., astr. Soc., 18,* pp. 1-32.

Milsom, J. S., 1970: Wooklark Basin, a minor center of sea-floor spreading in Melanesia. *J. geophys. Res., 75,* pp. 7335-7339.

Minear, J. W. & Toksöz, M. N., 1970a: Thermal regime of a downgoing slab and new global tectonics. *J. geophys. Res., 75,* pp. 1397-1419.

Minear, J. W. & Toksöz, M. N., 1970b: Thermal regime of a downgoing slab. *Tectonophysics,* 10, pp. 367-390.

Morgan, W. R., 1966: A note on the petrology of some lava types from east New Guinea. *J. geol. Soc. Aust., 13,* pp. 583-591.

Nicholls, I. A. & Ringwood, A. E., 1973: Effect of water on olivine stability in tholeiites and the production of silica saturated magmas in the island-arc environment. *J. Geology, 81,* pp. 285-300.

Oxburgh, E. R. & Turcotte, D. L., 1970: Thermal structure of island arcs. *Bull. geol. Soc. Amer, 81,* pp. 1665-1688.

Page, R. W. & McDougall, I., 1970: Potassium-argon dating of the Tertiary f1-2stage in New Guinea and its bearing on the geological time scale. *Amer. J. Sci., 269,* pp. 321-342.

Ringwood, A. E., 1974: Petrological evolution of island arc systems. *J. geol. Soc. London, 130,* pp. 183-204.

Ripper, I. D., 1975: Some earthquake focal mechanisms in the New Guinea/Solomon Islands region, 1963-68. *Bur. Miner. Resour. Aust. Rep. 178.*

Rittmann, A., 1953: Magmatic character and tectonic position of the Indonesian volcanoes. *Bull. Volcanol., 14,* pp. 45-58.

Robinson, G. P., Jaques, A. L. & Brown, C. M., 1974: Explanatory notes on the Madang geological map. *Geol. Surv. Papua New Guinea Rep. 74/13.*

Ryburn, R. J. Mackenzie, D. E. & Johnson, R. W., in prep: Geology of New Britain. *Bur. Miner. Resour. Aust. Bull.*

Stern, C. R. & Wyllie, P. J., 1973: Melting relations of basalt-andesite-rhyolite-H2O and pelagic red clay at 30 kb. *Contr. Mineral. Petrol., 42,* pp. 313-323.

Thompson, J. E. & Fisher, N. H., 1965: Mineral deposits of New Guinea and Papua; and their tectonic setting. *Proc. 8th Comm. Min. Metall. Congr., 6,* pp. 115-148.

Tilley, C. E., 1950: Some aspects of magmatic evolution. *J. geol. Soc. London, 106,* pp. 37-61.

Wiebenga, W. A., 1973: Crustal structure of the New Britain-New Ireland region. *in* Coleman, P. J. (Ed.), *The Western Pacific: island arcs, marginal seas, geochemistry,* Univ. W. Aust Press. pp. 163-177.

Wyllie, P. J. (Ed.), 1973: Experimental petrology and global tectonics. *Tectonophysics, 17,* part 3.

ERUPTIVE HISTORY OF MANAM VOLCANO, PAPUA NEW GUINEA

W.D. PALFREYMAN and R.J.S. COOKE

Bureau of Mineral Resources, P.O. Box 378, Canberra City, A.C.T. 2601.

Geological Survey of Papua New Guinea, Volcanological Observatory,
P.O. Box 386, Rabaul, Papua New Guinea.

ABSTRACT

Manam, an imposing basaltic stratovolcano rising 1725 m above the sea, lies 12 km off the north coast of mainland Papua New Guinea. It was probably first recognised as an active volcano during the mid-sixteenth century, but until the 1870s recorded sightings of the island were very infrequent; since then probably all the major eruptions have been recorded. Manam is clearly one of the most active volcanoes in the country and after it was brought under constant surveillance in 1957, it became apparent that the volcano is more-or-less constantly active. A compilation of all available accounts of past activity shows that the longest period of quiescence during the last 100 years is 9 years. A few large eruptions have occurred at the main crater but most strong activity takes place at the southern crater. The larger eruptions are typically of powerful strombolian character, with frequent but usually brief lava flows and occasional nuées ardentes of basaltic composition. Heavy falls of ash and scoria result from paroxysms, typically of only a few hours' duration, but often recurring after a few days to a few weeks.

INTRODUCTION

Manam, 12 km from the north coast of mainland Papua New Guinea (Fig. 1), is one of the most active volcanoes in the country (Fisher, 1957). Frequent strombolian eruptions have been recorded, many involving lava flows, and the volcano's longest repose period during the last one hundred years appears to have been only about nine years. The occurrence of nuées ardentes during some of its eruptions is well established (Taylor, 1963). Lavas produced during recent eruptions have been olivine-bearing quartz tholeiite basalt and low-silica andesite (Table 1). Some 5000 people live in a dozen or so village groups around the coast, and at least some of these people are in considerable danger during the most violent eruptions.

Full-time volcanological studies commenced in Papua New Guinea in 1937. An observatory was established in Rabaul in 1940, and was re-established in 1950 after a break following the Second World War. Volcanological observations commenced at Manam in June 1957 when, following a period of activity earlier in the year, an observation post was established by G.A.M. Taylor. Almost continuous observations were made by Taylor from this and from another, temporary, post during the violent eruptions of 1957 to 1960. A well-equipped permanent volcanological observatory, planned by Taylor, was established on the island in 1966 (Fig. 2).*

This paper surveys the known eruptive history of Manam, also referred to in earlier years as Manumudar, Mammamur, Manamur, Vulcan Island, or Hansa Island. The compilation of this account has involved the searching of explorers' journals dating from before the time of European settlement (1870s), of explorers', naturalists' and administrators' accounts from the period of German administration (till 1914) and from various (mostly non-scientific) sources from the period of Australian administration, up until 1951. Many of the older German

*See Myers (this volume). Ed.

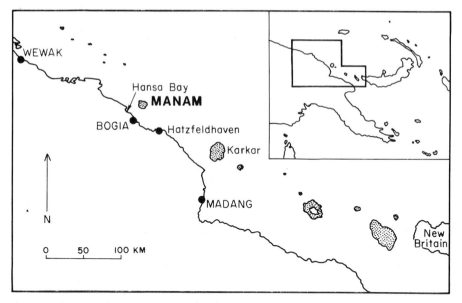

Fig. 1: North coast of mainland Papua New Guinea, showing Manam and other volcanic islands (stippled).

texts are not generally available in Australia and the descriptions of eruptions at Manam have been translated and included in the following account. To give a balanced picture of activity, descriptions (mostly originally in English) of activity from the period 1914 to 1951, have also been included although these texts are generally more accessible. Until the 1870s, observations had been possible at only very long intervals, and their main value lies in their indication of the character of activity, rather than as contributions to a connected eruptive history. Since then, observations have been intermittent, although it is probable that at least the major events have been reported. In recent years it has become clear that eruptive activity of some sort is more-or-less continuous. Much of the activity consists only of minor ejections of dust, occasional weak glows, and minor rumblings. However, periods of vigorous strombolian eruption occur at intervals, and, for the period since the 1950s, only activity of this type is covered in this paper. A detailed study of the activity since 1957 is in progress, and will be reported elsewhere.

TOPOGRAPHY

Manam is an impressive island stratovolcano, with a basal diameter of about 25 km, and a height of about 2800 m (1725 m above sea level). It appears to be very symmetrical at a distance, but a closer view shows that its flanks are strongly dissected and that it has a complex inner 'core' (Fig. 3). The lower parts of the island, to about 1250 m above sea level, are at the planeze stage of erosion (Kear, 1957), four planezes of similar dimensions being separated by four major radial, amphitheatre-headed valleys in the northeast, southeast, southwest, and northwest quarters (Fig. 2). Rising above this level is the central eroded core buttressed by a dense network of flows and dykes as seen exposed in the rugged headwalls and upper sidewalls of the valleys.

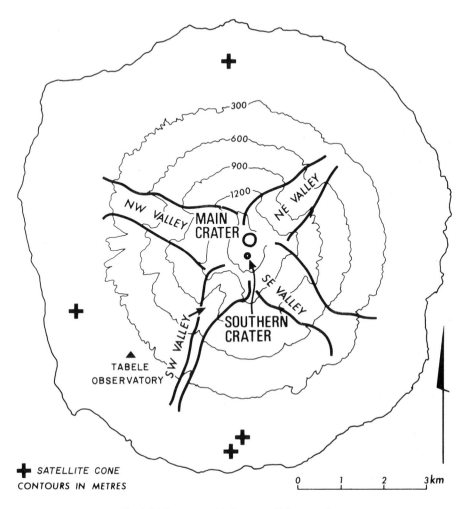

Fig. 2: Main topographic features of Manam volcano.

Recent activity has occurred chiefly at two craters: the main crater, at the head of the northeast valley and partly breached towards the northeast, and the southern crater, perched high near the summit at the head of the southeast valley. The position of the southern crater seems to shift about within a small area from eruption to eruption, and sometimes there may be two or three active vents in the vicinity. Although there are four parasitic cones (Fig. 2), historical activity is known only from the summit.

Recent activity is rapidly building up an apron of ejecta about the flanks of the central core. The apron is most strongly developed in the northwest and southeast valleys, less so in the northeast valley, and hardly at all in the southwest valley (Fig. 3). The apron consists of large talus cones built up against the valley headwalls; lava flows are incorporated in the talus in both the northeast and southeast valleys, at least. Nowhere has the apron yet built up sufficiently to overflow on to the outer planezes of the island, although eruptions in the late 1950s filled-in the head of the southeast valley (Taylor, 1963).

The amount of apron material present can be accounted for by comparatively recent activity (perhaps 500-1000 years). Imaginary removal of this young material would reveal the deeply eroded skeletal core of the island, and studies of erosion rates on shield and stratovolcanoes (McDougall, 1964, 1971; Ruxton & McDougall, 1967) suggest that this degree of dissection may have needed at least tens of thousands of years to develop. Detailed geological studies of Manam are in progress, but as yet there is no stratigraphic history available to determine if the topographic features result from two or more separate periods of volcanic activity.

EARLY EUROPEAN OBSERVATIONS

Manam may have been first observed by European navigators, and identified as an active volcano, during the mid-sixteenth century. A volcano was observed in this area during the voyage of de Retes in 1545 (Wichmann, 1909), although its identity cannot be positively established. Certainly, a volcano is marked on a European map of 1601 (Sharp, 1960, Pl. 1) in this vicinity; possibly it really lies among the islands just south of Manus Island (300 km to the northeast).

Schouten and le Maire observed Manam in eruption on 6-7 July 1616 (Spilbergen, 1906); their map is detailed enough to allow positive identification of the volcano, which they describe as emitting flames and smoke.

Tasman also saw Manam in eruption on 21-22 April 1643; it 'burned steadily directly out of the top with flames of fire' (Sharp, 1968). Again, the volcano can be positively identified from contemporary maps of the region. Tasman made a sketch of an erupting volcano, but its identity is not clear, as he also saw Karkar (Fig. 1) erupting at about the same time.

Dampier observed Manam in 1700: 'On Tuesday the 2d of April, about 8 in the Morning, we discovered a high peeked Island to the Westward, which seem'd to smoak at its Top. The next Day we past by the North-side of the Burning Island, and saw a Smoak again at its Top; but the Vent lying on the South-side of the Peek, we could not observe it distinctly, nor see the Fire.' Dampier's map and sketch of the volcano both unmistakably indicate Manam (Williamson, 1939).

In a fanciful account of his voyage along the coast near Manam in 1830, Morell (1832) claimed to have seen seven active volcanoes, five of which were actually erupting, and the other two smoking. Six of these were islands, and the other was said to be in the coastal ranges of the mainland, where no youthful volcanoes are now known; only four recently active volcanic islands are known in the area described by Morell. In spite of the doubt about the veracity of the account, circumstantial detail of the only eruption actually described seems to suggest Manam: 'In the direction of north-north-east from this cape [stated to be at $4^\circ59'$S, $145^\circ16'$E; that is, 50 km inland from the present coast] is a small volcanic island, lying about six leagues from the mainland, which was in full blaze . . . the flames ascended upwards from the lofty summit of the isolated mountain, at least one thousand feet; while the red burning coals of pumice-stone were carried to the north-west . . .'

OBSERVATIONS SINCE THE 1870s

1877. An eruption was observed between 11 and 13 November 1877 by von Miklucho-Maclay (1878) and was first seen from the vicinity of Karkar. 'During the whole night one could watch the flickering of the red light . . . The fire occurred at intervals of several minutes and lasted each time for only about ½ to 2 minutes.' Early on 12 November, 'the mountain-top was covered with a smoke-and-cloud cap above which a high, nearly vertical column of smoke towered up' and in the afternoon, from a distance of about 60 km, von Miklucho-Maclay saw 'three different columns of smoke which rose from the crater and, as its southern rim was a little lower, one could see the new eruption cone out of which rose two strong columns of smoke, while the third one, being a bit lower, rose at the southwest rim of the crater. From time to time great quantities of white smoke emerged. They were thrown out at similar intervals as the flickering of the light during the previous night had appeared. The three columns of smoke combined to form a huge whitish-grey cumulo-stratus cloud. As it grew darker, I saw, several times, mighty fork-lightnings flash through the darkening clouds. The white smoke clouds, periodically thrown out, changed into columns of fire which flared up in the same way as they had done the day before.' In a footnote von Miklucho-Maclay stated that the eruption had been noted in the logbook of the schooner *Flower of Jarrow* on 29 October 1877.

1885. Finsch (1888) observed the island from a distance: 'A rosy glow surrounded the crater wall and in the collapsed western vent could be seen clearly a fiery glow like a burning chasm which appeared more intense after nightfall, until the mountain was, stage by stage again covered by cloud.' Sapper (1917) noted that this observation was made during May 1885. Later in his account, though probably referring to the same eruption, he described the 'mighty roar' and the 'enormous white masses of cloud' emanating from the volcano.

1887-88. Grabowsky (1895) described an apparently more violent eruption in June 1887 which was clearly visible from Hatzfeldhaven, 35 km distant, on the mainland (Fig. 1). 'Weak tremors were observed several times while the double craters (Auroka and Ujap) of the volcano Manamur . . . were continuously active, and in the night from 27 to 28 June 1887 a violent eruption with a large lava flow took place . . .'

Similar activity was seen by Zöller (1891) who was stationed at Hatzfeldhaven in 1888. 'During the day one sees only a column of smoke whilst in the evening the gleam of the embers from the outbreaks of flames which are repeated at intervals of a few seconds gives much more variety to the scene. At times one saw molten lava flowing down to half way down the mountain.' Sapper (1917) made the comment 'apparently glowing ash', in relation to the lava flow. This is possibly correct as from the distance of Hatzfeldhaven it might be difficult to distinguish lava flows from hot ash avalanches or nuées ardentes.

1889-95. Sapper (1917) quoting Schleinitz (1889) stated that Manam was active during 1889. However, Schleinitz only remarked generally on activity.

Zdekauer (1899) is quoted by Sapper (1917), who gave his source as Hammer (1907), as having 'observed sparks of fire and lava flows' sometime during 1889.

However, Hammer stated that Zdekauer 'noted lava producing eruptions which occurred every eight minutes.' As Zdekauer's book (Zdekauer, 1899) was not seen by us, neither the reliability of the observation nor the date of the eruption could be checked.

According to Tappenbeck (1901) activity continued at Manam until about 1895: 'Until the mid 1890s the crater of Vulcan Island was highly active and ejected lava almost continuously.' Subsequently: 'Vulcan crater quietened down and showed no sign of its inner activity for long periods of time.' Tappenbeck equated the change in activity with the eruption of Karkar in 1895.

1898-1900. No further activity was reported until some time between 1898 and 1900 when Pflüger (1901) noted that 'The summit is jagged and produces two mighty columns of smoke.'* Von Hesse-Wartegg (1902) also noted the twin smoke columns. He further stated that 'on the northern slopes can be seen the glow of the fresh lava streams which stem from the latest eruption.'*

1904. Pöch (1908) made the following observations from Potsdamhafen (on the mainland opposite Manam; now abandoned) on volcanic activity at Manam during October 1904: 'On 26 October, at 2 p.m., a very high column of smoke suddenly appeared above the summit of the volcano ... At about 4 p.m. the cloud became larger and darker and one could see a shower of ash falling into the sea to the west of the island ... After sunset a powerful column of fire could be seen at the summit ... One could hear the rumblings and rifle-shot-like detonations. I could quite clearly see through my binoculars the fire column which issued from the westerly vent from time to time, always to be followed by a detonation. From the second and easterly crater vent the same weak but steady smoke column persisted just as before the eruption. This vent, though very close to the westerly one, was not at all affected by the eruption ... I could further see through the binoculars that large glowing boulders fell to the west of the fire column, and a lava stream rolled down the western valley. The lava had set fire to a forest and one could see burning trees or the glowing, burnt-out trunks. The lava reached the foot of the mountain and came to rest where the angle of slope diminished; it did not reach the sea. It was said that villages were destroyed. Next morning, 27 October, tremendous explosions occurred, but at 11 a.m. activity ceased. The westerly vent then smoked less than before the eruption. There were no more eruptions or tremors as long as I was at Potsdamhafen (till 25 November).' An illustration of the volcano by Pöch on 26 October shows a tall ash column rising to 2000 m or more above the summit. Two other, much shorter accounts, of this eruption (Pöch, 1907 a & b) state that the lava flow reached the sea. Pöch was told that eruptions had occurred at Manam regularly in the past and that the last eruption prior to his visit had been during April or May 1904.

1909. Under Sapper's entry for 1909 is, 'strong smoke development'; the source is given as Friederici ms. This reference was not seen by us so it could not be checked. G. Häntke (pers. comm., 1974) noted that 'explosions and lava' occurred during 1909; no source was given.

1911. Manam was visited briefly in December 1911 by Scholz (1912) who

---

*Sapper (1917) gave the dates of these observations at the publication dates of each book. However, it is possible that both accounts are of the same event.

made the following observation: 'On the evening of my arrival [12 December] it was active and ejected smoke and flaming fire. Every now and then there occurred deep rumblings inside the mountain and once a strong and persistent earth tremor was felt. Wide furrows on the slopes mark the paths of lava flows towards the sea. In one particular place, about a year ago, the volume of lava built up a kind of knoll.'

1913. Behrmann (1917), on his return from a voyage up the Sepik River in September 1913, noted that 'the easternmost crater ejected a brown smoke column. About every 10 minutes an eruption occurred concurrently with an underground thunder, the ejected material falling back into the crater. At night the reflection of fiery lava masses could be seen.' He stated that there had been several eruptions during the course of his expedition.

1917. A brief explosive eruption following a strong earthquake occurred sometime during 1917. Stanley (1923) stated that the eruption lasted a few minutes only and that 'boulders 3 or 4 cubic yards in mass were hurled to a height of 2000 feet.'

A much more violent eruption during 1917 is implied by a report in the mission journal, *Steyler Missionsbote* (1921-22). There, the quite violent eruption of December 1920 (see below) is described and compared in magnitude with one which took place during 1917.

1919. The eruption of 11 August 1919 was one of the most violent ever reported; Stanley (1923) gave the following account derived from a missionary who observed it. The eruption was preceded by several small earthquakes. 'At about 2 p.m. large volumes of steam and black vapour rising in the form of a vortex formed about the apex and soon obscured the sun, but the glow from the crater illuminated the lower clouds. A spasmodic rumbling at the base of the mountain commenced, and soon after 2 o'clock lava commenced to flow down the Eastern side and in about 5 minutes reached the sea near (B) causing the water to boil furiously, and huge volumes of steam and black vapour to rise to a great height, which covered the whole mountain and surrounding country . . . At 2.55 p.m. the red steam of lava could be clearly seen, but ceased to flow, and dense grey-coloured clouds of vapour were emitted from the crater. When the rumbling ceased a grey-brown halo of dust encircled the mountain, which gradually spread for miles over the mainland . . . At 3 p.m. the South-east wind cleared the Western centre of the island, but the lower Western portion was black with dense vapour and dust clouds. At 3.3 p.m. the largest crater was alone in eruption, the lava continuing to flow from the Eastern edge for a short distance. At 3.45 p.m. the lava ceased to flow, and only a small quantity of black vapour rose from the crater. The villages of Zogari and Josa [on the west coast of the island] were covered with dust, scoria, and hot fragments of vesicular lava.' Further on in his account Stanley noted: 'At night, it was possible to read a newspaper 8 miles away by the reflection of the lava on the clouds.' The point (B) mentioned is shown on Stanley's sketch map at the foot of the southeast valley.

1920-22. A slightly less violent eruption commenced in December 1920. A report in *Steyler Missionsbote* (1921-22) dated 19 December 1920 states: 'Opposite our station [Monumbo, near Potsdamhafen] lies the fire-ejecting mountain

Manam ... For a fortnight, it has been ejecting a fiery column many metres high above which hang black smoke clouds. Out of the clouds, twice, ash has rained for many hours. Other times it ejects glowing lava out of its fiery vent which rolls, destroying everything, down the mountainside and into the sea.' Another description is given in a further passage. 'Day and night one can hear for many hours its underground rumblings. It ejects glowing lava house-high which rolled into the sea as a fiery stream.'

The eruption probably continued until at least March 1921 as Stanley (1923) recorded an eruption at that time 'damaging a few gardens and extruding some lava, but it was not as violent as that in 1919.' Stanley (1923) also reported that 'the crater was in partial eruption, emitting huge volumes of steam and black dusty vapours which darkened the whole of the North-western portions of the island' during his visit in 1922 (the precise date is not reported).

1925. G.A.M. Taylor (unpublished ms) recorded that an eruption took place during 1925. He was informed by a local resident that it was possible to read a newspaper at Dugomur plantation on the mainland opposite the island by the night-time glow from incandescent lava being ejected from the volcano.

This eruption is not recorded by either the *Official Handbook of the Territory of New Guinea* (published 1936), or the *Pacific Islands Year Book 1935* both of which state that the last big eruption was in March 1921. However, Höltker (1942), drawing on Mission archives, stated that between 1917 and 1926 six large eruptions were observed, one of which presumably was that mentioned by Taylor.

1926-28. G. Häntke (pers. comm., 1974) informed us that strombolian activity occurred between March 1926 and February or March 1928. However, the original source of the information is not known.

1933-34. The anthropologist Camilla Wedgwood was on Manam between January 1933 and February 1934. She wrote (Wedgwood, 1934): 'it is not uncommon to hear the sound of volcanic explosions and to see a red glow at night in the sky above the mountain top. Sometimes, too, a thick brown-black cloud of dust rolls down over the upper slopes, and during the latter part of my stay there was a very brief shower of fine, glowing cinders.' Höltker (1942) also mentioned that he has photographs showing smoke plumes in 1932 and 1934.

1936-39. Höltker (1942), an anthropologist, who was stationed on Manam between July 1936 and April 1939, stated that, between the beginning of October and the middle of December 1936, 'the volcano was subjected to numerous large paroxysms. Six large eruptions followed one another at intervals of five to seven days.' He stressed that 'mostly the eruptions began towards evening, and lasted with only short interruptions into the early hours of the morning. On only one occasion did an eruption take place in the morning, and this was on 11 October 1936.' He described their general pattern as follows: 'In the late afternoon the usual smoke plume became darker and denser. On could notice a peculiar unrest in the mountain; earthquakes of different strengths and underground rolling. One could already see in the sky a red glow above the crater (the east crater was active). The rolling became stronger and stronger until the first eruption began as projections of glowing ejecta rose up hundreds of metres, and one could distinguish plainly large glowing blocks of lava. These glowing sheaves now played up and down like a fountain, as if the driving force were sometimes stronger, sometimes weaker. In between, the whole sky was covered with heavy black clouds of ash. Such an outburst lasted for several minutes, and sometimes up to a quarter of an hour. Then the glowing fountain sank back beyond the crater rim, the glow of the lava above the open crater remained, and the smoke clouds rose higher until after a short time fresh rolling in the interior of the mountain heralded a new outburst. Each outburst followed the previous one fairly frequently in the first hours of the night, at intervals of about 10 to 15 minutes. Towards morning the pauses became longer until after a last big outburst the volcano again became quiet, and only the usual smoke plume remained above the summit.' Then followed 'a pause of 6 to 7 days in which, throughout the day, the usual smoke plume, and at night the glow of the lava, could be seen, but no eruption followed'.

Höltker did not describe a nuée ardente, but in one of his published photographs, taken on 11 October 1936, he identified 'a glowing cloud descending' the southeastern valley. This identification appears to be accurate although details are not particularly clear in the photograph. On the same day: 'Out of the clouds there fell on us, not very thickly but yet continuously, ash and little glowing pieces of scoria . . . From time to time small and large lapilli poured down on to the corrugated iron roof of our house . . . After every outburst there was a layer of ash on the leaves of the taro plants in our garden, up to a centimetre thick, and more'.

A report by the captain of a vessel (data file, Rabaul Observatory) states that a 'heavy lava flow' was observed at 7 p.m. on 18 October 1936, ceasing at 8.20 p.m., and that 'terrific eruptions' occurred between 6.30 p.m. and 10 p.m. on 24 October, accompanied by lava and by the ejection of incandescent blocks 1000 to 1200 metres into the air.

Fisher (1939) gave dates for peaks in the 1936 activity at 17, 25, and 31 October and stated that 'A similar outbreak occurred on 15 March 1937.' Höltker (1942), however, noted that, after a lull from December 1936, eruptions began again in May to June 1937, the renewed activity taking the form of quiet outpourings of lava. He noted that 'After a couple of months it [the lava] could be observed from Bogia Harbour on the mainland on almost every evening; sometimes it was stronger, sometimes weaker . . . During these fiery outpourings glowing fragments bounced and rumbled down the hill; these were probably the larger lava blocks . . . This activity on Manam, with outpourings of lava at night, continued almost unbroken during the following years to the end of my stay in New Guinea [April 1939].' Höltker further stated that two craters were active during his stay on the island. From the descriptions, these can be readily identified with the present main and southern craters. The main crater was active in 1936, and the southern crater was 'constantly and intensely active'.

Another report concerning activity in January and February 1938 (*Rabaul Times*, 1 April 1938) notes that 'the crater on the south side, which is beneath the main one has been very active, shooting lava into the air hundreds of feet', and that the volcano had been in eruption 'night and day' for some time.

A correspondent to the *Pacific Islands Monthly* of June 1938 described activity at an unknown date early in 1938. 'The occasional wisps of steam and smoke normally seen issuing from the crater have given place to an almost constant eruption-cloud, ever-changing in shape and size with the wind and intensity of the eruption. Down the southern side of the cone may frequently be seen a sloping streak of steam, rising from flowing lava. Apparently the lava has not yet reached more than half-way from the crater to sea-level but it is being watched with interest for further developments. Rumblings and explosions have been heard on the mainland more than 12 miles away.'

1946-47. There appears to be no report of volcanic activity during the period of the Second World War. Best (1956) reported: 'Towards the end of 1946 an eruption even more severe than that of 1919 commenced and continued intermittently until September 1947. During this eruption a large lava flow spilled out from below the crest on the southeast side of the cone, and flowing down a broad avalanche valley, bifurcated near its outer extremity and overran the coast in two places . . . Thick deposits of ash and lapilli covered most of the island, rendering unproductive many of the native gardens.'

1956-66. Strombolian activity occurred in every year during this period, although there were distinct phases of activity which could be described as separate 'eruptions', particularly in 1957-58 and 1960. During these two periods the most violent events recorded at Manam took place.

Reynolds (1957) described the first phase of activity, between about 8 December 1956 and February 1957. During January, ash-free strombolian ejections took place at the southern crater to heights of about 300 m, at an average rate of six or seven per minute. One or more lava flows were extruded, and avalanche phenomena are described, some of which can be interpreted as small nuées ardentes. The explosive activity was strongest in late December 1956 when strong aerial concussions were occurring. In an unpublished account G.A.M. Taylor (1958) gave a summary of the course of events following the resumption of similar activity in mid-1957. A phase of activity in May-July 1957 'was almost identical in nature' with the previous phase, with the exception that the main crater became involved, producing incandescent lava (presumably as lava jets, and not flows) and a little ash. 'The next major phase in the eruption began early in October with rapidly fluctuating spasms of explosive activity which quickly culminated on the 18th in a strong eruption. Heavy nuées ardentes swept down the southern flanks . . .' After this, 'gas and ash emission greatly increased, coarse cinder fragments fell on coastal areas for the first time and a more mobile lava flow descended the southern slopes . . . On 4th December the volcano began to rumble continuously and the sounds were louder than any heard before. Two days later villages on the eastern coast received a heavy fall of ash and cinder blocks. The largest blocks, measured 2″ by 3″ — some of them penetrated the roofs of houses. More powerful activity occurred on the next day, 7th December, when a number of nuées ardentes descended the south-eastern valley, and dust and lapilli fell over the northern and western sectors of the island. The eruption reached a peak on the next day, 8th December. The early morning rumbling and ash emission changed at 10 a.m. to spasms of continuous roaring and the voluminous ejection of ash and cinders. For the greater part of the afternoon the northern and western sectors of the island were blacked out by heavy clouds of ash . . . On this day further nuées descended the south-eastern valley', and 'one of them entered the sea'.

The whole population of Manam was evacuated from 10 December. Activity continued at a reduced level after the 8th, although 'rhythmical explosive jets continued from the summit vents until the 13th, and nuées continued to be expelled from the southern vent', the last of them on Christmas eve. Fluctuations occurred in the intensity of activity over the next month.

'On the night of January 9th [1958] brilliant lava fountaining originated from the main crater . . . At 1600 hours on the 10th of January an enormous cloud rose more than 20,000 feet above the summit with loud roaring and rumbling. This outburst expelled a heavy nuée ardente into the north-east valley', after which 'a lava flow then pured over the north-eastern rim . . . Heavy ash emission followed this eruption' and 'on the 14th a more powerful eruption poured a much larger lava flow into the northwestern [sic, northeastern] valley', which reached to 'within half a mile of the coast'. Heavy falls of ash and scoria followed, and a lull in activity commenced on the 19th. 'At 0605 hours on 25 January, the inactive southern vent started silently emitting ash and vapour. An hour and a half later the main crater opened up. At 0800 hours nuées ardentes were expelled on to the southeastern flank' and 'for the next five hours the column was fed by incessant explosions roaring and rumbling from the summit vents . . . Heavy nuées ardentes were expelled' apparently into all four valleys. 'Most of the western side of the island received a deposit of coarse scoria which was 5 to 6 inches thick along the coast.' Another major phase occurred on 3 February, which also laid down about 15 cm of scoria on the west coast. During this period 'voluminous quantities of fragmental and molten material were discharged and powerful nuées swept down the southeastern flanks'. Although some observations suggested 'that this eruption was more intense than its predecessor . . . no evidence was found that nuées had descended into the other three radial valleys'.

Yet another major phase commenced at 1415 hours on 4 March and was the most sustained event on record. It lasted for more than 24 hours and was similar to the previous one although 'effusive activity was greatly emphasised and explosive phenomena were correspondingly reduced. Some nuées descended the southeastern flanks during the initial stages' and heavy scoria falls occurred on the eastern side of the island. 'The most prominent feature of the activity was a heavy lava flow which descended from the southern crater', and entered the sea on 6 March.

Only occasional minor activity occurred after this, until strombolian activity recurred at the southern crater at the end of June 1958, lasting into August. Brief strombolian ejections also took place at the southern vent in June and July 1959 (data file, Rabaul Observatory).

Taylor (1963) summarised the next main eruptive phase, most of which occurred at the main crater between December 1959 and August 1960. He stated that 'a new phase of eruptive activity began from the main crater with Strombolian explosions of increasing intensity', and that 'this rising trend of activity reached a climax on 17th March, 1960 . . . At 0900 hours an explosive outburst greatly augmented the size and height of the eruption column and at the same time nuées ardentes were discharged from the crater. For the next two and half hours gas-charged masses of hot fragmental material were successively poured over the low northeastern rim of the crater to descend at high velocity into the valley below . . .

Fig. 3: View towards the southwest valley of Manam, showing a small explosion column rising from the southern crater, and the surface of 1958 nuee ardente deposits (foreground). Photograph taken 8 May 1963 by G.A.M. Taylor.

Heavy lava outpouring immediately followed this explosive activity with incessant lava fountaining from the crater . . . This effusive activity continued with only short intermissions until the end of May'.

Strombolian activity occurred again at the southern crater from September 1960, and data files at Rabaul Observatory show that explosive activity occurred frequently at the southern crater during the next few years (e.g. Fig. 3) with minor strombolian activity and probable lava flows occurring in every year from 1960 to 1966 (cf. Branch, 1967). However, no more activity of this kind occurred at the main crater.

1974-75. Only minor activity occurred at either crater after April 1966, but in 1974, on 23 May, a brief spasm of lava jetting at the southern crater took place. A few days of lava jetting at the main crater from 1 June 1974 led into a prolonged phase of southern crater strombolian activity which lasted until about mid-August. During this period, nuées ardentes occurred at two periods, 5-6 June and 29 June, followed in each case by lava flow, in the southeast valley. Five short bursts of large-scale activity, each lasting 1-3 hours, took place between 29 June and 10 July. Several heavy falls of ash and scoria were dropped on the west coast during these periods, accumulating to a depth of about 5 cm.

A brief recurrence of southern crater lava jetting in September, and a more prolonged recurrence of intermittent activity from late October, followed. A lava flow again entered the southeast valley at the commencement of the latter phase.

Activity died away again early in January 1975. The 1974-75 activity is summarised in greater detail by Cooke et al. (this volume).

DISCUSSION

Nuées ardentes and lava flows

Nuées ardentes of basaltic composition (Table 1; Taylor, 1963) appear to be a regular part of the eruptive phenomena at Manam. A nuée can be recognised in the description by Stanley (1923) of the 1919 eruption, and nuées were identified during the 1936, 1957-58, 1960, and 1974 eruptions. Although they are less clearly identifiable in the descriptions of other eruptions, nuées and not lava flows may have been the phenomena observed overnight on 27-28 June 1887 and in 1904, and very probably also in December 1920. The implication of high speed, of rolling motion, and the general destruction, all suggest nuées ardentes. Although there is little information available, it is probable that at least the larger nuées at Manam would be incandescent at night, creating the illusion of a lava flow; J. Holbrook (pers. comm., 1974) observed daylight incandescence at the base of the 29 June 1974 nuée from a vessel offshore. Certainly, the only historical lava flows to be seen on Manam which have entered the sea are those of 1947 and 1958.

Table 1. Chemical analyses of rocks produced by post-1957 eruptions of Manam volcano*.

	1	2	3	4	5
SiO2	53.90	53.05	52.70	51.10	53.08
TiO2	0.38	0.35	0.37	0.58	0.35
Al2O3	16.60	13.90	17.00	16.65	14.84
Fe2O3	4.90	6.07	4.45	4.98	5.96
FeO	3.90	3.99	4.40	4.00	3.81
MnO	0.17	0.16	0.17	0.16	0.16
MgO	6.70	8.60	6.80	6.80	7.37
CaO	10.50	10.68	10.75	10.60	10.70
Na2O	2.40	2.57	2.50	4.25	2.61
K2O	0.69	0.90	0.66	0.88	0.83
P2O5	0.12	0.25	0.11	0.19	0.19
H2O+	0.06	—	0.08	—	0.07
H2O-	0.07	—	—	—	0.03
CO2	0.05	—	0.02	—	—
Total	100.44	100.52	100.01	100.19	100.00

1. Lava from southern crater, January 1957 (Morgan, 1966, specimen 6)
2. Lava from main crater, 1957/58 (Morgan, 1966, specimen 9)
3. Lava from unknown crater, 1962 (Morgan, 1966, specimen 19)
4. Lava from southern crater, April 1964 (Branch, 1967)
5. Nuée deposit from southern crater, October 1957 (Morgan, 1966, specimen 8)

* For an analysis of a Manam lava flow erupted in 1974, see Cooke et al., this volume, Table 2.

Paroxysmal pattern

It is clear from the foregoing descriptions that the strongest activity occurs only in brief spurts. If observations from passing vessels are excluded, and only reports derived from longer periods of observations are examined, specific dates are often nominated as eruptive peaks — for example 27-28 June 1887; 26 October 1904; 11 August 1919. In 1936, Höltker (1942) described the typical form of 'numerous large paroxysms' which 'followed one another at intervals of 5 to 7 days'. Each of these lasted, by implication, for less than one night. Taylor (1958) described a similar pattern for the eruption of 1957-58. Furthermore, Cooke et al. (this volume) point out that there were five striking pulses of stronger activity, on 29 June and 1, 3, 6, and 10 July 1974, each of only short duration (1-3 hours).

Although there is some tendency for activity to appear to cluster about the middle and end of the year, the paroxysms seem to have occurred in almost every month.

Identification of active craters

Many of the earlier descriptions are not sufficiently detailed to allow the active crater to be identified. However, where identification is possible the active crater is usually the southern one. Apart from the notable main crater activity in January 1958 and in 1960, the only occasion on which large-scale main crater activity seems to be indicated is in 1898-1900, when 'the glow of the fresh lava streams' could be seen on the northern slopes (von Hesse-Wartegg, 1902). Pending more detailed study of eruptive activity since the 1950s it is not possible to deduce any clear-cut pattern of interplay between the southern and main craters.

ACKNOWLEDGEMENTS

This paper is published with the permission of both the Director of the Bureau of Mineral Resources, Canberra, and of the Chief Government Geologist of the Geological Survey of Papua New Guinea, Port Moresby. Helpful discussions were had with R.W. Johnson and D.H. Blake, and the great assistance given by E.H. Feeken and W.B. Dallwitz in making translations is acknowledged.

REFERENCES

Behrmann, W., 1917: Der Sepik (Kaiserin-Augusta-Fluss) und sein Stromgebiet. Mitteilungen aus den deutschen Schutzgebieten. *Ergänzungshaft Nr 12*, 69 (local translation).

Best, J.G., 1956: Investigations of recent volcanic activity in the Territory of Papua and New Guinea. *8th Pac.Sc. Cong., Proc. 2*, pp. 180-204.

Branch, C.D., 1967: April 1964 eruption of Manam volcano. *Bur. Miner. Resour. Aust. Rep. 107*, pp. 27-35.

Finsch, O., 1888: Samoafahrten — Reisen in Kaiser-Wilhelms-Land und Englisch-Neu-Guinea, 296, 367. Leipzig (local translation).

Fisher, N.H., 1939: Report on the volcanoes of the Territory of New Guinea. *Terr. N. Guin. geol. Bull. 2.*

Fisher, N.H., 1957: Catalogue of the active volcanoes of the world, including solfatara fields — Pt V, Melanesia. *Int. volc. Ass.*, Naples.

Grabowsky, F., 1895: Der Bezirk von Hatzfeldthafen und sein Bewohner. *Petermanns Mitt. 41*, pp. 186-189 (local translation).

Hammer, K.L., 1907: Die geographische Verbreitung der vulkanischen Gebilde und Erscheinungen im Bismarckarchipel und auf den Salomonen. Giessen (local translation).

Hesse-Wartegg, von, E., 1902: Samoa, Bismarckachipel und Neuguinea drei deutsche Kolonien in der Südsee, 42. Leipzig (local translation).

Höltker, G., 1942: Meine Beobachtungen über die Vulkantätigkeit in Kaiser-Wilhelms-Land (Neu Guinea) 1936-1939. *Z. dtsch. geol. Ges., 94*, pp. 550-560 (local translation).

Kear, D., 1957: Erosional stages of volcanic cones as indicators of age. *N.Z. Sci. Techn., B, 38*, pp. 671-682.

McDougall, I., 1964: Potassium-argon ages from lavas of the Hawaiian Islands. *Geol. Soc. Amer. Bull. 75*, pp. 107-128.

McDougall, I., 1971: The geochronology and evolution of the young volcanic island of Reunion, Indian Ocean. *Geochim. Cosmochim. Acta 35*, pp. 261-288.

Miklucho-Maclay, N. von, 1878: Uber vulkanische Erscheinungen an der nordostlichen Kuste Neu Guineas. *Petermanns Mitt. 24*, pp. 408-410 (local translation).

Morell, B., 1832: *A narrative of four voyages to the South Sea, etc*, pp. 462-463. Harper, New York.

Morgan, W.R., 1966: A note on the petrology of some lava types from east New Guinea. *J. geol. Soc. Aust., 13*, pp. 583-591.

Pflüger, A., 1901: Smaragdinseln der Südsee, 204. Bonn (local translation).

Pöch, R., 1907a: Travels in German, British and Dutch New Guinea. *Geog. J. 30*, 609.

Pöch, R., 1907b: Über meine Reisen in Deutsch-Britisch-und Niederländisch-Neu Guinea. *Zeit. der Gesellschaft für Erdkunde*, pp. 150-151 (local translation).

Pöch, R., 1908: Reisen an der Nordküste von Kaiser Wilhelmsland. *Globus. Bd. XCIII Nr. 10*, pp. 150-151 (local translation).

Reynolds, M.A., 1957: Eruption of Manam Volcano, Territory of New Guinea, December 1956 — February 1957. *Bur. Miner. Resour. Aust. Rec.* 1957/43 (unpubl.).

Ruxton, B.P. & McDougall, I., 1967: Denudation rates in northeast Papua from potassium-argon dating of lavas. *Amer. J. Sci. 265*, pp. 545-561.

Sapper, K., 1917: Melanesische Vulkanzone. *in* Katalog der geschichtlichen Vulkanausbruche Schr. Wiss. Ges. Strassburg, 27, pp. 204-215 (local translation).

Schleinitz, A., 1889: Beschreibung der Nordküste von Kaiser Wilhelms-Land von Kap Cretin bis zu den Leguarant-Inseln. Nachrichten über Kaiser Wilhelms-Land und den Bismarck-Archipel, 87. Berlin (local translation).

Scholz, W., 1912: Der Kaiserliche Bezirksamtmann Dr Scholz in Friedrich Wilhelmshafen. *Amtsblatt fur das Schutzgebiet Neuguinea, Jahrgang 4 Nr 1*, pp. 36-7 (local translation).

Sharp, A., 1960: *The Discovery of the Pacific Islands*. Oxford.

Sharp, A., 1968: *The Voyages of Abel Janszoon Tasman*. Oxford.

Spilbergen, J., 1906: The East and West Indian Mirror. *Hakluyt Society*, 2nd series, 18.

Stanley, E.R., 1923: Report on the salient geological features and natural resources of the New Guinea Territory. *Rep. on the Territory of New Guinea, 1921-22*, pp. 52-53.

Tappenbeck, E., 1901: Deutsch Neuguinea, 12. Berlin (local translation).

Taylor, G.A.M., 1958: Notes on the current eruption at Manam. *Bur. Miner. Resour. Aust. Rec.* 1958/67 (unpubl.).

Taylor, G.A.M., 1963: Seismic and tilt phenomena preceding a Pelean type eruption from a basaltic volcano. *Bull. Volc., 25*, pp. 5-11.

Wedgwood, C.H., 1934: Report on research in Manam Island, Mandated Territory of New Guinea. *Oceania IV*, 377.

Wichmann, A., 1909: *Nova Guinea. Vol. 1*, 26. Leiden.

Williamson, J.A., (Ed.), 1939: *A voyage to New Holland (by William Dampier)*. London, Argonaut Press.

Zdekauer, 1899: Über die Sundainseln nach Neuguinea und dem Bismarckarchipel.

Zöller, H., 1891: Deutsch-Neuguinea und meine Ersteigung des Finisterre-Gebirges. *Union Deutsche Verlags Gesellschaft*, pp. 32, 160 (local translation).

VOLCANIC HISTORY OF LONG ISLAND, PAPUA NEW GUINEA

E.E. BALL and R.W. JOHNSON

Department of Neurobiology, Research School of Biological Sciences, Australian National University, P.O. Box 475, Canberra City, A.C.T. 2601

Bureau of Mineral Resources, P.O. Box 378, Canberra City, A.C.T. 2601

ABSTRACT

Long is a Quaternary volcanic island at the southern margin of the Bismarck Sea. It consists of two stratovolcanoes at either end of a caldera complex containing a freshwater lake, Lake Wisdom. The outer flanks of the island are covered by a thick pyroclastic mantle which, from geological evidence, the accounts of early explorers, and stories of the islanders, is thought to have been deposited during a major eruption sometime during the first half of the period 1700-1827 A.D. This eruption probably coincided with a period of cauldron subsidence. Subsequent volcanic activity in the southern part of Lake Wisdom built up Motmot Island whose eruptions have been recorded in 1953-55, 1968, and 1973-74.

INTRODUCTION

Long Island is a Quaternary volcanic complex off the north coast of mainland Papua New Guinea (Fig. 1). It was named in 1700 A.D. by the explorer William Dampier who described it as a 'long Island, with a high Hill at each End'. The island consists of two extinct stratovolcanoes at either end of a freshwater caldera lake, Lake Wisdom, and an active volcanic centre whose eruptions since the Second World War have built up an island, Motmot, in the southern part of the lake.

Long, also known as Arop (or Ahrup), is one of several volcanic islands off the north coast of mainland Papua New Guinea (Fig. 1). These islands form part of a 1000 km-long chain of volcanoes along the southern margin of the Bismarck Sea, between the Schouten Islands in the west and Rabaul in the east (see Johnson, this volume). Long Island, Crown Island, and Hankow Reef form a line which intersects the general trend of the volcanic chain at about 30°, and Long and Crown rise from the same northwest-trending submarine ridge (Fig. 1). The western part of the volcanic chain is associated with a belt of seismicity that defines the boundary between the Indo-Australian plate in the south and the 'South Bismarck' plate in the north (Johnson & Molnar, 1972). During the last 15-20 years, intermediate-focus earthquakes (between 150 and 230 km deep) have been especially frequent immediately south of Long Island, where they define an almost vertical zone (Denham, 1969; Johnson et al., 1971; Curtis, 1973).

Except for some information given by Fisher (1939, 1957), Taylor (1953, 1956), and Best (1956), little was previously known about the geological history of Long Island. However, geologists from the Central Volcanological Observatory (Rabaul) who have visited the island to investigate the post-war volcanic activity of Motmot, have been struck by the evidence for a major recent eruption on Long Island. G.A.M. Taylor, in particular, noticed the young, widespread, volcaniclastic deposits on the island and, hearing islanders' stories which were consistent with the

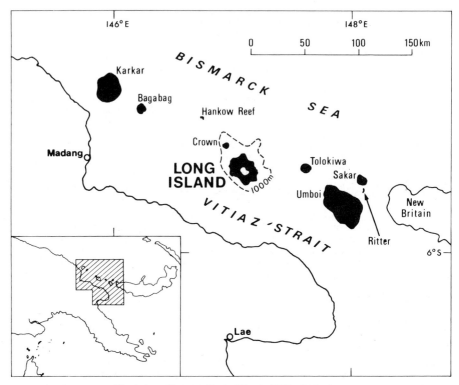

Fig. 1: Locality map. Dashed line is 1000 m isobath.

geology, concluded that a catastrophic volcanic eruption had devastated the island about one or two centuries before (Taylor, 1953; pers. comm. 1971).

GENERAL INFORMATION

Long Island is in fact nearly round (Fig. 2); but when viewed from the sea from many directions it appears to consist of two peaks connected by a long, low-lying isthmus, and it is therefore understandable why Dampier named the island as he did (see Fig. 5 and later discussion). The maximum diameter of Long Island is about 30 km, and the total area is about 425 km².

Mount Reaumur*, which dominates the north end of the island, is an eroded stratovolcano with three main peaks, the highest of which reaches 1304 m above sea level (Fig. 3). The other stratovolcano is Cerisy Peak which rises to 914 m above sea level at the south end of the island. Both volcanoes flank a larger, central volcanic complex whose centre collapsed to form the steep-sided caldera now occupied by Lake Wisdom. The two stratovolcanoes are older than the last major eruption from the central volcanic complex, which produced an extensive pyroclastic mantle on Long Island. In addition to Mount Reaumur and Cerisy Peak, smaller satellite hills, which were probably minor eruptive centres, are

_____

* Although the peak was named after the French physicist R.A.F. de Reaumur, recent maps show the spelling "Reumur" or "Reamur".

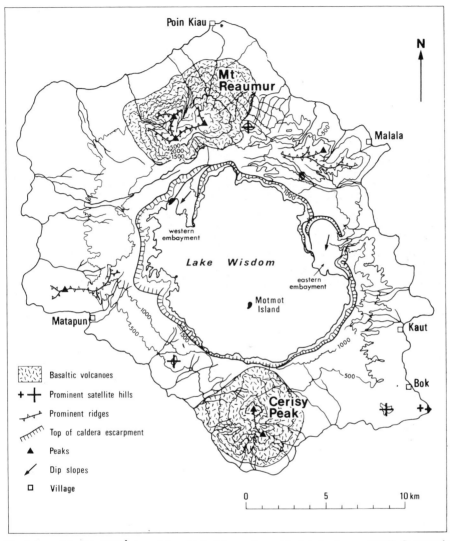

Fig. 2: Long Island in 1972, adapted from United States Army 1 inch to 1 mile map (1943). Contours in feet.

present immediately east of Mount Reaumur, 6 km northwest of Cerisy Peak, and south of Bok village (Fig. 2).

The surface of Lake Wisdom is about 190 m above sea level, and most of the lake is surrounded by caldera wall cliffs between 200 and 300 m high (Fig. 3). The lake is nearly circular; it has a maximum width of about 11.5 km, and covers about 95 km². Using a fathometer, Ball & Glucksman (in prep.) found that much of Lake Wisdom has depths greater than 300 m. Repeated soundings in one of the deeper parts of the lake gave depths in the vicinity of 360 m. A water sample from just above lake bottom at this depth was fresh, indicating that the caldera is probably sealed from the sea. The lake level shows annual fluctuations of 1-1.5 m.

Fig. 3: Southern flank of Mount Reaumur and northeastern wall of the Lake Wisdom caldera, viewed northwards from the western shore of Lake Wisdom showing part of the western embayment (cf. Fig. 2).

Lake Wisdom has no surface drainage, but it probably feeds springs which emerge from the outer flanks of the island. The most important springs join to form a stream which originates below a low point on the eastern rim of the caldera, and which is the only one on the island that flows to the sea year-round. In 1972, following a period of six months without rain, this stream was still running. According to the islanders, it started flowing during the 1930s (Taylor, 1953). Another stream, about half-way between the first one and Malala village, also carries substantial volumes of water, but disappears underground before reaching the sea. Other springs have been reported to emerge at or near the coast on the south and west sides of the island (D.A. Wallace, I. Hughes, pers. comms., 1974, 1975).

The recent major volcanic eruption greatly affected the flora, fauna, and human population of the island. Today, Long Island is fully vegetated, but the forest below 900 m has the appearance of being quite young. It is certainly more open than the forests on many of the nearby islands. However, judging from the accounts of visitors to Long Island throughout this century (Coultas, 1933-35; Evans, 1939; Taylor, 1953), who also commented on the youthful appearance of the forest, it appears likely that the lowland vegetation has reached a stunted steady-state caused by the extreme porosity of the soil (cf. Taylor, 1953). The montane forest above 900 m, however, which is almost always in clouds, is similar to that found on the peaks of adjacent volcanic islands (Diamond, 1974).

The faunas of Long and the adjacent islands are in general so poorly known that it is difficult to say whether Long is missing species that would be expected to be present. However, Diamond (1974) has found that Long has far fewer montane bird species, and more 'supertramp' species, than would be predicted on the basis of its size and elevation, and this may indicate that the island was devastated in the not-too-distant past. The human population presents a similar picture, as Long has been either uninhabited or only sparsely inhabited since 1700 (Dampier). Since about 1930, however, the population has been increasing rapidly, reaching 697 in 1972.

CENTRAL VOLCANIC COMPLEX

Rocks produced from the central volcanic complex, and deposited before the major eruption that formed the pyroclastic mantle of Long Island, are exposed in the walls of the Lake Wisdom caldera and, to a much lesser extent, on the northeastern coast of the island. The base of the pyroclastic mantle has not been identified in the caldera wall, but it seems that at least the lower half of the wall consists of materials laid down by earlier eruptions of the central complex.

Parts of the eastern and western sides of the caldera wall have been examined, and consist mainly of volcaniclastic deposits, particularly air-fall deposits, but also minor lava flows and some possible nuées ardentes deposits. Many of the deposits appear to have been reworked.

On the northeast and west sides of the island two prominent ridges extend outwards from the northeast and west sides of the caldera, and maintain a more-or-less constant elevation before dropping steeply to the coast (Fig. 2). Both ridges may represent outlying centres of volcanism which have been modified by deposition of later volcaniclastic deposits.

It is difficult to assess the form of the central volcanic complex prior to the formation of the present-day caldera. The caldera has obliterated the centres of eruption, which could have had any, or a combination, of the following forms: a single central crater; a widespread group of craters; a row of craters, or a fissure (perhaps on the line between Mount Reaumur and Cerisy Peak); a caldera, or calderas, smaller than the present-day caldera; etc. However, the complex was probably unlike the large stratovolcanoes found elsewhere along the southern margin of the Bismarck Sea, whose flanks slope at up to about 35^0 (e.g., Manam — Fisher, 1957; Ulawun—Johnson et al., 1972). Because lava flows appear to be extremely rare in the Lake Wisdom caldera walls, it is thought the central complex probably had the form of a large pyroclastic volcano, whose flanks sloped at perhaps no more than about 10^0 (cf. 'ash flow volcanoes' described by Macdonald, 1972). In profile, therefore, the complex may have been low-lying and, depending on the form and distribution of its eruptive centres and on the rate at which reworked clastic deposits were redeposited at its periphery, the profile could have been more-or-less flat-topped.

Five rocks from the central complex have been chemically analysed (Morgan, 1966; Johnson, in prep.). Three are basalts (53% SiO_2, or less), and two are low-silica andesites (more than 53, and up to 57% SiO_2).

MOUNT REAUMUR AND CERISY PEAK VOLCANOES

Mount Reaumur is an eroded, steep-sided volcano, and forms the highest point on Long Island (Fig. 3). Its base is surrounded by thick pumiceous materials deposited mainly by nuées ardentes and lahars (see below). The upper part of the

Fig. 4: Deposits of the pyroclastic mantle on the northeast coast of Long Island. The lower two-thirds of the cliff are made up of deposits of the 'middle unit', which are overlaid by bedded 'upper unit' deposits containing accretionary lapilli; the 'basal unit' is not exposed here. Cliffs are about 5 m high.

volcano consists of a series of ridges which define two watersheds, each drained by a stream valley which reaches the coast (Fig. 2). Porphyritic lava flows and intercalated scoriae and lapilli crop out in the eastern stream up to at least 720 m above sea level.

Cerisy Peak is a similar steep-sided and deeply dissected volcano. Stream courses radiate from a 2 km-long ridge at the summit of the volcano, and at the northern end of the ridge is the highest point which, when viewed from the south and southwest, has a dome-shaped profile. Loose boulders in a stream on the southwest side of the volcano are thought to have been derived from minor intrusions which may be exposed at higher levels.

The rocks of both volcanoes are basalt. Of four samples chemically analysed, three are olivine tholeiite and one is a quartz tholeiite (Johnson, in prep.).

PYROCLASTIC MANTLE

Poorly consolidated volcaniclastic deposits are exposed along most of the coastline of Long Island (Fig. 4). They cover the surface of the central volcanic complex, constitute the upper part of the caldera wall, and probably mantle most of Mount Reaumur and Cerisy Peak volcanoes.

The pyroclastic mantle shows a range of thickness and lithology, but overall it appears to consist of three parts: 'basal' and 'upper' well-bedded units separated by a 'middle' unsorted unit containing tree trunks and branches. This sequence suggests that a series of explosive eruptions preceded a catastrophic eruptive phase, which was followed by a period of waning explosive events.

Basal unit

The lower bedded unit crops out at the base of cliffs along many parts of the coast, but the bottom of the unit has not been observed. The maximum measured thickness is about 4 m at a point on the west coast where the sequence consists of internally bedded and unbedded lapilli and ash layers, one of which contains uncharred leaf remains. Taylor (1953) described a similar sequence from the east coast of the island: 'Numerous plant fossils and imprints left by standing trees' were contained in the basal part of this sequence, which was overlain by a layer of accretionary lapilli, a lapillus layer, another accretionary lapillus layer, and a 'chocolate tuff containing marine fossils'. This sequence suggests initial subaerial deposition followed by subsidence and submarine deposition and later uplift. Layers of accretionary lapilli and coral fragments are present in the basal unit at other coastal exposures.

Middle unit

The middle unit is between 3 and 5 m thick on the coast (Fig. 4). Unlike the basal unit, it is unsorted and unbedded, and in places shows crude columnar jointing. The middle unit consists mainly of pumice lapilli, accretionary lapilli, glassy angular lava chips, and fragments of crystalline lava; these are enclosed in a light-coloured earthy matrix. At numerous localities, both carbonised and uncarbonised fallen tree trunks and branches are present, particularly near the base of the middle unit.

The middle unit was almost certainly deposited by pyroclastic flows, but at several localities it is difficult to judge whether the deposits are those of nuées ardentes or of lahars. However, the deposits containing accretionary lapilli and uncarbonised wood were probably laid down by cold lahars, or slurries, rather than by hot nuées ardentes, which are more likely to disintegrate accretionary lapilli and char vegetation. Exposures of the pyroclastic mantle in the western stream draining Mount Reaumur (Fig. 2) show thicknesses of over 30 m (similar values were reported by Taylor (1953) from the east flank of Long Island), and provide evidence for at least two superincumbent flow-units.

Deposits of the middle unit are found only on the lower slopes of Long Island. They appear to be absent from the upper parts of Mount Reaumur and Cerisy Peak volcanoes, and possibly also from other high points on the island. During their movement down the outer flanks of the central volcanic complex, the pyroclastic flows moved around obstacles. Mount Reaumur, for example, is completely surrounded by middle-unit deposits; and the satellite hill southwest of Bok, and the small hills east of it, may have been offshore islands and sea stacks before the deposits were laid around them. Unless immediately removed by the sea, the pyroclastic mantle, and especially deposits of the middle unit, may have extended the coastline of Long Island, particularly on its southeastern and northwestern sides. However, if the east coast was extended, later erosion and subsidence must have taken place because the remains of coastal settlements buried by the middle unit are now below mean sea-level and have been almost entirely removed by the sea (I. Hughes, pers. comm., 1975).

Upper unit

The upper bedded unit occupies the high part of many coastal cliff sections (e.g. Fig. 4), but is absent in other places, presumably having been removed by erosion. At most exposures the unit is only about 1 m thick, or less, but at a few localities the thickness is more than 2.5 m. The upper unit consists of well-bedded lapilli, and commonly contains well-formed accretionary lapilli up to 1.5 cm in diameter. These accretionary lapilli are also found in the soil cover which extends up to the rim of the caldera. No wood fragments appear to be present in the upper unit.

The following evidence indicates that rain storms accompanied deposition of the pyroclastic mantle: cross-bedding and infilled stream channels, particularly in the basal and upper units; the presence of accretionary lapilli, formed when fine volcanic ash aggregated during rain storms; and the presence of mudflow deposits.

Two vesicular clasts from the pyroclastic mantle have been chemically analysed. They have the highest silica contents (56.7 and 58.5%) of all the analysed rocks from Long Island, and although not necessarily representative of the entire pyroclastic mantle, they suggest that the major eruption produced the most fractionated rocks on Long Island.

CALDERA AND ITS ORIGIN

Lake Wisdom caldera is 13.5 km from northwest to southeast and 11 km from southwest to northeast, and has an area of about 120 km^2 (Fig. 2). It is slightly larger than, for example, Hargy and Dakataua calderas in New Britain (Fisher, 1957; Branch, 1967; Lowder & Carmichael, 1970), and Crater Lake, Oregon (Williams, 1942). The highest parts of the caldera wall are in the north and west, where the rim is at least 490 m above sea level (Fig. 3). The lowest point on the caldera rim is a breach in the eastern wall, at least 30 m above lake level.

The caldera has two prominent embayments on its western and eastern sides (Fig. 2). The western embayment is defined by two peninsulas which project into the lake (Fig. 3). The northern peninsula has a steep escarpment on its eastern side, and the western part appears to be the southwest-dipping constructional surface of the central volcanic complex. The embayment seems to have formed by partial collapse of a separate arcuate segment, the northern end of which remained attached to the northern caldera wall. Water depths in this embayment are mostly less than 150 m (Ball & Glucksman, in prep.). In contrast, the eastern embayment is not flooded by the lake (Fig. 2); its forest-covered base dips southwards, and no high cliffs are preserved at lake-level.

Volcaniclastic beds in the southern peninsula of the western embayment, and in cliffs west and northwest of the eastern embayment, dip inwards towards the caldera. These inward dips could be due to mantling of parts of the caldera escarpment, but it is more likely they are the result of downwarping which accompanied collapse of both embayments. Elsewhere in the caldera wall similar beds appear to dip away from the caldera.

The pyroclastic mantle of Long Island closely resembles in lithology, thickness, and extent the pyroclastic mantles associated with many calderas, throughout the world, in which a close genetic relationship is implied, or can be proven, between cauldron subsidence and the eruption of copious pyroclastic materials (cf. Williams, 1941; McBirney & Williams, 1969). In the field, however, it is extremely difficult to identify the precise geological relationship between the eruptions and subsidence. It is impossible to say, for example, if the eruptions took place from central vents, from caldera ring faults, or from ring fractures associated with the caldera; if cauldron subsidence took place before, during, or after the eruptions; or if the subsidence was caused by deep-seated withdrawal of magma, by gravitational collapse into a less-dense body of magma, by subsidence into a chamber evacuated by a catastrophic eruption, or by the initial formation of cone-sheet fractures (by doming) and subsequent subsidence. On Long Island it is also not certain that the Lake Wisdom caldera was formed entirely during the period of deposition of the pyroclastic mantle. Indeed, the abundance of volcaniclastic materials in the lower parts of the caldera wall could be taken to indicate the contrary — that a *series* of subsidence events took place, each associated with a major pyroclastic eruption, and that only the last of these produced the caldera as it is seen today.

AGE OF THE LAST MAJOR ERUPTION

Several lines of evidence can be used in an attempt to date cauldron subsidence and deposition of the pyroclastic mantle. These include direct geological observations and carbon-14 dating, the accounts of early explorers, and stories told by the islanders.

Geological evidence

The young age of the pyroclastic mantle is suggested by the poor consolidation of the deposits and by the fresh appearance of the uncarbonised wood remains, which seem to have undergone little alteration since their burial. G.A.M. Taylor also obtained carbon-14 dates from two samples of wood collected in the early 1950s — one from a carbonised tree trunk from the base of the middle unit on the east coast, the other an uncarbonised sample from a clastic layer in the eastern part of the caldera wall. Both samples gave dates of not older than the beginning of the 19th century (unpublished BMR data). More recently, however, I. Hughes (pers. comm., 1975) obtained two carbon-14 dates of 1720 ± 75 and 1750 ± 65 A.D. on separate charcoal and charred wood samples from the middle unit on the northwest coast. Three carbon-14 dates from coastal habitation sites showed that the island had previously been inhabited for at least 800 years, though not necessarily continuously.

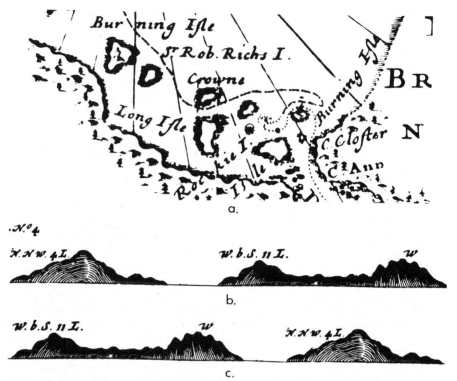

Fig. 5: Facsimiles of diagrams drawn by William Dampier during his voyage in 1700 A.D. (Williamson, 1939). (a) Detail from chart showing Dampier's course (cf. Fig. 1); note in particular the elongate ('long') outline of Long Island (cf. Fig. 2). (b) Profiles of islands in Table 13, sketch no. 4, of Dampier. (c) The islands of *b* transposed to give a consistent change in bearings across the diagram; the left-hand island is Long, and the right-hand one is identified as Tolokiwa.

Accounts of explorers

The earliest recorded description of Long Island was by William Dampier in 1700*.

> The 31st in the Forenoon we shot in between 2 Islands, lying about 4 Leagues asunder; with Intention to pass between them. The Southernmost is a long Island, with a high Hill at each End; this I named *Long Island*. The Northermost is a round high Island towering up with several Heads or Tops, something resembling a Crown; this I named *Crown-Isle*, from its Form. Both these Islands appear'd very pleasant, having Spots of green Savannahs mixt among the Wood-land: The Trees appeared very green and flourishing, and some of them looked white and full of Blossoms. We past close by *Crown-Isle*, saw many Coco-nut-Trees on the Bays and the Sides of the Hills; and one Boat was coming off from the Shore, but return'd again. We saw no Smoaks on either of the Islands, neither did we see any Plantations; and it is probable that they are not very well peopled. We saw many Shoals near *Crown-Island* and Riffs of Rocks running off from the Points, a Mile or more into the Sea. My Boat was once over-board, with Design to have sent her ashore; but having little Wind, and seeing some Shoals, I hoisted her in again, and stood off out of Danger.

* Dampier's *'Voyage to New Holland'* was originally published in two parts — the first in 1703, and the second in 1709. These works have since been republished many times, under different editorships. The editions consulted for this paper are those of Masefield (1906) and Williamson (1939).

Dampier also made profile drawings of the islands he saw along the north coast of mainland Papua New Guinea. Although the drawings are unlabelled, and no cross-reference is made with the text*, analyses by Reche (1914) and R.J.S. Cooke (pers. comm., 1974) indicate that Long Island should be shown in Table 13, sketches 4 and 5, of Dampier's book. In sketch 4 (Fig. 5b) the bearings are not consistent with the way the islands are drawn. However, when the sketch is redrawn according to the bearings (Fig. 5c), the new arrangement is consistent with the islands being Long and Tolokiwa (R.J.S. Cooke, pers, comm., 1974).

Dampier did not land on Long Island, and his sketch map (Fig. 5a, showing no caldera) is simply a coastal outline. The profile illustrated in Figure 5b & c, however, shows the interior as low-lying and flat-topped, which, as previously discussed, does not necessarily indicate the presence of a central caldera at that time, but does exclude the presence of any large central cones.

Dumont D'Urville (1833) sailed past Long Island in August 1827, named the two main peaks, and observed that 'The ground in the vicinity of the shore appeared more arid than all the other islands . . . we saw neither coconut trees nor any trace of inhabitants'.

In 1884, Finsch (1888) visited Long Island, which he described as being heavily wooded throughout, although showing more undergrowth-covered areas than Karkar Island to the west (Fig. 1). He noted two or three small settlements along the coast of Long. Later reports also agree that the island was fully vegetated, although three observers — Coultas (1933-1935), Evans (1939), and Taylor (1953) — all suggested that the vegetation indicated the possibility of a recent eruption.

Stories of islanders

Stories concerning the eruption of Long Island have been collected from the local people on many different occasions. Unfortunately, the stories are not consistent. Coultas (1933-1935), for example, stated:

> According to native legends, Ahrup was at one time a large active volcano, much higher than Tolokiwa, and with a large population. Eventually an eruption occurred which blew the cone completely out of the center of the island, throwing out hot stones and lava and killing the people, with the exception of one woman who escaped in a canoe to the mainland of New Guinea where her descendants are supposed to be living now.

Taylor (1953) wrote:

> Recent investigation by A.D.O. [Assistant District Officer] Parish suggests that the eruption was of comparatively recent origin as stories of the escape from Arop are still current among natives of the surrounding islands. It seems evident that some very alarming warning phenomena preceded this eruption as a considerable number of natives appear to have escaped from the island before the catastrophic eruption took place.

> Mr. Parish believes that the Siassi Island people originally come from Long Island, and has found, on the harsher parts of the neighbouring New Guinea coast, settlements of natives who are also evacuees. One group, he believes, settled near Lutheran Anchorage on northern Umboi but were subsequently wiped out by the 1888 eruption of Ritter Island.

* According to Williamson (1939, viii), this absence of a close correlation between the illustrations and the text is due 'to the fact that Dampier was at sea when the second part of his book was printed and was dead when the second edition appeared in 1729'.

Stories concerning the recolonisation of the island are another potentially useful source for dating the eruption. Data collected by several investigators indicate that some of the people now living on the island are fifth and sixth generation islanders, and the islanders claim their ancestors returned to the island at a time when the vegetation was just starting to become re-established. However, as pointed out by T. Harding (pers. comm., 1972), dating by this technique can be uncertain, because many Melanesian cultures tend to maintain their genealogies to only a certain number of generations, and then drop out additional generations between themselves and the founding ancestors.

Conclusions

The above evidence indicates that the major eruption took place sometime during the first half of the eighteenth century. To our knowledge no observations of Long Island were recorded between 1700 (Dampier) and 1827 (Dumont D'Urville) and this is the only record-free interval in the post-1700 period long enough for revegetation to have taken place following an eruption without the post-eruption devastation being noted.

The relationship of this major eruption to the time of formation of the Lake Wisdom caldera is less clear. It is possible that cauldron subsidence was a short-lived event that took place at the same time as the eruption; but the possibility cannot be excluded that the caldera has had a long and complex history consisting of several periods of subsidence and associated eruptions. In either case, the central volcanic complex of Long Island may never have had a high central peak; or if it did, the peak had disappeared by 1700 A.D.

POST-CALDERA ACTIVITY

The first well-documented records of volcanic activity in Lake Wisdom are aerial photographs taken in 1943, which show a horseshoe-shaped, low-lying island crater about 3 km from the southern shore of the lake. The photographs show material from the active crater discharging into the lake. However, after eruptions in 1953, islanders who had formerly been unable to provide information about previous eruptions informed Best (1956) that there had been minor eruptions within the lake in 1933, 1938, and 1943. According to N.H. Fisher (pers. comm., 1972) no island was visible from the caldera rim in 1938 (cf. Fisher, 1939). Taylor (1953) and J. G. Best visited Lake Wisdom in August 1952, but saw no sign of volcanic activity, and from lake-level were unable to see traces of any island.

In May 1953, volcanic activity was reported from the same site in Lake Wisdom (Best, 1956). An island was created consisting of two contiguous craters which joined to form a ridge about 400 m long, 100 m wide, and 30 m high. According to Taylor (1956), volcanic activity continued periodically until 7 January 1954. Incandescent ash was emitted from the same site on 5 June 1955, and activity is said to have reached a climax on 13 June before terminating abruptly (Fisher, 1957).

Except for fumarolic activity observed in 1961 (unpublished data, Central Volcanological Observatory), the island appears to have remained quiescent until 1968. During this interval, wind and waves eroded the island until only three small islets remained. In March 1968, a new period of explosive activity began, and

Fig. 6: Motmot Island from the southwest in November 1969, showing the smaller, un-named island west of its southern end.

created an island about 300 m long by 180 m wide (G.W. D'Addario, pers. comm., 1972). Further activity took place, and by November 1969 a second small island could be seen just above water level, separated from the main island by a channel about 1 m deep (Fig. 6). On all subsequent visits this second island could not be seen although a shoal was still present below lake level.

Conditions on Motmot in November 1969 were described by Bassot & Ball (1972) and conditions up to November 1972 have been summarised by Ball & Glucksman (in press). The situation in November 1972 is described in Figure 7. The bathymetric map of Ball & Glucksman (in prep.) suggests that Motmot caps a slightly elongated cone whose northeast-trending axis extends southwestwards towards a submarine ridge running out from the caldera wall. Another submarine ridge trends east-southeastwards from the peninsula on the west side of the lake (Fig. 1) as far as the Motmot cone. This ridge has a high area about half-way between the caldera wall and Motmot. It is unknown if these submarine ridges are the result of post-caldera volcanic activity or if they represent submerged parts of the pre-caldera central volcanic complex.

Ball & Glucksman (in press) measured water temperatures at various points around the circumference of Motmot in 1969, 1971, and 1972. Shifts in the position of the thermometer by only a few centimetres caused a change of more than 20^{0}C in the recorded temperature, but the following generalised trends could be inferred from the data: (1) a general cooling during the interval 1969-1972; (2) rising temperatures in the crater pond, and on the east side of the island north of the

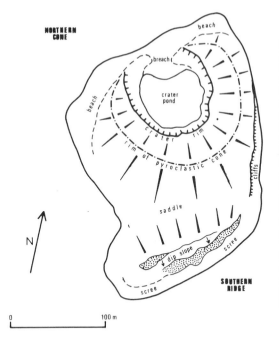

Fig. 7: Sketch-map of Motmot Island, Lake Wisdom, in November 1972. The southern part of the island consists of a northeast-southwest ridge, about 23 m high, made up of poorly consolidated ash and lapilli beds that dip southwards at about 30°. Bedding is especially prominent in those parts of the ridge marked by the stippling. The ridge is surrounded by scree slopes. The northern part of the island is a pyroclastic cone containing a crater, which is partly filled by a pond. The cone consists of bedded ash, lapilli, bombs, and blocks. The crater rim was continuous as late as October 1971, but aerial photographs taken in March 1972 showed that the crater wall had been breached, and that the pond was connected with Lake Wisdom. This sketch shows that by November 1972 the breach had been partly sealed again by beach deposits that form a fan at the edge of the crater pond.

saddle, in 1972. A decrease in the island's circumference at the water line, from 892 m (in 1969) to 845 m (1971) to 814 m (1972) was also noted.

The only activity observed between 1969 and 1972 was the opening of a vent, about 1.5 m in diameter, in the crater wall on the eastern shore of the pond between October 1971 and November 1972. This hole was probably caused by a gas or steam explosion; no molten material appears to have been expelled. Motmot was again intermittently active between May 1973 and February 1974, producing lava flows (see Cooke et al., this volume).

Two rocks formed during the 1968 eruption of Motmot (Johnson, in prep.) and four formed in 1973-74 (Cooke et al., this volume) have been chemically analysed. All are tholeiitic basalts.

CONCLUSION AND ACKNOWLEDGEMENTS

This paper provides a broad overview of the volcanic evolution of Long Island. It is a preliminary account which we hope will stimulate further, more detailed, geological work on the island.

We thank I. Hughes and R.J.S. Cooke for their criticisms of the draft manuscript and for supplying some of the factual data, and gratefully acknowledge reviews by referees D.H. Blake and J.G. Jones. The paper is published with permission of the Director of Bureau of Mineral Resources, Canberra.

REFERENCES

Ball, E.E. & Glucksman, J., in prep.,: A limnological survey of Lake Wisdom, Long Island, Papua New Guinea.

Ball, E.E.,& Glucksman, J., in press: Biological colonization of Motmot, a recently-created tropical island. *Proc. R. Soc. Lond. B.*

Bassot, J.M., & Ball, E.E. 1972: Biological colonization of recently created islands in Lake Wisdom, Long Island, Papua New Guinea, with observations on the fauna of the lake. *Papua New Guinea Sci. Soc. Proc. 1971, 23,* pp. 26-35.

Best, J.G., 1956: Investigations of recent volcanic activity in the Territory of New Guinea. *8th Pac. Sci. Cong., Proc., 2,* pp. 180-204.

Branch, C.D., 1967: Short papers from the Volcanological Observatory, Rabaul, New Britain. *Bur. Miner. Resour. Aust. Rep. 107.*

Coultas, W.F., 1933-1935: Unpublished journal and letters of William F. Coultas, v. IV, Whitney South-Sea Expedition, October 1933-March 1935.

Curtis, J.W., 1973: Plate tectonics and the Papua-New Guinea-Solomon Islands region. *J. geol. Soc. Aust., 20,* pp. 21-36.

Denham, D., 1969: Distribution of earthquakes in the New Guinea-Solomon Islands region. *J. geophys. Res., 74,* pp. 4290-4299.

Diamond, J.M., 1974: Colonization of exploded volcanic islands by birds: the supertramp strategy. *Science, 184,* pp. 803-806.

Dumont D'Urville, J.S.C., 1833: *Voyage de decouvertes de l'Astrolabe execute por ordre du Roi, pendant les annees 1826-27-28-29 sous le commandement de M.J. Dumont D'Urville. Capitaine de Vaisseau: Historie de Voyage.* v. 4 Paris.

Evans, G. 1939: The characteristic vegetation of recent volcanic islands in the Pacific. *Kew Bulletin (1939),* pp. 43-44.

Finsch, O., 1888: *Samoafahrten — Reisen in Kaiser Wilhelms-Land und Englisch-Neu Guinea in den Jahren 1884 u 1885 an Bord des Deutschen Dampfers "Samoa".* pp. 188-189. Ferdinand Hirt und Sohn, Leipzig.

Fisher, N.H., 1939: Report on the volcanoes of the Territory of New Guinea. *Terr. N. Guin. geol. Bull. 2.*

Fisher, N.H., 1957: *Catalogue of the active volcanoes of the world, including solfatara fields. Part 5, Melanesia.* Int. volc. Assoc., Naples

Johnson, R.W., in prep.: Late Cainozoic volcanoes at the southern margin of the Bismarck Sea, Papua New Guinea.

Johnson, R.W., Mackenzie, D.E., & Smith, I.E., 1971: Seismicity and late Cenozoic volcanism in parts of Papua-New Guinea. *Tectonophysics, 12,* pp. 15-22.

Johnson, R.W., Davies, R.A., & White, A.J.R., 1972: Ulawun volcano, New Britain. *Bur. Miner. Resour. Aust. Bull. 142,* 42 pp.

Johnson, T., & Molnar, P., 1972: Focal mechanisms and plate tectonics of the southwest Pacific. *J. geophys. Res., 77,* pp. 5000-5032.

Lowder, G.G., & Carmichael, I.S.E., 1970: The volcanoes and caldera of Talasea, New Britain: geology and petrology. *Bull. geol. Soc. Amer. 81,* pp. 17-38.

Macdonald, G.A., 1972: *Volcanoes.* Englewood Cliffs, Prentice-Hall Inc.

Masefield, J. (Ed.), 1906: *Dampier's voyages.* London, Grant Richards.

McBirney, A.R., & Williams, H., 1969: A new look at the classification of calderas (abst.). *Intern. Assn. Volcanology and Chem. Earth's Interior, Symposium on volcanoes and their roots,* Oxford Univ. Supplementary abstracts.

Morgan, W.R., 1966: A note on the petrology of some lava types from East New Guinea. *J. geol. Soc. Aust, 13,* pp. 583-591.

Reche, O., 1914: Dampier's Route langs der Nordkuste von Kaiser-Wilhelms-Land, *Petermanns Geographische Mitteilungen., 60,* pp. 223-225.

Taylor, G.A.M., 1953: Notes on Ritter, Sakar, Umboi and Long Island volcanoes. *Bur. Miner. Resour. Aust. Rec. 1953/43* (unpublished).

Taylor, G.A.M., 1956: Australian National Committee on Geodesy and Geophysics. Report of the sub-committee on Vulcanology, 1953. Review of volcanic activity in the Territory of Papua-New Guinea, the Solomon and the New Hebrides Islands, 1951-53. *Bull. Volcanol., 18,* pp. 25-37.

Williams, H., 1941: Calderas and their origin. *Univ. Calif. Berkeley Pub. Geol. Sci. 25,* pp. 239-346.

Williams, H., 1942: The geology of Crater Lake National Park, Oregon, with a reconnaissance of the Cascade Range southward to Mount Shasta. *Carnegie Inst. Washington Pub., 540,* 162 pp.

Williamson, J.A., (Ed.), 1939: *A voyage to New Holland (by William Dampier).* London, Argonaut Press.

STRIKING SEQUENCE OF VOLCANIC ERUPTIONS IN THE BISMARCK VOLCANIC ARC, PAPUA NEW GUINEA, IN 1972-75

R.J.S. COOKE, C.O. McKEE, V.F. DENT, and D.A. WALLACE

Geological Survey of Papua New Guinea, Volcanological Observatory,
P.O. Box 386, Rabaul, Papua New Guinea.

ABSTRACT

Eruptions in 1972-75 at six volcanoes in the Bismarck volcanic arc, Papua New Guinea, form a striking space-time cluster. Five of these volcanoes are the only ones known to be active in a 420 km-long segment of the western part of the arc, and all five were in eruption within an eight-month period in 1974. Two of them, Karkar and Ritter, erupted after repose periods of about 80 years, and the eruptions at Langila were the strongest known there in at least 100 years. The effusion of lava flows at Motmot volcano (Long Island) and Karkar volcano had not previously been observed. Lava flows occurred also at Ulawun, Manam, and Langila volcanoes, although lava flows at Bismarck arc volcanoes other than Manam have been rare during the twentieth century. Eruptions at the five volcanoes other than Ritter were dominantly of strombolian type, although some non-strombolian phenomena occurred. A strong increase in seismic activity commenced in late 1970 in the Adelbert Range area, on the mainland adjacent to Karkar and Manam volcanoes, although no increased seismic activity was evident in the parts of the arc adjacent to Long, Ritter, and Langila volcanoes. The Adelbert Range activity comprised two aftershock series whose apparent fault planes may represent a pair of conjugate shear fractures consequent on the north-northeast compression existing in the area, and whose locations are influenced by the presence of Karkar and possibly Manam volcanoes. Apart from this possible association of local seismic and volcanic events, there is no known regional tectonic event which could account for the 1972-75 sequence of eruptions. All eruptions of the 1972-75 sequence, except those at Karkar, are briefly described.

INTRODUCTION

Seven volcanoes in Papua New Guinea erupted during the three years July 1972 to June 1975, some of them more than once. Six of the volcanoes — Ulawun, Langila, Ritter, Long, Karkar, and Manam — lie in the Bismarck volcanic arc (Johnson et al., 1973), the chain of volcanoes along the southern margin of the Bismarck Sea (between Rabaul and Vokeo; Fig. 1). Only ten volcanoes in the Bismarck arc have definite written records of eruption (see below), and this conjunction of activity at six of them represents a marked space-time clustering of eruptions. In fact, five of the six occupy adjacent positions along the arc, if only these historically active volcanoes are considered.

The seventh erupting volcano was Bagana, on Bougainville Island in the arc of the Solomon Islands (Fig. 1). It was active throughout 1971 to 1975; lava was extruded continuously, but brief periods of explosive activity occurred in March 1971 and April 1975. This eruption will not be treated here, but up-to-date information is provided by Bultitude (this volume).

This paper aims to discuss first the features which characterise the present 'cluster' of eruptions in the Bismarck arc, and to compare it with other apparent sequences in the incompletely known volcanic history of Papua New Guinea. Possible causes of the sequence are considered, but as studies are still in progress, this part of the paper is necessarily brief and speculative.

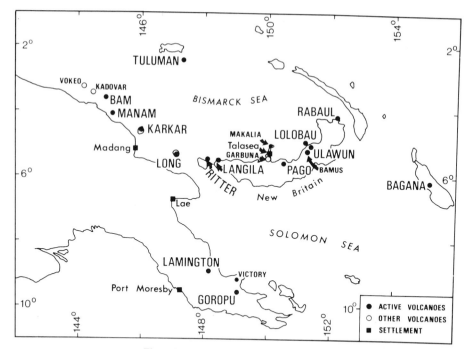

Fig. 1: Papua New Guinea locality map.

The second part of the paper presents summaries of all eruptions in the 1972-75 Bismarck arc sequence except those at Karkar, which are treated in an accompanying paper (McKee et al., this volume), and discusses some aspects of the eruptions.

PART 1: ERUPTION SEQUENCES IN BISMARCK VOLCANIC ARC

Eruptive histories

For the time before the 1870s, there exist only rare reports of volcanic eruptions in Papua New Guinea because of the absence of a literate population in the area. Since European settlement began, in the 1870s, the record of larger eruptions is probably nearly complete, although some smaller eruptions may have been missed, even up to the present day, at the more remote volcanoes.

A diagrammatic representation of the eruptive history of Bismarck arc volcanoes during the last 100 years is given in Figure 2. The source of much of this information is Fisher (1957), with data updated and corrected from material held at Rabaul Observatory. Eruptive histories of several other Papua New Guinea volcanoes are included in Figure 2.

Ten of the Bismarck arc volcanoes in Figure 2 have definite records of eruption, and Makalia appears, on the basis of a local legend, to have erupted at an indefinite date late last century. Several other volcanoes, not included in Figure 2, are believed to have either been active within the last few hundred years or to be capable of erupting again; these include Kadovar, west of Bam, and Garbuna and

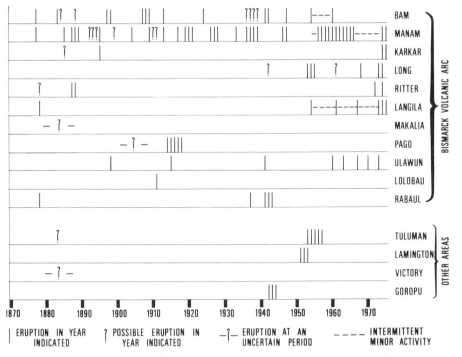

Fig. 2: Eruptive histories for some Papua New Guinea volcanoes, 1870-1975.

Bamus in the central New Britain group of volcanoes. Apart from these, several unconfirmed reports suggest that submarine activity may have occurred at other locations in the Bismarck arc; these are neglected in the present discussion.

Characteristics of the 1972-75 sequence

Eruptions in the present sequence, or 'cluster', have taken place between October 1972 and the time of writing, July 1975 (Fig. 3). In this period, six Bismarck arc volcanoes have erupted; in fact all six were active in a single twelve-month period from October 1973 to October 1974, and in only eight months, from February to October 1974, the five volcanoes Langila, Ritter, Long, Karkar and Manam, which lie successively along 420 km of the arc (neglecting extinct volcanoes), were all active.

The greatest number of Bismarck arc volcanoes active in any few-year period in the past 100 years was five, and no more than two adjacent volcanoes have been active at such times (Fig. 2). The only earlier marked 'cluster' of eruptions since 1950, when fairly detailed observation of Papua New Guinea volcanoes commenced, was that during the 1950s, when seven volcanoes erupted, including Bagana; however, only four of these were in the Bismarck arc (Fig. 2).

The present sequence is characterised not only by the number of erupting volcanoes, and by the spatial relations between five of them; the characters of several of the eruptions are notable relative to previous known histories of these volcanoes. For instance (for fuller details, see below):

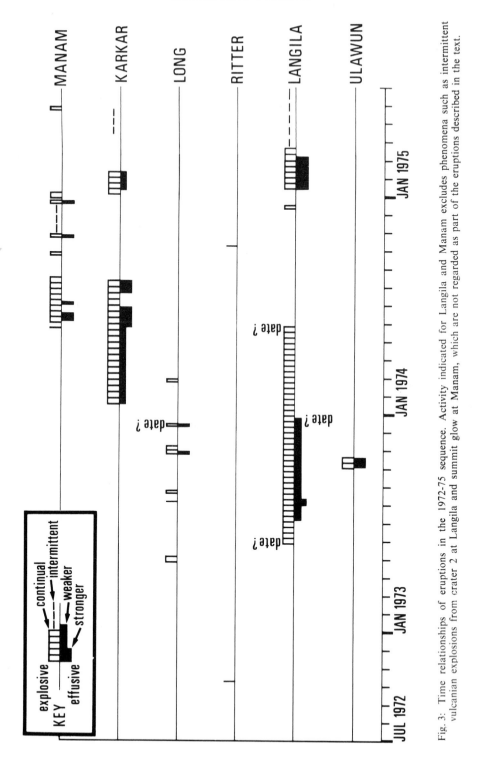

Fig. 3: Time relationships of eruptions in the 1972-75 sequence. Activity indicated for Langila and Manam excludes phenomena such as intermittent vulcanian explosions from crater 2 at Langila and summit glow at Manam, which are not regarded as part of the eruptions described in the text.

1. The two effusive eruptions at Langila volcano were on a markedly larger scale than any eruption known there since 1878, at least.

2. Lava flow occurred at Motmot volcano (Long Island) for the first time since the first observation of Lake Wisdom, in the 1930s (although underwater activity may have taken place unrecognised).

3. Karkar and Ritter volcanoes both became active after long repose periods, 79 and 84 years respectively (again, underwater activity at Ritter may have gone unrecognised), and a large volume of lava was extruded at Bagiai cone (Karkar).

4. Eruptions at Manam and Ulawun volcanoes were not remarkable compared with the last previous eruptions at these volcanoes. However, the 1970 and 1973 Ulawun eruptions together mark a striking change in the volcano's known mode of activity, as neither lava flows nor nuées ardentes had been observed there before 1970.

During the present century, lava flows have been common at Manam; otherwise, lava effusion at Bismarck arc volcanoes has been rare, being known only at Lolobau in 1911(?), Pago in 1914-18(?), and Langila in 1961 and 1967. Since 1970, lava has flowed at Ulawun in 1970 and 1973, Langila in 1973 and 1975, Motmot (Long Island) in 1973 (twice), Bagiai (Karkar) in 1974 (twice) and 1975, and on four occasions at Manam in 1974.

Of the seven volcanoes from Langila westwards known, or believed, to be 'active' (i.e. not extinct) only the two westernmost, Bam and Kadovar, have not erupted during the current sequence. Of the seven volcanoes to the east of Langila which are known, or believed, to be 'active' only Ulawun *has* been active during the current sequence, and as it has erupted at intervals of only about three years since 1960 (Fig. 2), its association with the other five volcanoes in the sequence may not be significant. Perhaps the outstanding feature of the present sequence is actually the association of eruptions at the five volcanoes in the western part of the arc (corresponding to the 'western' arc of Johnson, this volume).

Regional stress

G.A.M. Taylor frequently emphasised a 'regional stress' factor, evidenced by the occurrence of strong tectonic earthquakes, in his studies of volcanic eruptions in Papua New Guinea (Taylor, 1958, 1963, 1966; Taylor et al., 1957).

Cooke (1975) examined the time correspondence between regional seismicity (as characterised by earthquakes with magnitude M greater than or equal to 6¼) and volcanism in the western part of the Bismarck arc for the period 1947-74. He found that there was no truly regional increase in seismic activity preceding or accompanying the 1972-75 sequence of eruptions. There *was* a significant increase in shallow and intermediate-depth seismicity near Manam and Karkar volcanoes, commencing in 1970, about 3.3 years before the start of the 1974 Karkar eruption. The last apparent increase in seismic activity in the parts of the western arc corresponding to the positions of Long, Ritter and Langila volcanoes took place in 1963, about 9.5 years before the 1972 Ritter eruption which commenced the 1972-75 sequence; this apparent increase was produced by a single large earthquake. Other apparent increases in seismic activity in parts of the western arc in 1947, 1951, and 1959 do not seem to have been related to a general upsurge in volcanic activity.

Everingham (1974) presented a cumulative seismic moment plot for seismic activity in the New Britain/north Solomon Islands region, which includes the eastern part of the Bismarck arc. This plot was based on earthquakes this century with magnitudes M greater than or equal to 7, and shows an apparent cyclic pattern with strong increases in activity just before 1920, in the mid-1940s, and in the early 1970s. The two earlier increases seem to have been *preceded* by increased volcanic activity in the eastern part of the Bismarck arc (Fig. 2). The 1970s increase in activity is due to two magnitude M8 earthquakes in the Solomon Sea in July 1971; the results of Everingham (1975a) indicate that the first of these was associated with the Solomon Islands arc, but the second (26 July) with the New Britain structure. The only known volcano-related events in the eastern part of the Bismarck arc corresponding with this seismicity increase apart from the two eruptions at Ulawun volcano, are a magnitude M6 earthquake directly beneath Pago volcano on 19 July 1971, and an increased level of volcano-seismic activity in the Rabaul caldera which began in late 1971 (unpublished data).

G.A.M. Taylor also investigated apparent luni-solar influences on volcanic activity in the references quoted above. These aspects, both in relation to the present sequence and to previous activity, are under study by D.A. Wallace. The most striking feature of the present sequence in this regard seems to be the simultaneous occurrence of strong pulses of activity at the adjacent volcanoes Manam and Karkar in early June 1974, just before the solstice.

Tectonic interpretations and the 1972-75 sequence

Many authors have attempted to interpret the tectonics of the Bismarck volcanic arc (e.g. Denham, 1969; Johnson & Molnar, 1972; Curtis, 1973; Krause, 1973; Johnson, this volume). While all agree in general terms on the interpretation of the eastern part of the arc (a zone of subduction of a Solomon Sea plate beneath New Britain), detailed interpretations of the western part vary, although all agree it is a zone of compression.

From these studies there are clear indications of westward continuation of the New Britain Benioff zone beneath the volcanoes Langila, Ritter, and Long. In this latter area, no strong increase in seismic activity occurred later than about 9.5 years before the first eruption at these volcanoes during the 1972-75 eruption sequence. There is no evidence at present for a Benioff zone underlying Karkar and Manam volcanoes but, as stated above, there was a marked increase in seismic activity in this area, commencing about 3.3 years before the first eruption at these volcanoes during the 1972-75 sequence. Yet the approximate simultaneity of rise of magma at four of these five volcanoes (and assumed at the fifth, Ritter) suggests tectonic continuity between them.

The earthquakes mainly responsible for the increased seismic activity near Karkar and Manam volcanoes comprised two aftershock series. One, the 'Madang' earthquake series, commenced on 31 October 1970 (Everingham, 1975b), and the other, here called the 'Josephstaal' earthquake series, commenced on 18 January

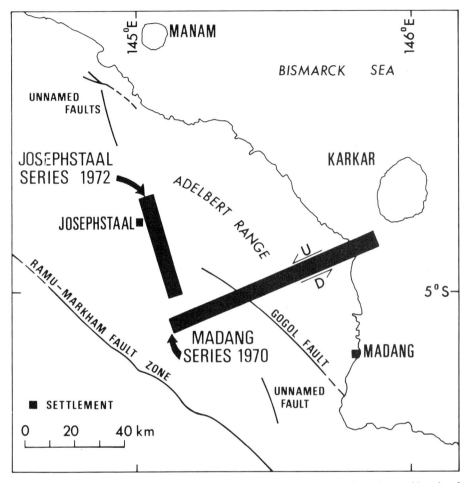

Fig. 4: Adelbert Range area locality map. Thick bars indicate approximate orientations and lengths of aftershock zones of the earthquake series identified. Faults are from 1:1,000,000 map, Geology of Papua New Guinea (1972).

1972. Each series revealed a structural trend different from the dominant west-northwest trend in the area: about 070° for the Madang series, and about 350° for the Josephstaal series.

Figure 4 shows the orientation and length of the aftershock zone (determined for larger aftershocks) of the Madang series (data from Bulletins of the International Seismological Centre, October to December 1970), the directions of relative motion on this assumed fault plane determined from focal mechanism studies (Everingham, 1975b), and the orientation and length of the aftershock zone (again, determined for larger aftershocks) of the Josephstaal series (data from Bulletins of the International Seismological Centre, January to March 1972). Focal mechanism data are not available for the Josephstaal series. Focal depths for events in both series were shallow, averaging 30-40 km. The assumed fault planes seem to bear some relation to the positions of the volcanoes, particularly that of the

Madang series to Karkar, and possibly that of the Josephstaal series to Manam (Fig. 4).

The direction of compression in this area is about north-northeast (Johnson & Molnar, 1972) and the two fault planes together call to mind a pair of conjugate shear fractures associated with this compression. Concentration of stress in an area close to the existing Ramu-Markham fault zone (Fig. 4) would be implied by such a model, suggesting that this fault zone would be the plate boundary in the Adelbert Range area. The actual locations of the shear fractures may have been imposed by existing weaknesses associated with the volcanoes. Compressional stresses would be relieved by shear fracturing rather than by overthrusting, because of the continent/island-arc type of collision envisaged here by Johnson & Molnar (1972), in which the two masses would have similar buoyancies and neither would be overthrust. One may speculate that the unknown Josephstaal series fault mechanism was a mirror image of the Madang series fault mechanism, reflected in the compression axis.

Present-day tectonic continuity from Manam and Karkar to the more easterly volcanoes is implied by the 1972-75 eruption sequence, suggesting that Manam and Karkar still have access to magma generated from a lithosperic slab, which further to the east is associated with northward subduction. The relief of compressional stresses in the Adelbert Range area by shear fracturing is in general agreement with the apparent absence of active northward subduction in the area. Presumably magma generation can continue from an inactive part of a slab by processes leading to partial melting, such as release of volatiles, provided that active subduction ceased sufficiently recently.

The tectonic event presumed to be responsible for initiating 'simultaneous' rise of magma at four, perhaps five, volcanoes in the western part of the Bismarck volcanic arc remains unidentified. The Madang and Josephstaal earthquake series seem unlikely causes for such a regional effect.

PART 2: ACCOUNTS OF 1972-75 ERUPTIONS

Ulawun

Introduction

Ulawun, a 2350 m-high stratovolcano, is one of the most imposing of the active volcanoes of Papua New Guinea. All historical eruptions have taken place at the summit. Eruptive activity was first recorded there in 1700 (Williamson, 1939). Strong explosive eruptions were reported in 1898 (Sapper, 1917) and 1915, and minor eruptions in 1941, 1960, 1963, and 1967 (Johnson et al., 1972). In 1970 another strong eruption occurred, basically of strombolian character, although nuées ardentes were erupted for the first time in Ulawun's recorded history, as were lava flows (Johnson et al., 1972). The sequence of eruptions 1960 to 1970 seemed to be cyclic in nature, and to show a trend of increasing intensity.

1973 Eruption

The 1973 eruption was similar to that of 1970 (Johnson et al., 1972) although it lasted only 14 days compared with 28 days in 1970; lava flows and nuées were again produced. The account which follows is based on nearly continuous observations commencing shortly after the eruption was reported and continuing until its end. The observation post was about 10 km northwest of the summit, and was equipped with a portable seismograph.

Late in the afternoon of 4 October, an aircraft pilot reported incandescence and traces of brown

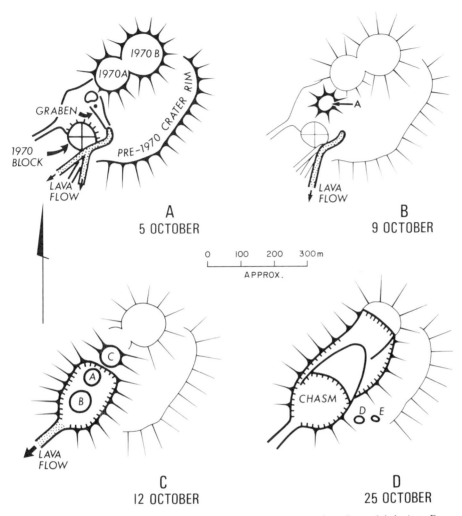

Fig. 5: Summit topography at Ulawun volcano during the 1973 eruption. Crater labels A to E are prefixed 1973 in text. See Johnson et al. (1972, Fig. 4) for contour map of Ulawun.

smoke in the summit area of Ulawun. Shortly after dusk on the same day, an incandescent lava flow descending the southern flank of Ulawun was observed from a plantation about 45 km to the south, and glow in the summit area was seen from near the foot of the volcano.

An aerial inspection early on 5 October showed that brightly incandescent lava flows were descending the southern flank of Ulawun from a source lying in a graben-like feature on the flank of a terminal cinder cone formed in 1970 (1970A; Fig. 5A and Plate 1). The lava flows were incandescent more than 1 km from their source, and movement of the lava from the source could be readily detected from the air. A newly formed circular pit was present higher up the slope in the graben-like feature, southwest of crater 1970A. This pit was emitting white vapours, but no explosions were seen and there was no trace of an ash column.

By morning on 6 October small explosions were occurring at intervals of 10-15 s, giving rise to a lightly ash-laden vapour column reaching about 150 m above the crater. Incandescent lava jets were observed overnight on 6th, and similar activity was continuing during an aerial inspection on 7 October.

The sources of the explosions lay in two closely adjacent vents in approximately the position of the new pit observed on 5th. The character and motion of the lava flows was unchanged; in addition, a small amount of lava appeared to have flowed into the pre-1970 crater from a source in about the position of the later vent 1973D (see below).

Explosive activity increased strongly after 7 October. Aerial inspections on 8 and 9 October showed that a dense, dark-brown ash column was pouring from crater 1973A (Fig. 5B), formed by coalescence of the two vents observed on 7th. The column rose 600-900 m above the summit, and bright incandescence was visible in its base every few seconds. By 11 October, the ash column reached 1800-2000 m above the summit, and the already high level of volcanic tremor had increased strongly, doubling in amplitude in about 15 hours. At the peak of this tremor, at 0615 LT on 11 October a nuée ardente was discharged down the south-southwest slopes. Two other nuées erupted on 11th, at 0848 and 1925 LT, following the same path. The last of these was the biggest, reaching about 3 km from the crater, but much of their energy was dissipated against a 150 m-high scarp which lay in their path (Johnson et al., 1972, Fig. 6).

Although volcanic tremor weakened gradually from about midday on 12 October, explosive activity continued to intensify, being strongest about 13-14 October. At this stage, almost continuous lava jets reached 500-600 m above the summit, accompanied by continual rumblings and detonations. Aerial inspections during this period showed major changes to the summit topography (Fig. 5C); a second crater (1973 B) was present in approximately the position of the 'raised block' of the 1970 eruption (Johnson et al., 1972). This had become active on 11 October, and was apparently the source of the nuées. A chasm opening to the south-southwest slopes was also present, and a new lava flow issued from its foot. The earlier lava flows had ceased, possibly on 10 October. A third crater (1973C, Fig. 5C) north-northeast of the others began to erupt about midnight on 11 October; its early activity was marked by heavy ash emission.

Similar activity continued until the early hours of 19 October, when all eruptive activity and volcanic tremor ceased abruptly. The final few days of the eruption had shown some decline in visible activity (with minor fluctuations in intensity), but the activity remained vigorous to the end. Crater 1973C was inactive from early on 18th, but two new small craters formed at about that time (1973D and 1973E, Fig. 5D), from which small-scale explosive activity was observed. A small lava flow in the pre-1970 crater, observed after the eruption, probably originated at 1973E at about this time.

Collapses of parts of the new pyroclastic structures at the summit took place between 19 and 25 October. Since the eruption Ulawun has shown only copious emission of white vapours, as it had done before the eruption. No gas odours were detected at any stage during the eruption.

The later lava flow moved over the top of 1970 lava in the western valley (Johnson et al., 1972, Fig. 6), and halted about 5 km from the summit, 1 km above the terminus of the 1970 flow and about 670 m above sea-level. About 10^7 m^3 of lava was extruded, as aa flows. From aerial observations of the flow source, it seems probable that effusion rates were of the order of tens of cubic metres per second early in the eruption. The flow nose in the southwest valley is about 8-9 m thick and 70-80 m wide.

Langila

Introduction

Langila comprises a group of young craters, with an extensive apron of recent lava flows extending about 5 km north and east from the craters, on the northeast flank of Talawe volcano (Fig. 6). A large eruption seems to have occurred at Langila in 1878, judging from the seemingly exaggerated account by Powell (1883), but no definite information on later activity is available until eruptions resumed in 1954 (Taylor et al., 1957). Since then, intermittent vulcanian eruptions from crater 2 (Fig. 7) have occurred frequently; in fact some activity of this type has taken place in almost every year since 1954. Crater 3 (Fig. 7) was created in 1960, and a small lava flow was extruded from it shortly afterwards. Crater 3 was again active in 1967, and produced another small lava flow. Both flows were subsequently buried by pyroclastic deposits.

The remoteness of the volcano and, until recently, poor physical communication, has meant that probably only the larger-scale activity has been reported. There is uncertainty as to the identification of the active crater in many reports.

Plate I: Oblique aerial photograph (camera tilted) of lava flow source, Ulawun volcano, 5 October 1973. Width of the source is about 3 m. The identity of the fin-like monolith (upper left corner) produced during the 1970 eruption, is not known. Photograph by R.J.S. Cooke.

Plate 2: Explosion at crater 3, Langila volcano, 22 July 1973. Incandescent lava surface is about 10 m across. Photograph by R.J.S. Cooke.

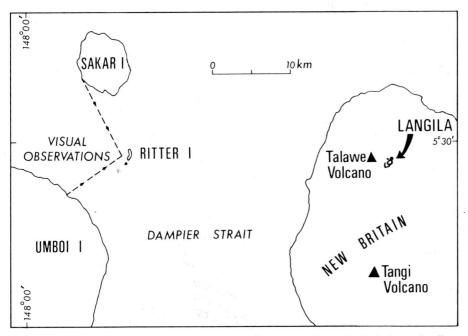

Fig. 6: Locality map of Langila volcano and Ritter Island. Intersection of visual observation lines indicates approximate site of 1974 eruption of Ritter.

1973-74 Eruption

An eruption took place between May 1973 and May 1974, distinct from the 'normal' pattern of intermittent vulcanian activity from crater 2. Rhythmic mixed strombolian-vulcanian ejections occurred at crater 3, accompanied between July and December 1973 by lava effusion to form a large flow, comparable in size to some of the flows comprising the recent apron. Lava also rose into the bottom of crater 2 at an early stage of this eruption but remained inactive there.

The following account of the 1973-74 eruption is based on continuous observations from 15 July to 22 August 1973 and from 9 to 16 November 1973, and on regular aerial inspections and information supplied by local residents. The observation post used was about 8 km north of the craters at the Cape Gloucester airstrip.

An increase in activity at Langila was noted by local residents from late May 1973. A red glow was

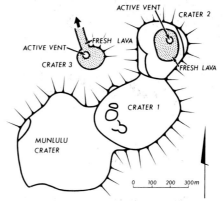

Fig. 7: Topography of Langila craters during July to December 1973.

observed from one of the craters on 3 June, frequent rumblings from the volcano were heard from mid-June onwards, and some small explosion clouds were seen. Most of this early activity probably occurred at crater 3, and the peak was apparently reached about 24-27 June. On 9 July the presence of a new lava flow originating from crater 3 was first noticed and from about this time, explosive activity again became stronger.

An aerial inspection on 15 July confirmed the early reports. Explosions were taking place at crater 3, which was filled with lava, and this lava had overflowed about 600-700 m to the north; crater 2 contained a mound of lava with a vigorous fumarole on its top surface (Fig. 7). Both craters are known to have been empty of lava on 2 May, although photographs from the air by a private photographer show the beginning of rise of lava into crater 2 on an unknown date in May.

During the continuous observations in July and August 1973 there were no changes in the pattern of activity at crater 3, although minor fluctuations in intensity occurred. The typical pattern consisted of roughly periodic phases of explosiveness (Plate 2) separated by lulls. The explosive phases took place at intervals of 10-30 minutes, and contained a complex sequence of events of varied character. They started with mushroom-like explosions of ash and lava fragments, accompanied by strong detonation sounds; lava fragments reached 250 m above the vent. Although they were brightly incandescent at night, daylight incandescence in the ejections was rarely observed, even from a range of only about 300 m. Similar explosions also occurred within explosive phases, as did impulsive blasts of vapour, periods of continuous outpouring of vapour, and periods of regularly repeated puffs of vapour at up to two per second, each type of activity having a characteristic sound. Explosive phases declined in intensity later in the eruption, and by May 1974 they consisted generally of weak and single events, at increasingly long intervals.

Lava flowed from crater 3 for about 150 days, probably until sometime in December 1973. The flow nose advanced until late August or early September, reaching its final length of about 1.9 km after 50-60 days; thereafter the flow showed only minor lateral spread but a marked increase in thickness. The final thickness was 25-30 m in many parts of the flow, and about 5×10^6 m^3 of block lava was produced. For about the first 20 days, until the end of July, the effusion rate was apparently steady at about 0.2 m^3s^{-1}, but for about the next 10 days more rapid spread of lava occurred, and the effusion rate probably reached 0.5-1.0 m^3s^{-1} before declining again. At night, faint incandescence could often be seen at the flow surface just below the region of overflow from the crater.

Crater 2 activity during the eruption consisted chiefly of almost continuous vapour emission although fairly regular fluctuations in its intensity were observed. Two moderately large explosions and a few small ones occurred at crater 2 during 38 days' continuous observation in July-August 1973. The period 17-21 July was unusual, as a period of explosiveness on 17 July was succeeded by a phase of high-pressure emission of white vapours on 18-20 July, during which the crater produced a steady roar, deafening at close quarters. The noise was prominent even at the observation post. Activity declined to 'normal' vapour emission on 21 July. Intermittent glow above crater 2 first occurred on the night of 18 July, and was then observed every night until full-time observation ceased on 22 August.

Small vulcanian ash ejections were occurring at crater 2 several times per day on average in October-November 1973. These were inaudible at the observation post. By 17 February 1974 the lava mound in crater 2 had been destroyed, possibly by strong explosions within the previous few days, and its place taken by a funnel-shaped vent. Small explosions were occurring frequently in this vent during an aerial inspection on that date.

Crater 2 remained unchanged in form during the remainder of 1974, although occasional moderate explosions occurred, probably representing 'normal' crater 2 eruptive activity. No explosions at crater 3 were reported from June to November 1974, although steady emissions of blue and white vapours continued there.

Gas odours were detected only weakly and rarely during close range inspections. Hydrogen sulphide was noted on three occasions in July to early August and once in November 1973, and sulphur dioxide about mid-August.

1974-75 Eruption
Crater 3 became active again in December 1974, and another large lava flow was extruded from it

in January and February 1975. This account of the 1974-75 eruption is based on continuous observations from 15 January to 6 February 1975, and from 23 to 27 June 1975, supplemented by regular reports by local residents and occasional aerial inspections.

Crater 3 produced occasional explosions, apparently of vulcanian type, from early December 1974. Explosive activity at crater 3 increased significantly on about 8 January, and became similar in character to 1973-74 activity at this crater. From mid-January to mid-February, the intensity of the new eruption was probably slightly higher than that in 1973-74. Lava was overflowing to the north by mid-January, and this flow was active for about 50 days. Its final length was about 5 km, and its volume about 8×10^6 m³, giving an average effusion rate about four times as great as that of the 1973 flow. In early March, activity at crater 3 became more vulcanian. Explosions were infrequent, but were strong, with loud rumblings. The character of activity at crater 2 also changed at this time, from occasional vapour emission to occasional strong vulcanian explosions. Towards the end of March the frequency of explosions increased at both craters, from a combined rate of several per day to several per hour; explosions from crater 3 were about three times as frequent as those from crater 2. Similar activity is continuing at the time of writing (July 1975). The cone around crater 3 is now larger than at any earlier period of its short history.

Ritter Island

Introduction

Ritter Island (Fig. 6) is the uninhabited remnant of a formerly much larger island which was destroyed by a catastrophic eruption in 1888 (Sapper, 1917). There were many casualties in the region, mostly caused by large-scale tsunamis accompanying the event. The eruption presumably involved some caldera collapse, although the form and dimensions of the conjectured submarine caldera are unknown. No evidence of craters or thermal areas now exists on the island. The only other volcanic activity known at Ritter Island before 1972 occurred in 1700 (Williamson, 1939), 1793 (Rossel, 1808), and 1887 (Chalmers, 1887), with a possible occurrence in 1878 (Powell, 1883).

Brief eruptions again took place in the sea off Ritter Island on 9 October 1972 and 17 October 1974; the events were markedly similar, each lasted only a few hours, and they took place at almost the same time of day. Neither event was witnessed by volcanologists. Accounts of these eruptions have been prepared from stories obtained from villagers on neighbouring islands.

1972 Eruption

R.A. Davies gave the following account* based on information collected 17-18 days after the event: 'Earthquakes were felt at about 0530 (9 October) and rumbling explosions could be heard. Sufficient daylight showed that explosions were taking place at Ritter Island. The explosive outbursts occurred at the rate of one or two per minute, resembled the detonation of large bombs, and sent clouds of black and white smoke into the air. The height of the explosion clouds must have been over 400ft (by comparison with the height of Ritter) and a strong rumbling was heard frequently. The main activity appears to have been offshore (to the west) about half-way along the length of Ritter, but smoke was also seen rising from near the extremities of the island . . . . . The activity ceased fairly abruptly sometime after 0730, though "smoke" continued to rise from the eruption site for two days. No smoke was seen after 11 October'.

Additional information concerning the 1972 activity was obtained during an investigation of the 1974 eruption; a small tsunami was observed at Umboi Island, but not at Sakar Island (Fig. 6) during the 1972 activity, and villagers said variously that there had been three to six individual explosions.

The earthquake swarm accompanying the eruption was recorded at seismograph stations throughout Papua New Guinea between about 0530 and 0740 LT. The events were felt at distances out to about 35 km from the eruption site, and maximum felt intensities on adjacent coasts at distances of about 12 km were estimated at about MM 4. About eight of the clearest individual events can be confidently located in the Dampier Strait region (Fig. 6), although location accuracy does not permit precise identification of Ritter Island as the source.

*Note on Investigation 73-008 of the Geological Survey of Papua New Guinea, entitled "Ritter Island — 1972 activity".

1974 Eruption

The following account was prepared from information collected 2-4 days after the event. At about daybreak, a single earthquake was felt. A little later, a series of frequent (possibly continuous) tremors commenced which gradually increased in strength. At this stage active landslides were observed at several places on Ritter Island. Minor rumbling noises commenced. A climax in noise and tremor culminated in an explosion bursting from the sea to the west of the island, accompanied by a loud 'bomb-like' noise. After a brief lull a second, larger explosion followed, again preceded by rumblings and tremor. The explosion columns were described as black at the bottom and light-coloured higher up; they apparently reached to several times the height of the island (possibly to 400-500 m). No rocks or incandescence were observed in the column.

The sea was observed to dome up at the site of these explosions immediately before one of them, and sea disturbances continued at the site afterwards. A small tsunami was observed shortly after one of the explosions, although there were conflicting stories as to which one. The sea retreated at both Umboi and Sakar Islands and came back after a few minutes; the total rise and fall was not greater than that of a normal tide (possibly 0.2-0.5 m). Soft rumbles continued for a few minutes after the second explosion, and then all activity ceased. The 1974 explosion site was estimated to be 600-900 m west of the southern half of Ritter Island (Fig. 6).

The earthquake swarm accompanying this activity was very similar to that of 1972; events were recorded throughout Papua New Guinea between 0530 and 0650 LT. They were felt out to about 30 km radius, and maximum intensities were again estimated at about MM 4 at the closest sites. However, most informants said that the felt earthquakes in 1974 were not as strong as those of 1972. This is confirmed by seismic records of the swarms. The station at Talasea was not operating during the 1974 activity, so 1974 events cannot be as confidently located as those of 1972. Arrival times at other close stations are in accord with the origin being in the vicinity of Ritter Island.

Long Island

Introduction

Long Island contains a large caldera lake, Lake Wisdom, but was not recognised as an active volcano until 1943 when an aerial photograph showed a small crater just above lake level.* This crater was not present in 1938 or 1952 (Ball & Johnson, this volume). Surtseyan activity (Walker & Croasdale, 1971) occurred at the site intermittently during 1953-55 (Best, 1956; Fisher, 1959) and in 1968, leading to the erection of subaerial ash cones which formed a small island. An observation of strong vapour emission at the volcano island during 1961 (data file, Rabaul Observatory) suggests the possibility that an eruptive episode had just occurred there, as normally there were little or no prominent vapour emissions except in association with eruptive phases.

The active volcano island is here called Motmot (meaning 'island' in the local language), although it has sometimes been referred to as Mani. Recent bathymetric surveys of Lake Wisdom (E.E. Ball and J. Glucksman, pers. comm.) show that the volcano stands in water at least 360 m deep.

The remoteness of Long Island, irregular communications, absence of a population with a direct view of Motmot, and the brevity and intermittency of the known eruptive phases, have led to poor knowledge of the details of activity, and indeed to doubt that all recent eruptive phases are even known. Eruptive activity recurred in 1973-74, in five main phases; two of these were observed, in part, by volcanologists.

Biological and geological investigations (Bassot & Ball, 1972; Ball & Johnson, this volume; E.E. Ball and J. Glucksman, pers. comm.) monitored the condition of Motmot in each of the years 1969-74. Features noted which were possibly premonitory to the 1973-74 eruption included formation of fissures and of a small pit (Fig. 8A) in 1972, and apparent rises in temperature at parts of the shoreline between 1969 and 1972.

*See Ball & Johnson, this volume, Fig. 2, for a sketch map of Long Island. Ed.

Fig. 8 (opposite): Topography of Motmot volcano island (Long Island) before and after each phase of the 1973-74 eruption. In the text, crater labels B1 to B7 (parts C and D) are prefixed 1973, and A2 (part F) is prefixed 1974.

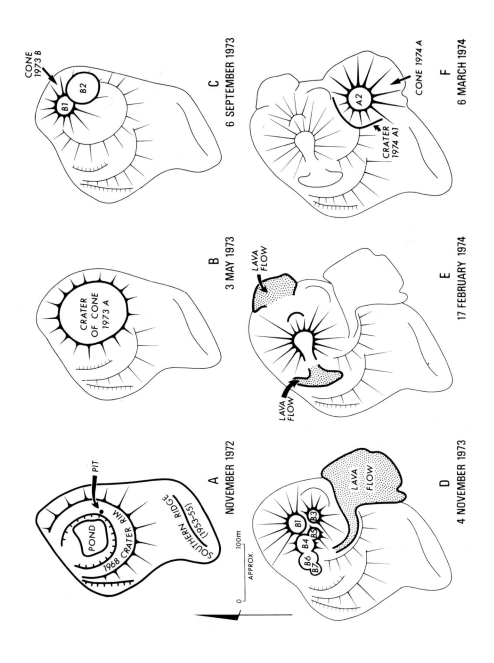

CONE 1973 B

B1 B2

C

6 SEPTEMBER 1973

CRATER OF CONE 1973 A

B

3 MAY 1973

PIT

POND

1968 CRATER RIM

SOUTHERN RIDGE (1953-55)

A

NOVEMBER 1972

0 100m APPROX.

CONE 1974 A

A2

CRATER 1974 A1

F

6 MARCH 1974

LAVA FLOW

LAVA FLOW

E

17 FEBRUARY 1974

B1 B3 B5 B4 B6 B7

LAVA FLOW

D

4 NOVEMBER 1973

April-May 1973 Eruption

The first positive evidence of a new eruption is in photographs of Motmot taken from an aircraft on 27 April 1973, which show a new ash cone, 1973A (Fig. 8B). Only weak vapour emission is visible in the photographs, but observers in the aircraft reported brown smoke (pers. comm. R. Creagh, who also reported secondhand information that wisps of smoke had been seen there on 19 April). The date of formation of the new cone is unknown.

Vigorous surtseyan activity was observed from the air at the 1973A and 1968 craters on 1 and 2 May 1973. Volcanologists noted that eruptions on 2 May were taking place about every 30 s, with ash reaching 50-150 m, and a vapour column to about 600 m above lake level. No incandescence was seen in the column, steam was not evolved when ejected rocks landed in the lake, and no gas odours could be detected. This phase had ended by 3 May.

August 1973 Eruption

Further eruptive activity was reported from passing aircraft on 5 August, and between 19 and 21 August 1973. Motmot had grown in an easterly direction, and incandescence at the base of explosions on the eastern side of the island was noted during the latter period.

Volcanologists observed the results of this activity during an aerial inspection on 6 September. No eruptions were observed, but a new cone, 1973B (Fig. 8C), had been erected within 1973A, obliterating its northeastern rim. 1973B was conical and had a small summit crater (1973B1); its different form, compared to the broad, low ash cone 1973A, suggests that it was a cinder cone. A larger crater (1973B2, Fig. 8C) was present on the eastern flank of cone 1973B; this had no visible associated pyroclastic deposits, and may have been formed by a single phreatic explosion after other eruptive activity had ceased. Extensive white sublimates were present inside and immediately around its rim.

October-November 1973 Eruption

Renewed eruptions occurred from 23-27 October, and from 1-3 November. The first indication was glow over Long Island, observed by the crew of a passing ship early on 23 October. The following account of the rest of this phase is derived from photographs and notes of activity made from 24 October by E.E. Ball (pers. comm.), from continuous observations by volcanologists from an observation post on the lake shore 4 km west of Motmot between 30 October and 7 November, and from several aerial inspections. A portable seismograph was operated at the observation post.

Activity during 24-25 October was characterised by ash-free jets of incandescent lava fragments from crater 1973B1, at intervals of 5-10 s and generally to heights of 10-20 m (occasionally 100-150 m) above the crater. These were accompanied by blue vapour which formed a thin column reaching about 300 m above lake-level. Continual noise sounded at close range 'like a steam train with periodic louder bangs' (E.E. Ball, pers. comm.). Ejected lava fragments were red-incandescent in daylight, and steam clouds were produced when they fell in the lake. A lava flow was observed to be entering the lake on the eastern side of Motmot (Fig. 8D) on the morning of 25 October. The flow was incandescent in parts, and profuse steaming occurred at its margin with the lake. The source of the flow appeared to be at the foot of the southwest flank of cone 1973B. A second crater (1973B3, Fig. 8D) commenced lava jetting during the night of 25 October. Ash and lava fragments were produced on 26th. Activity declined thereafter and ceased late in the afternoon of 27 October. Only vapour emission occurred between then and the morning of 1 November.

New activity commenced about dawn on 1 November with small and lightly ash-laden explosions from a new crater 1973B4, west of the others (Fig. 8D). Initially intervals between explosions averaged about one minute, but activity gradually built up to a phase of almost continuous dark ash emission late the same morning. Light ash falls occurred over the lake shores, and dark lava fragments could be distinguished in the ejections later in this phase. A lull preceded a period of similar activity in mid-afternoon. Late in the afternoon, the activity abruptly changed to noisy lava jetting, similar to that of 24-25 October. At first these jets reached only to about 10 m above the vent, but later in the evening they reached 50-60 m, and showers of incandescent fragments covered the cone after each ejection.

This type of activity continued at a reduced scale early on 2 November, but about midday the ejections contained dark lava fragments and more ash than before, and the noises were stronger. At about this time a graben-like feature could be seen down the west flank of cone 1973B, below the active crater. Late in the afternoon, ash-free jets of incandescent lava fragments again occurred, gradually becoming larger and more violent. During the evening these ejections reached up to 150 m above the vent, and took place at intervals of about 30 s. Activity declined overnight, and by morning of 3 November consisted of weak ash ejections at increasingly long intervals. Two new craters, 1973B6 and 1973B7 (Fig. 8D), were seen to form during the morning, in the graben-like feature. Another crater, 1973B5 (Fig. 8D), was visible although no activity was observed there. No explosive activity was observed after mid-afternoon on 3 November. Volcanic tremor of low amplitude was recorded by a seismograph at the observation post during the activity; its amplitude varied in step with the intensity of the visible activity.

Six craters associated with cone 1973B were active during the October-November activity (Fig. 8D); the base angle of the cone was 34°. The lava flow was aa, about 1.5 m thick, and its volume was about 5000 m³. The surface of the northern part of Motmot was strewn with scoriae, mostly irregular in shape, but with a small proportion of bombs up to 0.8 m across, including cow-dung, fusiform, and ribbon types (Macdonald, 1972). No gas odours were detected during the active periods.

December 1973 Eruption

Volcanologists noted during an aerial inspection on 17 February 1974 that a further eruptive phase had recently occurred. Two small new lava flows were present, at the foot of the northeastern and the western flanks of cone 1973B, and this cone had increased markedly in size, apparently from activity at crater 1973B4 (Fig. 8E). The dates of this active period could not be determined reliably, although some time in December 1973 is probable, as Glucksman (pers. comm.) had already observed these new features in mid-January 1974.

February 1974 Eruption

Another eruptive phase occurred from about 23 to 28 February 1974 (R. Creagh, pers. comm.). Volcanologists observed on 6 March that a small new cinder cone, 1974A (Fig. 8F), had been erected by this activity on the site of the first (October) lava flow, obliterating it. Slight traces were seen of a larger new crater (1974A1, Fig. 8F) surrounding the new cone, whose crater was labelled 1974A2. Voluminous vapour emission from the latter was the only activity observed.

Manam Island

Introduction

Manam is an impressive island stratovolcano, whose summit is about 1725 m above sea level. Its first recorded activity was in 1616, and it has a history of frequent eruptions, with a longest repose period of probably about nine years (Palfreyman & Cooke, this volume). In a sense, it could be regarded as continually active, as full-time observation carried out from an observatory there over the last 18 years has shown that occasional weak ejections of ash and summit glows are observed almost every year. The major 'eruptions', the most recent of which were in 1936-39, 1946-47, and 1956-66, consist essentially of intermittent strombolian activity; lava flows and occasional nuées ardentes take place during such eruptions. Phases of activity in 1957-58 and 1960 (Taylor, 1963) were the strongest known at this volcano, and caused the island to be wholly or partly evacuated.

Eruptive activity has occurred chiefly at two locations near the summit, the 'main crater', and the 'southern crater'.* The main crater is a large, stable feature some 300 m in diameter, and the southern crater is generally 70-100 m in diameter although topography in its vicinity changes from time to time. When observations have been sufficiently detailed to identify the active craters, most of the stronger activity has taken place at the southern crater; the main crater has produced large-scale activity only in 1958 and 1960. However, the main crater is the source of most of the other weak activity between 'eruptions', mentioned above. The effects of lava flows and nuées from these two craters are restricted because of channelling by four major radial valleys; southern crater activity affects the southeast, and to a lesser extent the southwest, valley, while main vent activity affects chiefly the northeast valley.

A fairly typical 'eruption' (see above) commenced in May 1974 after a 'repose' of about 8 years during which only fairly minor eruptive phenomena occurred. This new activity continued intermittently through the rest of 1974 into 1975. During 1974, four periods of lava flow occurred, and moderate nuées

*See Palfreyman & Cooke, this volume, Fig. 2, for a sketch map of Manam Island. Ed.

ardentes were erupted. The following brief account is based on full-time observations by the resident volcano observer, and additional continuous observations for most of the period 25 May to 4 August by visiting volcanologists, supplemented by occasional aerial inspections. The main observation post was Tabele observatory, about 4 km southwest of the summit.

Activity March-August 1974

The first reported activity of 1974 was at the main crater in early March. This consisted of occasional weak to moderate ash emissions and from late in March, occasional weak glows and rumblings. Little main crater activity of this type had taken place in 1972 and 1973, and none at the southern crater. After this time activity fluctuated, but overall showed a slight increasing trend. From about mid-April, weak glows were occasionally also observed at the southern crater.

On 18 May, moderate to strong ash emission commenced at both craters. Strombolian activity occurred at the southern crater for a few hours early on 23 May; the incandescent lava jets were markedly tall and narrow and were associated with brilliant lightning displays (J. Feeley, pers. comm.), an unusual feature for this type of activity (cf. Macdonald, 1972). During the next week, only white or lightly ash-laden vapour emissions occurred at the southern crater, and fluctuating moderate ash emission continued at the main crater. During an aerial inspection on 25 May, incandescent ejections were seen deep inside the main crater. Moderate glow could be seen above the main crater at night in this period.

Lava jetting at the main crater became visible overnight on 1 June, peaked during 2-3 June, and died away on 4th; thereafter its activity was confined to mainly white vapour emission, with occasional light ash content.

Strong fluctuating glow returned above the southern crater on the night of 31 May, and weak jetting commenced there on 2 June, becoming stronger from the night of 4th. About 35 small nuées ardentes erupted from the southern crater during 5-6 June, and travelled part-way down the southeast valley; for about a half-hour near midday on 5th, more-or-less continuous larger nuées were produced. Several more small nuées were observed on 11 June, again in the southeast valley. An aa lava flow from the southern crater commenced sometime on 5th, and had reached about 3 km down the eastern edge of the southeast valley and was still advancing when it was examined on 7 June. Four new flow units were present on 13 June parallel to and adjoining it on the south; one of these was still active on 13th. It is not clear whether these units represented the product of continuous effusion or of separate pulses of lava outflow. Lava flow had ceased by the next inspection on 21 June, although the amount of lava in the valley was considerably larger than on 13th.

Lava jetting at the southern crater continued uninterruptedly between about 4 and 16 June at intervals from 1-10 s, accompanied by moderately strong ash emission. Lava jets during this phase reached maximum heights of about 300 m above the crater. Early in this strong June activity, the active vent of the southern crater migrated from the summit ridge towards the head of the southeast valley, apparently reopening an older infilled vent. Short periods of thick, dark ash emission from the main crater were seen on 11 and 14 June.

Southern crater lava jetting became intermittent from 17 to 20 June, and was not seen at all between about 21-26 June, although ash was still ejected from the southern crater. A temporary build-up in ash emission took place from the main crater. Flashing arcs (visible acoustic shock waves; see Perret, 1950) accompanied lava jets on 20 June.

Southern crater lava jetting resumed on 27 June, and continued until about 22 July, usually accompanied by moderate to strong ash emission. During the first half of this period there were five striking pulses of stronger activity at the southern crater, each lasting no more than a few hours, on 29 June and 1, 3, 6, and 10 July. The first of these pulses culminated at about 1715 LT in the eruption of a moderately large nuée, which travelled about 4 km down the southeast valley. Immediately before the nuee, observers on the island reported an extraordinary increase in the intensity of ejections of ash and incandescent fragments (J. Feeley, pers. comm.). This first pulse and those on 1 and 3 July deposited up to 50 mm of ash, and scoria blocks up to 60 mm across, on the west coast of the island. The pulses on 6 and 10 July produced unusually high lava jetting, estimated at 600-700 m above the crater.

Extensive new but inactive lava flows were present along the eastern edge of the southeast valley on

6 July, and it is believed that they were produced between 1 and 4 July. These flows were about 2 m thick and more sheet-like than the earlier ones, as they had advanced over the fairly broad and smooth path of the nuée of 29 June. One narrow tongue of lava reached a point about 1 km from the coast.

Thick ash continued to accompany southern crater ejections until 13 July, but from about 14th, activity consisted typically of ash-free ejections of lava fragments, which were dark in daylight. The intensity of activity declined slightly during this time and lava jetting ceased between about 22 and 26 July, although some ash emission continued during the lull.

Weak lava jetting again occurred between about 27 July and 13 August, after which all eruptive activity ceased. Volcanic tremor also ceased at this stage, having been present since early May; its intensity had fluctuated broadly with the intensity of visible eruptive activity at the southern crater during this time.

Activity September 1974-June 1975

A resurgence of minor activity at the main crater occurred during the first half of September. Glow was observed there at mid-month for the first time since early June, and there was an increase in ash emission. From 21 to 27 September, the southern crater produced a weak and brief recurrence of strombolian activity with ejections of ash and lava fragments. Volcanic tremor reappeared on seismic records from 21st until the end of September.

A similar sequence of activity at the two craters took place in October. During the first half of the month the main crater showed increased ash emission, and glows at mid-month. The southern crater recommenced lava jetting on 21 October, accompanied by a recurrence of volcanic tremor. The October activity was stronger that that in September, and new lava flowed down the southeast valley on about 22-23 October. Activity at the southern crater declined at the end of October, and continued at low level until nearly mid-December; volcanic tremor ceased, having persisted at low to moderate intensity until nearly mid-December. The southern crater activity from the beginning of November consisted of alternate periods of a few days with weak ash emission and glows, and a few days with only vapour emission. Ash emission at the main crater was observed several times during November.

Volcanic tremor was not recorded for about a week before 19 December. A strong build-up of tremor began on 19 December, accompanied by lava jetting at the southern crater. Tremor and visible activity reached a climax in the late afternoon of 21 December, when a moderately large nuée was expelled from the southern vent into the southeast valley. A substantial flow of lava followed immediately after the nuée, in the same path. A few days of vapour emission ensued. On 27 December another phase of lava jetting and ash emission began at the southern crater, lasting until 4 January 1975.

Glow at the main crater was observed for a few days at the beginning of January and again in mid-February. Stronger activity occurred at the main crater for about two weeks in late April and early May, and included ash emission and weak ejections of lava fragments.

Weak ejections of ash and lava fragments took place at the southern crater for a few days about 20 to 23 May, although glow at the crater had been often observed earlier in the year.

Discussion of aspects of individual eruptions

Eruption types and intensities

Eruptions at all of the volcanoes except Ritter were dominantly strombolian, although there were variations in the detailed character of activity between one volcano and another. There were non-strombolian elements such as nuées ardentes during eruptions at two of the volcanoes, and a surtseyan phase preceded strombolian activity at Long Island.

In no case did a strombolian-type eruption commence other than quietly, even when a long period of dormancy had elapsed, such as at Karkar. A-type volcanic earthquakes are not known to have preceded or accompanied any of the strombolian-type eruptions, and periods of continuous volcanic tremor or explosion earthquakes constituted all the volcano-seismic activity during these eruptions (cf. Minakami, 1960). As estimated from records of volcanic tremor, Ulawun produced the most intense sustained activity, although Manam intensities reached similar levels during periods of a few hours; Karkar produced maximum intensities less than these by a factor of about two. Langila and Long Island eruptions were weak in comparison, with intensities lower than the others by a factor of 10 or 20.

It is noteworthy that during all the strombolian eruptions reported, gas odours were rarely detected; hydrogen sulphide was noticed weakly at Langila, Karkar, and Manam occasionally, and sulphur dioxide was weakly detected once at Langila and possibly once at Karkar.

Nature of the Ritter Island eruptions

During the 1972 eruption the clearest seismic events can be accurately correlated with acoustic signals recorded at hydrophone stations in the central Pacific (R.H. Johnson, pers. comm.). Acoustic signals from normal earthquakes in the Ritter Island region are generally not recorded at these hydrophone stations, and the character of the eruption signals suggests a very shallow origin for the events, possibly right at the ocean bottom. The unusual long-period character of the seismic events as recorded by seismographs supports the idea of shallow focus. Event commencements were emergent at all stations, so no conclusions can be drawn from the first motions about possible source mechanisms. The mid-Pacific hydrophone stations were not operating during the 1974 activity. Ocean-bottom sources of both explosive and caldera-collapse types can be envisaged to account for the seismic observations, but the existing data do not seem to clearly favour one explanation more than the other.

There are marked similarities in the patterns of the 1972 and 1974 swarms and in the actual event times, although the 1974 swarm is of shorter duration and contains fewer events. Table 1 lists the times of clearly recorded events at Lae, the closest seismic station, for both swarms.

Table 1. 1972 and 1974 Ritter Island seismic swarms*

8 October 1972		16 October 1974
19 30 15½		19 32 37½
19 42 53		19 45 20
19 52 27		
19 56 36 ⎫		⎧ 19 58 26½
20 06 45 ⎬ continuous		⎨ 20 33 50½
(20 53...) ⎭		⎩ (20 52 ...)
21 02 23½		
21 07 34		
21 16 44½		
21 40 45		

*Listed times are first arrivals at station LAT (Lae) in U.T., for clearly recorded events; times in brackets indicate end of period of continuous activity.

Nuées ardentes

The occurrences of nuées ardentes at Ulawun volcano in 1973 and at Manam volcano in 1974 add examples to the sparse literature on nuées at basaltic volcanoes (Taylor, 1963; Williams & Curtis, 1965; Johnson et al., 1972).

Each nuée during these eruptions was discharged onto a single narrow sector of the cone. All three nuées at Ulawun on 11 October 1973 were observed in profile during their descent; observation of their commencements indicated that they were expelled from the crater, and were not initiated by back-falling of ejecta from a dense eruption column. In fact the eruption column immediately before the first nuée was markedly less dense than it had been during the previous few days. None of these three nuées was expelled violently by remarkable vertical or lateral explosion; the influence of the 1970 'raised block' (Johnson et al., 1972), which is here taken to be a plug of solidified lava, does not seem to have led to laterally directed nuées of Pelée type (Macdonald, 1972). In each case, during a period of apparently normal activity, the lower parts of the summit ash column expanded towards the south, initially at an almost imperceptible rate. As the nuée gathered speed in its descent, its leading edge became visible at intervals as a spurt of ash rising above the flank of the volcano ahead of the main body of the developing nuée cloud; this apparent separation was merely a result of the observer's viewpoint. The intervening parts of the cloud rose into view from behind the flank, appearing to overtake one spurt as another was forming down-slope. The material comprising the second nuée was expelled in a brief pulse, as evidenced by the thick convoluted ash cloud having completely separated from the upper slopes while the nuée was still descending only 1.5-2.0 km from the summit.

An overflowing mechanism satisfies these observations, and readily explains the narrow sectors of the Manam and Ulawun cones to which the nuées were limited. The sectors were determined by the positions of the lowest parts of the crater rims, and the expulsions were weak enough to overflow only at the lowest points.

In a general way, the sequence of events associated with the discharge of nuées ardentes was similar at Manam and Ulawun, although the time scales were different. At Manam, a strong and short-lived increase in volcanic tremor, lasting no more than a few hours, accompanied nuées. At Ulawun, both in 1970 (Johnson et al., 1972) and in 1973, nuées also occurred at times of peak volcanic tremor, although the build-up period was at least a week and the decline lasted even longer. At both volcanoes, the nuées were expelled at times of powerful strombolian eruptive activity. As already noted, at Ulawun the intensity of the visible eruption built up over a period of days, and there was no outstanding activity immediately preceding a nuée, but at Manam, in at least one case, there was an extraordinary increase in intensity beginning about a half-hour before the nuée. Lava flow closely succeeded each phase of nuée generation at Manam during 1974, and a new flow commenced from the nuée vent at Ulawun shortly after the nuées of 1973. Observations were not sufficiently detailed to determine the interval between nuée expulsion and the onset of lava flow.

Petrology

Chemical and modal analyses and CIPW norms of representative lava

Table 2. Chemical* and modal analyses and CIPW norms of 1973-74 lava flow samples

Volcano	Ulawun	Langila	Long Island	Manam Island
Sample No.	74710020	74710016	74710012	74710026
Date erupted	about 9/10/73	about 14/7/73	about 25/10/73	June 1974
SiO_2	52.7	55.3	48.7	52.7
TiO_2	0.85	0.55	1.04	0.32
Al_2O_3	18.6	16.8	18.6	14.9
Fe_2O_3	3.35	3.25	4.80	4.85
FeO	6.35	5.75	6.55	3.90
MnO	0.16	0.18	0.20	0.17
MgO	4.55	4.45	4.65	8.75
CaO	10.6	9.15	12.5	11.0
Na_2O	2.40	2.60	1.96	2.35
K_2O	0.40	1.49	0.88	0.62
P_2O_5	0.10	0.18	0.17	0.09
H_2O+	0.05	0.08	0.04	0.04
H_2O-	0.03	<0.01	<0.01	0.01
CO_2	0.05	0.05	0.05	0.05
Total	100.19	99.84	100.14	99.75

CIPW norms

Q	7.37	8.65	2.37	4.62
or	2.36	8.83	5.19	3.67
ab	20.28	22.05	16.56	19.94
an	38.76	29.85	39.32	28.36
di ⎰wo	5.34	5.91	8.85	10.63
di ⎨en	3.03	3.40	5.31	8.33
di ⎱fs	2.08	2.24	3.08	1.12
hy ⎰en	8.28	7.70	6.26	13.52
hy ⎱fs	5.69	5.08	3.63	1.83
mt	4.85	4.72	6.95	7.05
il	1.61	1.05	1.97	0.61
ap	0.24	0.43	0.40	0.21
cc	0.11	0.11	0.11	0.11

*Volume percent** phenocrysts*

Plagioclase	30.5	31.0	31.5	20.0
Olivine	<0.5	0.5	1.5	4.0
Orthopyroxene	1.0	1.5	—	0.5
Clinopyroxene	1.5	9.0	5.5	15.5
Fe-Ti oxides	—	3.0	0.5	1.0
Total	33.0	45.0	39.0	41.0

* Supplied by Australian Mineral Development Laboratories, Adelaide.
** Determined by C.O. McKee.

samples from the 1972-75 group of eruptions at Ulawun, Langila, Long and Manam are given in Table 2; analyses of ten samples of eruptive products from Karkar are presented in an appendix to the accompanying paper by McKee et al. (this volume). Lavas produced in the Langila and Karkar eruptions are low-silica andesite. Ulawun, Manam, and Long Island lavas are quartz tholeiite, although the Long Island lava would be classed as olivine tholeiite after standardisation of the oxidation state of iron by the method of Irvine & Baragar (1971).

None of the lavas shows marked chemical contrast with products of other ecent eruptions of the same volcano, although the new Karkar lavas are slightly more basic than older ones. The only significant chemical variation during one of these eruptions took place at Karkar; this corresponds to the much greater volume of lava extruded at Karkar compared with the other eruptions. This variation, believed to be controlled by separation or addition of calcic plagioclase, suggests that appreciable time elapsed after magma generation, permitting fractionation; the time scale of this process is not known.

ACKNOWLEDGEMENTS

The authors thank: technical officers and assistants of Rabaul Observatory for various kinds of aid in making observations and collecting data; local residents near the active volcanoes for supplying information on activity; E.E. Ball for supplying observations and photographs of Long Island activity; Observer-in-Charge, Port Moresby Geophysical Observatory, for supplying seismic records of the Ritter Island seismic swarms; R.W. Johnson for discussions; K. Somerville for drafting illustrations; G. Brooker and E. Kent for typing the manuscript. The Chief Government Geologist, Geological Survey of Papua New Guinea, authorised publication of this paper.

REFERENCES

Bassot, J.M., & Ball, E.E., 1972: Biological colonisation of recently created islands in Lake Wisdom, Long Island, Papua New Guinea, with observations on the fauna of the lake. *P.N.G. Sci. Soc. Proc., 23,* pp. 26-35.

Best, J.G., 1956: Investigations of recent volcanic activity in the Territory of Papua and New Guinea. *Proc. 8th Pacific Sci. Cong., 2,* pp. 180-192.

Chalmers, J., 1887: *Pioneering in New Guinea.* London, Religious Tract Society.

Cooke, R.J.S., 1975: Time variations in recent volcanism and seismicity along convergent plate boundaries of the south Bismarck Sea, Papua New Guinea. *Bull. Aust. Soc. Expl. Geophys., 6,* pp. 77-78.

Curtis, J.W., 1973: Plate tectonics of the Papua-New Guinea-Solomon Islands region. *J. Geol. Soc. Aust., 20,* pp. 21-35.

Denham, D., 1969: Distribution of earthquakes in the New Guinea-Solomon Islands region. *J. geophys. Res., 74,* pp. 4290-4299.

Everingham, I.B., 1974: Large earthquakes in the New Guinea-Solomon Islands area, 1873-1972. *Tectonophysics, 23,* pp. 323-338.

Everingham, I.B., 1975a: Faulting associated with the major north Solomon Sea earthquakes of 14 and 26 July 1971. *J. Geol. Soc. Aust., 22,* pp. 61-69.

Everingham, I.B., 1975b: Seismological report on the Madang earthquake of 31 October 1970 and aftershocks. *Bur. Miner. Resour. Aust. Rep. 176.*

Fisher, N.H., 1957: Catalogue of the active volcanoes of the world, including solfatara fields — Pt V, Melanesia. *Int. Volc. Assoc. Naples.*

Fisher, N.H., 1959: Report of the sub-committee on volcanology, 1954-56. *Bull. Volc., 21,* pp. 153-161.

Irvine, T.N., & Baragar, W.R.A., 1971: A guide to the chemical classification of common volcanic rocks. *Can. J. Earth Sci., 8,* pp. 523-548.

Johnson, R.W., Davies, R.A., & White, A.J.R., 1972: Ulawun volcano, New Britain. *Bur. Miner. Resour. Aust. Bull. 142.*

Johnson, R.W., Mackenzie, D.E., Smith, I.E., & Taylor, G.A.M., 1973: Distribution and petrology of late Cenozoic volcanoes in Papua New Guinea. *in* Coleman, P.J. (Ed), *The Western Pacific: island arcs, marginal seas, geochemistry,* Univ. W.Aust. Press, pp. 523-533.

Johnson, T., & Molnar, P., 1972: Focal mechanisms and plate tectonics of the Southwest Pacific. *J. Geophys. Res., 77,* pp. 5000-5032.

Krause, D.C., 1973: Crustal plates of the Bismarck and Solomon Seas. *in* Fraser, R. (Ed), *Oceanography of the South Pacific 1972,* N.Z. Nat. Comm. UNESCO, Wellington. pp. 271-280.

Macdonald, G.A., 1972: *Volcanoes.* Englewood Cliffs, Prentice-Hall Inc., 510 pp.

Minakami, T., 1960: Fundamental research for predicting volcanic eruptions (Part 1). *Bull. E.R.I., 38,* pp. 497-543.

Perret, F.A., 1950: Volcanological observations. *Carnegie Inst. Washington Pub. 549.*

Powell, W., 1883: *Wanderings in a wild country.* London, Sampson, Low, Marston, Searle & Rivington.

Rossel, —, 1808: *Voyage de d'Entrecasteaux.* Paris, l'Imprimerie Imperiale.

Sapper, K., 1917: Katalog der geschichtlichen Vulkanansbruche Melanesien. *Schr. wiss. Ges. Strassburg, 27.*

Taylor, G.A.M., 1958: The 1951 eruption of Mount Lamington, Papua. *Bur. Miner. Resour. Aust. Bull. 38.*

Taylor, G.A.M., 1963: Seismic and tilt phenomena preceding a Pelean type eruption from a basaltic volcano. *Bull. Volc., 26,* pp. 5-11.

Taylor, G.A.M., 1966: The surveillance of volcanoes in the Territory of Papua and New Guinea. *South Pacific Bulletin.*

Taylor, G.A.M., Best, J.G., & Reynolds, M.A., 1957: Eruptive activity and associated phenomena, Langila volcano, New Britain. *Bur. Miner. Resour. Aust. Rep. 26.*

Walker, G.P.L., & Croasdale, R., 1971: Characteristics of some basaltic pyroclastics. *Bull. Volc., 35,* pp. 303-317.

Williams, C.E., & Curtis, R., 1965: The eruption of Lopevi, New Hebrides, July 1960. *Bull. Volc., 27,* pp. 423-433.

Williamson, J. (Ed) 1939: *A Voyage to New Holland* (by William Dampier). London, Argonaut Press.

1974-75 ERUPTIONS OF KARKAR VOLCANO, PAPUA NEW GUINEA

C.O. McKEE, R.J.S. COOKE, and D.A. WALLACE

Geological Survey of Papua New Guinea, Volcanological Observatory,
P.O. Box 386, Rabaul, Papua New Guinea.

ABSTRACT

Karkar volcano, dormant since 1895, erupted from Bagiai cone, on the floor of the inner summit caldera of the volcano, between about 14 February and 8 August 1974, and again from about 30 December 1974 to 10 February 1975. These eruptions were predominantly effusive, and accompanying explosive activity was chiefly of strombolian character. Weak explosive activity recurred at Bagiai in April and May 1975. Lava flow in 1974 occurred in two periods, for about 135 days from the beginning of the eruption to late June, and again for about 15 days in July and August. Only weak explosive activity was observed during a 25-day lull between these periods. The highest levels of activity were attained during the 40 days preceding the lull, and during the second effusive period, when lava outflow rates exceeded 100 m^3s^{-1}. Lava flowed again for about 30 days in January and February 1975. Altogether about $1.7 \times 10^8 m^3$ of low-silica andesitic lava, mainly aa, was extruded to form a compound sheet on the floor of the inner caldera, almost completely covering it. The greatest thickness of the sheet is probably about 50 m, but individual flow units are about 3-5 m thick. The main lava sources were distributed around the eastern and southern base of Bagiai cone. Systematic variation in the physical and chemical composition of the lava during the course of the eruptions may have resulted from separation of plagioclase.

INTRODUCTION

Karkar Island is an andesitic stratovolcano which lies about 16 km off the north coast of mainland Papua New Guinea (Fig. 1). The island is one of a chain of fifteen Quaternary volcanic islands which constitute the western part of the Bismarck volcanic arc (Johnson et al., 1973). It has a population of about 20,000 living around its coast and, to a lesser extent, in its foothills. The coastal margin and parts of the lower slopes of the volcano are under cultivation, mostly plantations of coconuts and cocoa, but the larger part of the island is covered in dense tropical rainforest.

Except for possible small-scale activity in 1962 (see below), the volcano had been dormant for 79 years prior to the large-scale eruption which lasted from February to August 1974. Large-scale activity resumed in January 1975, and minor activity is continuing at the time of writing (June 1975). These eruptions are part of the 1972-75 sequence of eruptions at volcanoes along the southern margin of the Bismarck Sea, which is discussed in an accompanying paper (Cooke et al., this volume).

This paper aims to provide a preliminary description of the eruptions; a complete report on all aspects of the eruptions is in preparation. A report on the chemical and mineralogical changes which occurred in the new lavas is appended.

STRUCTURE AND TOPOGRAPHY

G.A.M. Taylor prepared the first geological account of Karkar Island, based on his investigations in 1963 and 1965[*]

[*] Accounts of these investigations on Karkar, and of subsequent investigations by other geologists, are included in Record 1972/21 (unpublished) of the Bureau of Mineral Resources, entitled 'Geology and petrology of Quaternary volcanic islands off the north coast of New Guinea', by R.W. Johnson, G.A.M. Taylor, & R.A. Davies.

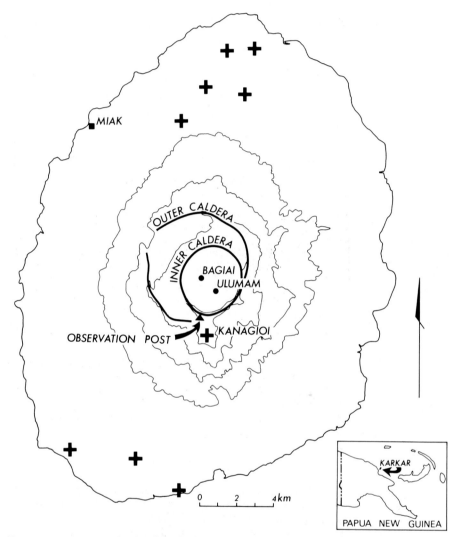

Fig. 1: Karkar volcano, showing form lines at approximately 300 m intervals. Crosses show parasitic cones. Dots are recently active volcanic centres.

Karkar Island is roughly elliptical in plan; its sea level dimensions are 25 km by 19 km, but its diameter on the sea-floor is probably 35-40 km. The island has the form of a truncated cone, with a double caldera near its centre. The flanks slope gently at less than 20°, and from a distance the volcano presents a fairly smooth regular profile. However, the topography is locally rugged, particularly in the south and east, owing to the development of deep gullies by radial erosion. Lava flows and pyroclastic deposits, including those of extensive pyroclastic flows, are exposed in stream and coastal sections.

Table 1. Analysis of gas condensate collected on Bagiai cone 22 April 1972*.

Temperature	pH	SO_4	Cl	Ca	Mg	Na	K	Li
		(in ppm)						
85°C	5.1	37	3	1.0	0.28	5.04	0.68	trace

* This analysis is included in Report 73/19 (unpublished) of the Geological Survey of Papua New Guinea entitled 'Gas condensate collection programme — progress report October, 1973', by I.H. Crick.

Focal mechanism and aftershock distribution studies of the 'Madang' earthquake sequence (main shock 31 October 1970, 1753 UT, 4.95°S, 145.68°E, depth 41 km, magnitude 7.0, about 45 km southwest of Karkar) defined a 130 km-long zone, trending east-northeast, which extended to the vicinity of Karkar (Everingham, 1975). Cooke (1975) finds possible grounds for associating the Madang earthquake sequence with the Karkar eruption, on the basis of correspondences between regional seismicity and volcanism in the western part of the Bismarck volcanic arc from 1947 to 1974, between longitudes 143¼°E and 149¾°E (see also Cooke et. al., this volume)*.

The 1974 eruption lasted from 14 February to 8 August, during which time lava flowed continuously except for about three weeks in June and July. Explosive activity, mainly of strombolian character, occurred throughout. Eruptive activity resumed about the beginning of 1975; lava flowed for about four weeks in January and February, accompanied by strombolian explosive activity. After a lull, explosive activity recommenced in mid-April, and at the time of writing (June 1975) minor eruptive phenomena are still occurring.

The eruptions were continuously monitored by Rabaul Observatory staff from an observation post on the caldera rim, 1.7 km south of Bagiai cone (Fig.1). A short-period vertical seismograph was operated at the observation post throughout.

Figure 2 illustrates the distribution of explosive and effusive vents and the lava fields of the 1974 and 1975 eruptions. All vent identifications in the eruption narratives below refer to Figure 2. Figure 3 is a simplified eruption diagram showing the trends in selected observed parameters during the eruptions. Quantitative information in the present paper such as lava volume and effusion rate, should be regarded as provisional.

The following eruption narratives have been divided into phases for convenience; the changes in character of activity in successive phases are gradational.

1974 Phase 1, 14-18 February: Early investigations

The first known events of the eruption were loud bangs heard on the south and east coasts of Karkar during 14 February. An aircraft pilot observed 'smoke' at the caldera rim during the afternoon of 15 February, and on investigating, found that explosions were occurring at Bagiai cone. Glowing lava ejections to 20-30 m above the vent, and a glowing lava flow issuing from the eastern base of Bagiai, were observed from the air on the morning of 16 February. The explosive vent was reportedly about half-way down the southeast side of Bagiai.

* In late 1970, Mr Taylor told me that he considered the 'Madang' earthquake and the shift in focus of the strongest fumarolic activity on Bagiai to be a prelude to eruptive activity from Bagiai. He did not speculate on when this predicted activity might take place. Ed.

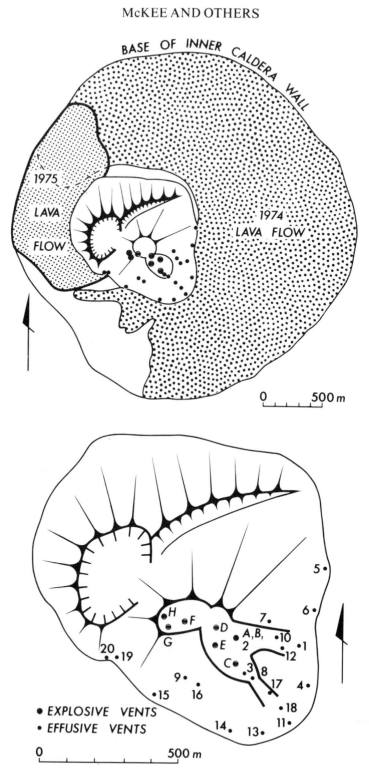

The authors first observed the eruption during an aerial inspection at mid-afternoon on 17 February. The details of activity were similar to those already described. Ejections containing either brightly incandescent lava fragments, or dark ash, or both, were occurring at intervals of 5-15 s from a small vent, A, and a mound of black cinders had already formed around the vent, which was located in the northern sector of the old southeast crater. Movement of the upper, incandescent part of the lava flow from its source vent, 1, could not be detected from the air.

Another aerial inspection on the morning of 18 February revealed similar activity, and in addition a small vent, 2, several tens of metres below vent A. Vent 2 was the site of occasional short-lived red glows and some minor vapour emission, but no actual explosions were observed there. A thin ribbon of faintly incandescent lava could be detected extending 10-20 m below this vent.

1974 Phase 2, 19 February—22 March: Steady mild activity

The observation post was established during the afternoon of 19 February. By that time unco-ordinated explosions of incandescent lava fragments ('lava jets') were occurring at three closely spaced vents — A, 2, and a new one, B, between them — building up the mound of fresh cinders. Lava was flowing more profusely from vent 2 and continued from the low-level source, vent 1. At this stage the ash content of most ejections was moderate to low, but by 25 February explosive activity consisted more of ash ejections than of lava jetting. From late February onwards, generally larger volumes of ash were emitted. During a peak of activity in the middle of March, vents A and B coalesced to form an enlarged vent A.

Lava issued from vent 1 at an apparently steady rate throughout this phase, and the resulting lava field expanded slowly towards the east and north. From the first week in March some lava was actively tunnelling beneath the congealed surface, reappearing as incandescent sources at the outer edges of the field.

Lava from vent 2 moved towards the south. Surges of lava effusion occurred at vent 2 at irregular intervals between 21 and 28 February, being occasionally preceded by distinctive projections of dark ash and incandescent fragments. A lava surface was visible in the vent on 23 and 25 February and was observed to swell before some explosions. After another series of lava surges in mid-March, associated with the then prevailing higher level of activity, flow of lava from vent 2 ceased on 17 March.

Early in the eruption, explosive events were prominent on the seismic records, but volcanic tremor became the dominant feature of the seismic activity after about 20 February.

The end of this phase was marked by weaker ejections of lava fragments, lower rates of ash emission, cessation of lava flow at the high-level vent 2, and weakening of seismic activity.

1974 Phase 3, 23 March—7 April: Weak activity

Explosive activity reached a low level late in March, although on 23 March 'flashing arcs' (visible acoustic shock waves, see Perret, 1950) were observed for the first time in the eruption. Ash emission weakened and yellow sublimates began forming around the explosive crater (A). By early April, the highest lava jets were reaching only about 100 m above the vent, and the ash content and frequency of ejections were markedly reduced. Lava continued to flow steadily in a major feeder to the north. The level of seismic activity was low.

1974 Phase 4, 8 April—18 May: Major new explosive vent formed

On 8-9 April, sound effects became louder and seismic activity increased. Strong aerial concussions were observed at the coast on 9 April. The peak of this higher level of activity was reached on 11 April, and seismic records for that day consist of more-or-less continuous tremor. Throughout 11 April the noise level was high and large volumes of ash were ejected from vent A. By the evening of 12 April explosive activity had declined slightly although dark ash clouds continued to be ejected, flashing arcs were common, and lava jets were reaching heights of about 220 m above the vent.

Fig. 2 (opposite): *Upper:* inner caldera, Karkar volcano, showing explosive and effusive vents, and lava fields of the 1974 and 1975 eruptions; topography on Bagiai cone before the 1974 eruption is sketched as three craters and an escarpment. *Lower:* Bagiai cone, showing details of summit topography in June 1975; 1974-75 explosive and effusive vents are indentified.

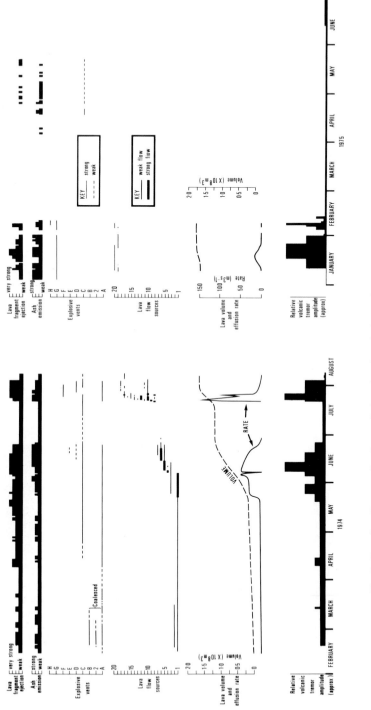

Fig. 3: Eruption diagram showing variations in selected observed parameters of the 1974-75 Karkar eruptions. Changes in cumulative lava volume and effusion rate are schematic, and figures are based on preliminary estimates.

Plate 3: Lava jet at crater C, Bagiai cone, Karkar volcano, mid-May 1974. Height of jet is about 180 m. Photograph by D.A. Wallace.

Plate 4: Explosion at crater C, Bagiai cone, Karkar volcano, mid-May 1974. Photograph by D.A. Wallace.

Explosive activity occurred briefly at the low-level effusive vent 1 in the pre-dawn hours of 11 April; for about 15 minutes small showers of incandescent lava were projected at a low angle from this vent. Early on 14 April similar explosive activity at vent 1 lasted about 2 minutes; 'choofing' sounds at intervals of about 1½ s, reminiscent of a steam locomotive, accompanied the low-angled blasts of lava fragments. Otherwise steady flow of lava from vent 1 continued, although the direction of the main feeder changed to south on 9 April.

Shortly after midday on 14 April a new explosive vent, C, was created at the southern base of the growing cinder cone about vent A. Coils of dark ash rose slowly and silently from it. The first emission lasted about 3 minutes; other similar emissions followed at intervals ranging from a few tens of seconds to more than 10 minutes. Small quantities of incandescent fragments could be seen near the base of these emissions by evening of the first day of this activity. At night, occasional flashes of lightning were observed in the ash clouds from vent C. Little noise accompanied ejections from this vent during its first two days of activity.

The activity at vent A at this time consisted of ejections of incandescent lava fragments at intervals of about 2-5 s, accompanied by loud explosive sounds and flashing arcs.

During the next 10 days conditions were similar, although the ash content of vent C ejections was gradually reduced and the amount of incandescent lava fragments increased. The main lava feeder from vent 1 flowed towards Ulumam cone from 21 April. Lava jetting at vent 1 was again observed for about 15 minutes near midnight on 25 April.

In late April and early May, explosive activity was at a moderate level, and only minor fluctuations in the level of volcanic tremor were recorded. Average intervals between ejections from vents A and C were respectively about 4 s and 1 minute. Vent A ejections reached about 150-200 m above the vent, and most had small ash content. There were exceptions, when dense, dark ash clouds were erupted, commonly without much noise and with almost no seismic signal. Several ephemeral vents around vent A were formed in early May, and either coalesced with it or were filled in.

Vent C produced larger explosions than vent A, many ejections reaching about 250 m high and spreading incandescent lava fragments in wide-angled jets. Ash content of vent C emissions was very high at times. On 6 May lava was visible in the western part of the crater around vent C; it had a dull red surface broken by brighter red lines at cracks in the crust. Some explosions originated there, but most were located in the eastern part of the crater.

Explosive activity at vent 1 recurred for a few minutes on the afternoon of 1 May; this consisted of vertical ejections of incandescent lava to about 20 m above the vent, in contrast to the almost horizontal projections on the earlier occasions.

An increase in explosive activity, and a slight change in its character, was evident from about 10 May. Lava jetting of greater frequency and of a more impulsive kind occurred at both explosive vents, accompanied by loud, sharp detonations. Volcanic tremor became larger and more continuous than it had been.

By the middle of the month lava jetting from vent A was virtually continuous with low ash content. Ejections at vent C (Plates 3, 4) continued to be larger than those from vent A although far less frequent, and were expressed seismically as discrete explosion shocks, superimposed on continuous tremor.

About midday on 18 May the northern and southwestern inner walls of the crater around vent A collapsed, considerably enlarging the crater.

1974 Phase 5, 19 May—4 June: Increased activity

Volcanic tremor increased to new high levels from about 19 May, and discrete explosion events almost disappeared from the record.

Between about 21 and 23 May, vent A activity consisted of an essentially continuous jet of particles directed towards the east at about 45° to the vertical; the jet returned to the vertical after 23 May. An aerial inspection on 23rd revealed a fissure-like feature in crater A, trending about northeast, which showed bright daylight incandescence along its length.

'Heave-ups' of lava in crater C were frequently observed during the period 20-25 May. A mound of lava would rise up to 5 m above the crater rim in a few seconds, and remain visible for periods ranging from a few seconds to two minutes before explosions destroyed it. A slight decline in the frequency and height of lava jets from vent C was noticed during the same period. A sluggish lava flow appeared from a new vent, 3, just below the southeast rim of crater C, early on 25 May.

Near the end of May ejecta were accumulating on the upper slopes of Bagiai so rapidly that scoria flows were produced; some of these almost reached the foot of the cinder cone on its southeast side. The average width of these flows was about 5 m and their maximum speed was about 1 km hr^{-1}.

The rate of lava flow from vent 1 increased strongly late in May, and the lava field spread rapidly. Flows reached the caldera wall along about 1 km of the southeast sector, completely surrounding and almost burying Ulumam cone. After about 8 June the main expansion of the lava field took place in the northeastern sector of the caldera floor.

In late May and early June, surges of lava and mild explosive activity occurred several times per hour at vent 1. This apparent explosive activity was often nothing more than turbulence at the lava source, although generally some obvious gas release was associated with this activity. At times gushing waves of lava up to 1 m high were observed, and occasionally small fountains of lava were produced. Activity of this type was not observed after 3 June.

Towards the end of this phase activity at vent A consisted of frequent ejections of lava fragments to heights of 200-300 m; flashing arcs were common. Greater ash content in the ejections was noticeable in early June.

1974 Phase 6, 5-30 June: Major peak in activity

Overall the activity was more vigorous from 5 June, corresponding to an increase in seismic activity to a new high level by 7th. Intervals between the showers of brightly incandescent fragments from vent A averaged only about 2-3 s, and the ash content became less. The ejections were accompanied by sharp and loud detonations, and aerial concussions were the strongest yet observed. Maximum heights of 500 m were reached by ejections from vent A, but those from vent C remained small, although more frequent from 5 June. Minor ejections of incandescent fragments were observed for a few hours at the otherwise effusive vent 3 in the afternoon of 6 June, and that night the vent 3 lava flow was glowing more brightly than before.

Lava effusion from vent 1 reached its highest rate near the end of the first week in June, when lava was bursting from the vent in great gushes and surges. However flow from this vent ceased abruptly on the morning of 7 June.

About midday on 7th, nearly continuous ejections of grey ash began at vent C, and the quantity of incandescent fragments in vent A ejections increased. An hour later a new lava source, 4, appeared at the southeast base of Bagiai. The effusion rate was high, but flow from this vent had ceased by the evening of 8th, apart from a brief resurgence on 11th.

On the night of 8 June, lava appeared from a new source, 5, at the eastern base of Bagiai. Lava flowed strongly from this vent, and reached the northwest caldera wall by 14 June.

During the first week of this phase, vent C was occasionally inactive for periods up to several hours long. On 12 June, summit explosive activity was noticeably weaker although vent A ejections were still occurring at about 4 s intervals, and volcanic tremor showed a corresponding decline.

Lava ceased flowing from vent 3 on 14 June, although explosive activity from the adjacent vent C was slightly stronger, occurring at intervals as short as 10 s. Lava began flowing from a new source, 6, at the eastern base of Bagiai on the night of 14th, joining with the flow from vent 5 to produce a major feeder to the northwest.

Yet another new lava source, 7, appeared about half-way up the eastern flank on the night of 17

Plate 5: Lava source 20, base of Bagiai cone, Karkar volcano, 16 January 1975. Width of source is about 3 m. Photograph by C.O. McKee.

June. The following morning some large masses of brown cindery material were carried down in this flow.

Two weakly explosive, ash-producing vents, D and E, were created near the western rim of vent A on 15 and 19 June, respectively. These vents had become inactive by the end of June.

Lava flow from vents 5 and 7 ceased on the night of 24 June, and from vent 6 on 26th, although weak incandescence at the latter on the night of 29th suggested a brief resurgence of flow. Explosive activity at vents A and C weakened considerably early on 25th, and the changes in visible activity corresponded to a marked decline and change in character of the seismic activity. Discrete explosion events were recorded in place of the declining but continuous volcanic tremor which had prevailed until then. The explosion events correlated with small to moderate ejections of incandescent fragments and variable amounts of ash from vent C, which took place at average intervals of about 2 minutes. During this period, infrequent ejections of off-white vapour occurred at vent A, gradually becoming weaker.

The most notable developments in the structure of the cone during this phase were the partial infilling of the old summit crater, pronounced increase in size of the cone about vent C, formation of a deep notch between craters A and D, enlargement of crater A to the east, and accumulation of lava and slumped cinder cone material to produce a prominent shoulder near the eastern base. Deposits of yellow sublimates in several places on Bagiai were first noticed on 25 June.

1974 Phase 7, 1-21 July: Weak explosive activity, no lava flow

During the first half of July, eruptive activity consisted only of small ejections from vent C at intervals not less than 3 minutes, generally similar in character to those at the end of June. However, early in the second week many ejections consisted purely of vapours, giving way from 9 July to more frequent dense, convoluted clouds of ash and lava fragments, although the intervals between ejections were similar.

Several arcuate fractures broadly connecting the western extremities of craters C and D were first noticed on 1 July; slippage amounted to about 1 m during this phase.

A low level of seismic activity persisted until about 14 July, after which a graded intensification of the seismic activity correlated well with the steady strengthening of visible explosiions in crater C. Vent C ejections had become richer in incandescent lava fragments and poorer in ash by 17th, and weak glow persisted at the crater between ejections. By 18th lava jetting was more frequent, and many of the ejections were prolonged. By 21 July vent C lava jetting was almost continuous and accompanied by loud, sharp detonations.

1974 Phase 8, 22 July—8 August: Major peak in activity, eruption ends

In the early morning of 22 July lava began flowing strongly from a source, 8, on the southeast flank of Bagiai. Vent A was reactivated, producing small jets of lava, while ejections at vent C continued at intervals of 2-3 s. A lava source, 9, appeared on the southwest flank of Bagiai, the first in that sector of the cone, soon after daylight. By mid-morning, billowing dark ash clouds towered over the volcano. During an aerial inspection at this time, another new lava source, 10, was observed on the eastern flank. Vent E was reactivated about midday and ejected dark ash clouds for several hours, although it was inactive by next morning.

From this point on, new lava sources became distinctly concentrated along radii of the cone, particularly towards the southeast. Sources 11 to 17 appeared successively between early afternoon on 22 July and the morning of 25th. Of the ten sources created during these three days, vents 10 and 12 in the eastern sector were the strongest; no more than about five of them were active at any one time. Outbreaks of lava from several different positions in the southeast sector were witnessed. One such outbreak late on 25th commenced as a dull red patch about 10 m across. Five minutes later this patch began to swell and glow more brightly, and with a slow heaving motion, viscous red-incandescent lava welled up from the vent with large chunks of similarly incandescent welded scoria. Within another ten minutes yellow-incandescent lava was gushing from the vent and spilling down the southeast flank, and large blocks of red-incandescent scoria were breaking from the walls of the vent. Other new vents behaved similarly, producing surges, fountains, and mild explosions of lava.

Undermining and collapse of cone materials in the eastern and southeastern sectors of lava flow led to the formation of 'chasms' in these areas overnight on 25 July. Lava was channelled by these chasms, and large irregular pieces of the cinder cone were carried along in the eastern flow. Weak flow of lava at vent 17 was the only activity remaining in the southeastern sector from mid-morning of 26th.

For the week beginning 22 July, ash emission from the volcano was strong, and ash falls over the western part of the island were heavy enough to break cocoa and coconut tree branches and to prevent outdoor work on plantations. Lightning in the ash clouds was common.

A new explosive vent, F, became active at the southern rim of the old summit crater of Bagiai about midnight on 24 July. Thick convoluted ash clouds were produced by the reaming-out of the new vent; these ejections were almost silent.

Activity overnight on 25 July and early on 26th was at a particularly high level. Explosions of ash and lava fragments from vents A, C, and F were strong and frequent. Brilliant quasi-continuous jetting of fluid lava continued at vent A in late July, although vents C and F were producing smaller and less frequent ejections of ash and lava fragments from 26th. Vent D was reactivated on the night of 27 July and produced dense ash and lava fragment ejections at high frequency for a day or two before a gradual decline commenced.

Seismic activity during this peak was in broad correspondence with visible activity. Strong tremor was present between 22 and 25 July; after 26th a decline was evident.

The first explosive vent to cease erupting was F, on 1 August, and vents A and D became inactive overnight on 3rd. Vent C ejected ash and incandescent lava fragments with varying intensity until the night of 7th. Lava ceased flowing in the east lava channel on 4th or 5th and ceased in the southeast channel overnight on 8 August.

Piecemeal collapse of the walls of craters A and C began on 5 August, causing dust clouds to be produced and exposing layers of partly welded cinder. Sublimate deposition became noticeable from 5 August. Moderate blue and white vapour emission and faint glow continued at a number of points during several weeks of observation after the end of the eruption. Visits to the cone did not enable the cause of the glows to be identified.

1975 eruption

Renewed eruptive activity probably started about 30 December 1974, when deep rumbling was first heard. From 5 January 1975, dark grey ash clouds were seen rising above the volcano.

Regular observations began on 11 January from the observation post established in 1974. Jets of brightly incandescent and fluid-looking lava fragments were being ejected at intervals of 3-5 s from a new vent, G, on the western flank of Bagiai cone, and reached heights of 150-200 m above the vent. Aerial concussions were frequently felt. Ash content was variable, and lightning occurred in ash clouds between 11 and 14 January. Strong volcanic tremor was recorded. Lava flowed at low to moderate effusion rates from new vents, 19 and 20, at the southwestern base of Bagiai (Plate 5), and weak explosive activity occasionally occurred at these vents. Lava flowed onto the western part of the caldera floor and eventually over-rode 1974 flows in the northwest.

The strongest activity was in the week 17-24 January. Eruptive activity ceased briefly on 3 February. Strong explosive activity recommenced on 5 February, but ended by 10 February after producing a new explosive vent, H, on 8th. Weak effusive activity recurred between 6 and 9 February. Weak glow was observed above the vents G and H during the next week.

Explosive activity resumed on 6 April, after a lull in which volcano-seismic activity was absent, or at least so weak as to be difficult to identify with confidence, although vapour emission continued; grey ash clouds were ejected spasmodically from vents G and H. Similar activity occurred on 9th. More frequent explosions, this time at vent C, started on 18 April, and roaring and rumbling sounds soon became pronounced. Small quantities of incandescent lava fragments were weakly ejected by some explosions. Seismic activity at this time was weak compared with that in January. Glow above vent C became visible on 27 April.

No clearly defined end to the 1975 eruption has yet occurred. Although visible explosive activity ceased on 22 May, roaring and rumbling persists at the time of writing (June 1975), blue and white vapour emission continues, and glow is still observed at vent C. Volcano-seismic events of low to moderate amplitude again became noticeable at the time of the April activity, and are still recognisable.

DISCUSSION

Available evidence suggests that the 1974 Karkar eruption commenced gradually, although perhaps a more violent beginning might have been expected after such a long dormancy. Possible precursory phenomena are discussed above, i.e. the change in the location of the main thermal activity on Bagiai cone, perhaps associated with the 'Madang' earthquake, the earthquake itself and aftershocks, the low Cl/SO_4 ratio of the gas condensate, and the slight apparent temperature increase of Bagiai fumaroles; even with hindsight, their real significance is not clearly established. There is no evidence of any precursory A-type volcanic earthquakes (Minakami, 1960), but as the nearest seismic station was then more than 100 km from the volcano their absence is not certain. If they did occur, none was reported felt, and they had ceased by the time seismic recording commenced at the observation post.

For the most part during the 1974 and 1975 eruptions, explosive activity was confined to the upper parts of Bagiai, while effusive activity occurred around the lower parts. The most noteworthy exceptions were lava flow for 27 days from the high-level vent 2 in February and March 1974, and mild explosions from the low-level vents 19 and 20 during January and February 1975.

Terminal explosive activity at Bagiai cone was located, until late July 1974, exclusively in the old southeast crater, the youngest of the pre-1974 craters, although an element of westward migration within the crater is suggested by the opening of vents D and E late in the first major phase. Effusive activity was restricted to the flanks east and northeast of this crater until the beginning of the second major phase. However, the long established and stable effusive vent 1 was replaced by several other nearby vents as the first major phase reached its peak, suggesting that this 'leakiness' of the cone may have been induced by increasing magma pressure, or the high level of volcanic tremor, or both.

A similar 'leakiness' was even more marked early in the second major phase. In this case, although the strongest outflow of lava continued eastwards, the leaky zone had rotated clockwise to include the southeast to southwest sector of the cone. In addition a radial distribution of some of the vents became apparent, leading to the formation of chasms by lava undermining and weakening the flanks of the cone, as described above. Although explosive activity continued in the old southeast crater during the second major phase, westward migration of vents extended to the old summit crater with the creation of vent F.

Activity in January-February 1975, both explosive and effusive, occurred exclusively on the western flank of Bagiai, confirming the migration tendency. However the minor explosive activity of April-May returned to the vicinity of the old southeast crater.

The 1974 Karkar eruption produced about 1.6×10^8 m³ of lava, and a small fraction of that amount as tephra; the 1975 activity added about another 0.1×10^8

m³ of lava. The pattern of effusion during the eruptions differed strikingly from that of some other large-volume effusive eruptions at volcanoes of other types such as Hekla (Thorarinsson, 1970) and Kilauea and Mauna Loa (Macdonald & Abbott, 1970). In these, effusion rates reach very high levels for only a few days at the beginning of eruption, but at Karkar about 95 days of steady low-rate lava flow preceded the first of the two major effusive phases in 1974, which lasted about 40 days. A complete halt in lava outflow then followed for about 25 days until the second major effusive phase which lasted about 15 days. In this phase the highest outflow rates of the eruption were reached, and *did* occur at or soon after the recommencement of effusive activity. Almost five months without reported eruptive activity preceded the 1975 effusive phase, which lasted about 30 days.

Most of the lava flows of 1974 and 1975 were of aa type although a minor amount of ropy pahoehoe texture was noted in the 1975 lava. The final lava field is very complex in structure, and is probably 40-50 m thick in parts, but individual active flow units were observed to be only 3-4 m thick (Fig. 4). An important feature of the developing lava field, first noticed in lava from vent 1 about 6 March 1974, was subsurface flow of lava, giving rise to incandescent lava 'sources' at the outer edges of the field. As there were no significant physical or chemical changes in the lava by this date (see Appendix), nor was there any reason to suspect a change in lava temperature, the onset of subsurface flow and its maintenance thereafter seem attributable to the fact that flow was proceeding on an essentially horizontal surface. This is indicated by the form of the lava field on 6 March 1974, when small and equidimensional overlapping lobes of fresh lava around the eastern base of Bagiai retarded direct access of new lava from vent 1 to some parts of the caldera floor. Subsurface flow is believed to have been mainly responsible for the presence of slabs of lava in the lava field, both in situ, and tilted and up-ended. Movement of lava beneath previously congealed skins of lava, where they had formed, led to fracturing and disruption of these surfaces.

Bagiai cone is now surfaced by brittle, angular blocks, some of which have weakly developed breadcrust surfaces, scoriae, and some rounded bombs.

Gas odours were rarely detected during these eruptions; hydrogen sulphide was noticed only at the end of the first major active phase of the 1974 eruption, and sulphur dioxide was never certainly detected, although an acrid gas observed near lava sources 19 and 20 during the 1975 eruption may have been sulphur dioxide.

More detailed analyses are required of many of the observations made during these eruptions, such as the volcano-seismic activity, the occurrences of explosive activity at the low-level effusive vents, and the relationship between, and variation in character of, activity at different vents. The complete report on the eruptions (in preparation) will cover these aspects.

ACKNOWLEDGEMENTS

The authors thank: technical officers and assistants of Rabaul Observatory for various kinds of aid, in particular V. Kaita, R. Pakat, A. Pidik, C. Matupit, J.

Fig. 4 (opposite): Post-eruption view of lava field, Karkar volcano, 13 September 1974. The crater rim of Ulumam cone is just visible, the remainder having been buried by lava. Photograph by R.W. Johnson.

Kuduon, J. Saun, and P. Daimbari, each of whom spent an extended period at the observation post; many local residents at Karkar for information and assistance, in particular Mr and Mrs R. Booth for their hospitality; R. W. Johnson for discussion; F. Simonis and W. Pearson for drafting illustrations; and G. Brooker for typing the manuscript. The Chief Government Geologist, Geological Survey of Papua New Guinea, authorised publication of this paper.

REFERENCES

Cooke, R.J.S., 1975: Time variations in recent volcanism and seismicity along convergent plate boundaries of the south Bismarck Sea, Papua New Guinea. *Bull. Aust. Soc. Expl. Geophys., 6,* pp. 77-78.

Everingham, I.B., 1975: Seismological report on the Madang earthquake of 31 October 1970 and aftershocks. *Bur. Miner. Resour. Aust. Rep. 176.*

Johnson, R.W., Mackenzie, K.E., Smith, I.E., & Taylor, G.A.M., 1973; Distribution and petrology of late Cenozoic volcanoes in Papua New Guinea. *in* Coleman, P.J. (Ed), *The Western Pacific island arcs, marginal seas, geochemistry,* Univ. W.Aust. Press, pp. 523-533.

Kunze, G., 1901: *Im dienst des Kreuzes auf ungebahnten Pfaden.* Verlag des Missionhauses, Bremen (3rd edition, 1925).

Macdonald, G.A., & Abbott, A.T., 1970: *Volcanoes in the sea.* Honolulu, Univ. Hawaii Press, 441 pp.

Masefield, J. (Ed.), 1906: *Dampier's Voyages.* London, E. Grant Richards, Volume 2, 624 pp.

Minakami, T., 1960: Fundamental research for predicting volcanic eruptions (Part 1). *Bull. E.R.I., 38,* pp 497-543.

Morell, B. Jr., 1832: *A Narrative of four voyages to the South Sea, North and South Pacific Ocean, Chinese Sea, Ethiopic and Southern Atlantic Ocean, Indian and Antartic Ocean from the year 1822 to 1831,* New York, J. & J. Harper.

Perret, F.A., 1950: Volcanological observations. *Carnegie Inst. Washington, Pub. 549.*

Sapper, K., 1917: Katalog der geschichtlichen Vulkanansbruche Melanesien. *Schr. wiss. Ges. Strassburg, 27.*

Sharp, A., 1968: *The voyages of Abel Janszoon Tasman.* Oxford, Clarendon Press, 375 pp.

Stoiber, R.E., & Rose, W.I., 1974: Cl, F, and SO2 in Central American volcanic gases. *Bull, Volc., 37,* pp. 454-460.

Thorarinsson, S., 1970: *Hekla, a notorious volcano.* Reykjavik, Almenna Bokafelagid, 62 pp.

Zöller, H., 1891: *Deutsch-Neuguinea und meine esteigung des Finisterre-gebirges.* Stuttgart.

Appendix. *Preliminary petrology*
by
R.W. Johnson, C.O. McKee, & R.N. England

Chemical and modal analyses and CIPW norms of ten rocks produced during the 1974-75 eruptions of Karkar are given in Table A, where the samples are arranged from left to right in order of eruption. Samples 1 to 8, and 10, are from lava flows, and sample 9 is an ejected block. The ten rocks are believed to represent the most important phases of the eruption.

The rocks are low-silica andesites. They are all especially porphyritic in plagioclase, but also contain some clinopyroxene, olivine, and iron-titanium oxide phenocrysts; sample 1 contains trace amounts of orthopyroxene phenocrysts. In order of eruption, samples 1 to 8 are progressively depleted in total phenocryst and plagioclase phenocryst contents (Table A). Samples 9 and 10, however, the youngest rocks, contain at least as many phenocryst as samples 6, 7, and 8, and more plagioclase phenocrysts than samples 7 and 8. Clinopyroxene phenocrysts are much more abundant, and olivine phenocrysts slightly more abundant in sample 9 than in the other nine samples. The groundmasses of all the samples contain glass, charged with iron-titanium oxide and clinopyroxene grains and plagioclase laths. The phenocryst and groundmass clinopyroxene is calcium-rich; no pigeonite or groundmass orthopyroxene has been identified.

Chemical analyses of older rocks from Karkar Island (Morgan, 1966; and ten unpublished analyses) suggest that low-silica andesite is the most common rock type on the volcano. In comparison, however, most of the low-silica andesites produced in 1974-75 are slightly more basic than most of the older rocks; they contain less silica and show higher Mg/Fe values. Sample 9 is the most magnesian and least siliceous of all the analysed rocks from Karkar Island. On the basis of total alkali:SiO_2 and K_2O:

Table A. Chemical analyses,[a] CIPW norms, and analyses of ten rocks produced during the 1974-75 eruptions of Karkar volcano.

	1	2	3	4	5	6	7	8	9	10
SiO_2	53.8	53.3	54.0	53.5	53.2	54.0	54.6	54.6	53.0	54.69
TiO_2	0.56	0.71	0.84	0.57	0.88	1.11	0.80	0.99	0.75	0.63
Al_2O_3	19.6	19.8	19.2	19.5	19.6	17.9	16.3	16.1	15.7	15.30
Fe_2O_3	2.45	2.60	2.70	2.85	2.45	2.80	3.75	3.40	4.15	3.31
FeO	6.30	6.00	6.30	5.95	6.15	6.95	7.05	7.60	6.15	8.15
MnO	0.16	0.15	0.16	0.16	0.16	0.17	0.19	0.20	0.19	0.22
MgO	3.15	3.10	3.20	3.10	3.15	3.60	4.00	4.05	6.15	4.15
CaO	10.3	10.6	10.2	10.6	10.7	9.90	9.25	9.00	10.4	8.66
Na_2O	2.50	2.50	2.55	2.50	2.50	2.65	2.70	2.75	2.20	2.97
K_2O	0.88	0.87	0.90	0.88	0.85	0.94	1.01	1.02	0.79	1.04
P_2O_5	0.16	0.15	0.17	0.16	0.16	0.17	0.20	0.19	0.14	0.20
H_2O+	<0.01	<0.01	<0.01	<0.01	<0.01	0.01	<0.01	0.01	0.02	0.06
H_2O-	0.06	0.01	0.01	0.06	0.05	0.03	0.05	0.07	0.18	0.02
CO_2	0.05	0.05	0.05	0.05	0.05	0.05	0.05	0.05	0.05	0.05
Total	99.97	99.84	100.23	99.88	99.89	100.28	99.95	100.03	99.87	99.44

CIPW norms										
Q	7.85	7.44	8.24	7.82	7.30	7.88	9.15	8.71	7.39	7.63
or	5.20	5.15	5.31	5.21	5.03	5.54	5.97	6.03	4.68	6.18
ab	21.16	21.18	21.52	21.18	21.18	22.36	22.86	23.27	18.67	25.28
an	39.70	40.31	38.20	39.46	39.81	34.09	29.41	28.59	30.73	25.51
di { wo	4.21	4.62	4.66	4.95	5.00	5.62	6.22	6.06	8.26	6.72
di { en	1.95	2.23	2.25	2.40	2.42	2.76	3.20	3.00	5.30	3.13
di { fs	2.22	2.31	2.33	2.46	2.50	2.75	2.86	2.94	2.42	3.51
hy { en	5.90	5.50	5.69	5.33	5.43	6.18	6.77	7.09	10.06	7.27
hy { fs	6.71	5.68	5.90	5.48	5.61	6.15	6.03	6.95	4.59	8.16
hy { mt	3.56	3.78	3.91	4.14	3.56	4.05	5.44	4.93	6.04	4.83
il	1.06	1.35	1.59	1.08	1.67	2.10	1.52	1.88	1.43	1.20
ap	0.38	0.36	0.40	0.38	0.38	0.40	0.47	0.45	0.33	0.48
cc	0.11	0.11	—	0.11	0.11	0.11	0.11	0.11	0.11	0.11

Volume [b] *percent phenocrysts*										
Plagioclase	37	34	33.5	31.5	29.5	19.5	12	9	14.5	18.5
Olivine	0.5	0.5	<0.5	0.5	1	0.5	0.5	1	2	0.5
Orthopyroxene	<0.5	—	—	—	—	—	—	—	—	—
Clinopyroxene	1.5	1.5	0.5	1	1.5	0.5	0.5	1	10	1.5
Fe-Ti oxides	<0.5	0.5	<0.5	<0.5	<0.5	0.5	<0.5	<0.5	<0.5	<0.5
Total	39	36.5	34	33	32	21	12.5	11	26.5	20.5

Rock type and locality description

1. Lava flow front, mid-late February, 1974, east base of Bagiai (74710001)[c]
2. Lava flow front, early March 1974, east base of Bagiai (74710002).
3. Lava flow front, late March 1974, northern part of caldera, about 1 km from vent (74710003).
4. Lava flow edge, early April 1974, east base of Bagiai (74710004).
5. Lava flow front, late May 1974, southern part of caldera, about 300 m from vent (7471005).
6. Lava flow front, early June 1974, southern part of caldera, about 1.2 km from vent (74710006).
7. Lava flow front, late July 1974, southern part of caldera, about 1.1 km from vent (74710008).
8. Lava flow front, late July-early August 1974, southern part of caldera, about 300 m from vent (74710009).
9. Ejected block, late July-early August 1974, base of Bagiai (74710010).
10. Lava flow front, mid-January 1975, west base of Bagiai (75710015).

[a] Supplied by Australian Mineral Development Laboratories, Adelaide
[b] Determined by C.O. McKee
[c] Bureau of Mineral Resources registered number.

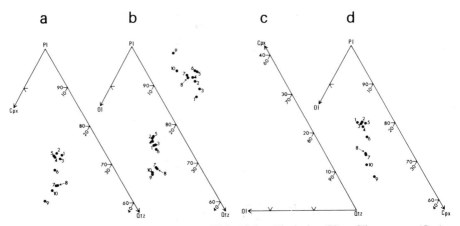

Fig. A. Normative pseudo-quaternary system Olivine(O1) — Plagioclase(Pl) — Clinopyroxene(Cpx) — Quartz(Qtz). Oxidation state of iron in all analyses standardised by the method of Irvine & Baragar 1971 — i.e. if Fe2O3 is greater than TiO2 + 1.5, sufficient Fe2O3 is converted to FeO so that Fe2O3 = TiO2 + 1.5) and recalculating volatile-free to 100 percent. Sample 9 is the most oxidised of the 1974-75 rocks.

SiO2 relationships, the rocks of Karkar could be classified as 'calcalkaline' although, as pointed out by Morgan (1966), some contain moderate amounts of total-iron similar to those of 'tholeiitic' rocks.

    In Figure A, the 1974 Karkar rocks are plotted in the normative pseudoquaternary system Olivine-Plagioclase-Clinopyroxene-Quartz. The four diagrams, a to d, are parts of the faces of the tetrahedron, and the sample points are projections from the interior of the tetrahedron onto each face. In Figure Aa,b,d, samples 1 to 5 fall in a cluster, but samples 6, 7 and 8, 10, and 9, form points progressively farther from the Pl apex, suggesting that plagioclase had a strong influence on the development of the series. This trend of depletion in normative plagioclase corresponds roughly with the depletion in plagioclase phencrysts mentioned above (cf. Table A). In Figures Aa,d, the point representing sample 9 is displaced towards the Cpx apex, and in Figure Ab it is displaced towards the O1 apex. These divergences correspond with the higher proportions of clinopyroxene and, to a lesser extent, olivine phenocrysts present in sample 9 (Table A).

    About one-hundred electron microprobe analyses show that the mean composition of analysed plagioclase phenocrysts in all ten rocks is An88, and that the mean plagioclase composition of each rock falls within the range An84-90. However, preliminary attempts to relate the change in composition of the rocks to the separation (or addition) or different amounts of plagioclase with an average composition of An88 have not been wholly successful, although there is little doubt that the evolution of the series has been controlled to a large extent by calcic plagioclase (reasons for the discrepancies will be discussed elsewhere).

    The available data do not establish which of the ten rocks, if any, is parental to the series. If the series represents a liquid line of descent from a parental composition represented by one of the older samples 1-5, the later part of the series evolved mainly by the successive separation of plagioclase, producing the compositions represented by samples 6, 7, 8, and 10. Alternatively, the series may have evolved from an alumina-poor parent (such as is represented perhaps by sample 10) by the addition of plagioclase, and the magmas were erupted in reverse order. Trends from compositions intermediate along the series, or from compositions not represented by any of the analysed rocks, are also possible. In any of these trends, sample 9 — the only rock not collected from a lava flow — is anomalous because, compared to the other samples, clinopyroxene and olivine, as well as plagioclase, appear to have influenced its position in the series.

References

Irvine, T.N., & Baragar, W.R.A., 1971: A guide to the chemical classification of common volcanic rocks. *Can. J. Earth Sci.*, 8, pp. 523-548.
Morgan, W.R., 1966: A note on the petrology of some lava types from east New Guinea. *J. geol. Soc. Aust.*, 13, pp. 583-591.

PUMICEOUS PYROCLASTIC DEPOSITS OF WITORI VOLCANO, NEW BRITAIN, PAPUA NEW GUINEA

D.H. BLAKE[1].

Division of Land Use Research, C.S.I.R.O., P.O. Box 1666, Canberra City, A.C.T. 2601

ABSTRACT

Witori is a low-angle volcanic cone with a central caldera, and is formed mainly of two groups of pumiceous pyroclastic rocks separated by a buried soil or erosion surface. Pyroclastic flow deposits predominate; most of these are not welded but some have moderately welded zones which show volcano-karst features such as fluted gullies. The pumice deposits are typical products of caldera-forming eruptions, the last of which probably took place about 2600 years ago, when the younger group of pumiceous pyroclastics was laid down.

INTRODUCTION

Pumiceous pyroclastic deposits are important constituents of several volcanoes in Papua New Guinea. However, such deposits have been little studied up to now, and it is the aim of this paper to draw attention to some of their features, as displayed on Witori volcano (Fig. 1), one of eleven Quaternary volcanoes in the Cape Hoskins area of New Britain (Blake & Bleeker, 1970).*

Witori volcano is a broad low-angle cone about 700 m high with a well preserved central caldera (Fig. 2), and was examined by the author briefly in 1968 and in more detail in 1970. The flanks of Witori are made up almost entirely of pyroclastic deposits formed predominantly of dacitic pumice, but some andesitic lava flows are exposed in the sides of the caldera and in valleys on the upper southwest flank. The petrography and geochemistry of these volcanic rocks were described by Blake & Ewart (1974). The present caldera, which was probably formed during a catastrophic eruption about 2600 years ago (Blake & McDougall, 1973), contains an active volcano, Pago, formed of dacitic lava flows, a remnant of a dacitic cumulodome, some pumice mounds, and, in the north and west, some narrow pumice ridges extending inwards from the caldera wall.

PUMICE DEPOSITS

The pumiceous pyroclastics consist mainly of non-welded flow deposits but also include partly welded flow deposits and air fall, alluvial, and possibly some laharic beds. They are well exposed along the upper part of the caldera wall, where they are about 70 m thick, and in incised valleys, some over 50 m deep, on the flanks. They are seen to overlie andesitic lava gravel on the east flank and highly vesicular andesitic lava flows and associated ash and agglomerate in the caldera wall and on the upper southwest flank: some of the andesitic ash between lava flows forms indurated layers thought at first to be densely welded pyroclastic flow deposits (Blake & Bleeker, 1970).

Stratigraphy

An older and a younger group of pumice deposits can be distinguished on the east and lower southwest flanks, separated from each other by a buried soil or an

[1]Present address: Bureau of Mineral Resources, P.O. Box 378, Canberra City, A.C.T. 2601

*See also the paper by Pain & Blong (this volume) on the tephras of some volcanoes in the Highlands of Papua New Guinea. Ed.

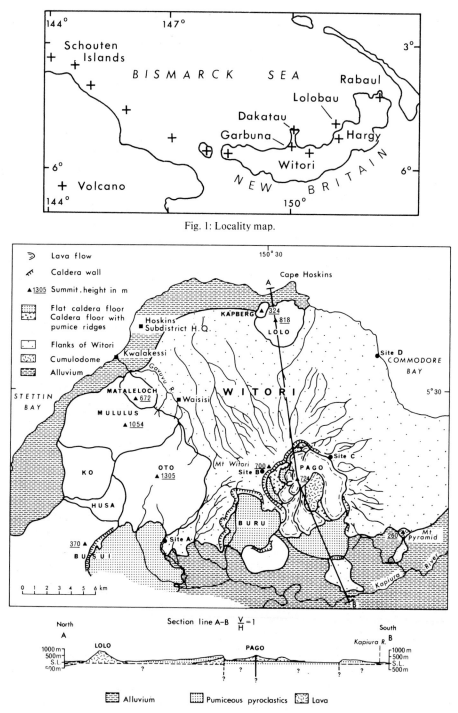

Fig. 1: Locality map.

Fig. 2: Witori and adjacent volcanoes, Cape Hoskins area, New Britain.

erosion surface. These two groups were probably deposited during two separate major eruptions. The buried soil between them is about 1 m thick, and changes colour downwards from dark to pale brown. It indicates a period of at least 100 years (P. Bleeker, pers. comm. 1974) during which no pumice was deposited, enabling soil formation to proceed unhindered. Charcoal from the buried soil on the southwest and northeast flanks (sites A and D, Fig. 2) has been dated at 2590 ± 300 years B.P., and this is considered to be the approximate date of the more recent of the two major eruptions and is when the present caldera is thought to have formed.

The buried soil and its associated erosion surface have not been identified on the northwest flank, where only the younger group of pumice deposits appears to be exposed, nor on either the north or south flanks. The north flank consists of pyroclastic flow and airfall deposits which may belong to either or both groups, and the remnant of the south flank preserved is formed mainly of airfall deposits possibly belonging to the older group. Most of the pumice on the south flank is somewhat weathered, in contrast to that on the other flanks.

The older group of pumice deposits on the east flank reaches a thickness of over 50 m and consists of a non-welded flow deposit made up of several flow units. It is overlain here by a layer of coarse airfall pumice 2 to 3 m thick representing the younger group.

The younger group of pumice deposits is well exposed at site A, on the lower southwest flank of Witori, where the following sequence is present (see also Fig. 3).

Top	Thickness (m)
8. Inaccessible and obscured by vegetation	about 10
7. Moderately welded zone of pyroclastic flow deposit, with gradational lower contact	about 3
6. Non-welded part of pyroclastic flow deposit made up of several thin flow units	6
5. Airfall pumice layer showing internal stratification	0.3
4. Non-welded pyroclastic flow deposit, a single unit filling a depression in the former land surface	up to 3
3. Airfall pumice layer, showing internal stratification, mantling former land surface	1.2
2. Buried soil containing charcoal (dated at 2590 ± 250 years) and artifacts consisting of small chips of obsidian (R.I. 1.486 ± 0.003)	1
1. Airfall pumice layers	up to 2

In the above sequence, the airfall pumice of unit 1 belongs to the older group and units 3 to 8 belong to the younger group. The moderately welded zone of the flow deposit, unit 7, can be traced northeast towards the caldera of Witori, but grades laterally into non-welded pumice less than 1 km to the southwest.

Another flow deposit with a welded zone is exposed on the northwest flank near Waisisi. This deposit is very similar to that at site A, and may be a product of the same major eruption.

Fig. 3: Pumiceous pyroclastic deposits at site A, lower southwest flank of Witori. Numbers 1-8 refer to sequence described in the text.

A compound non-welded pyroclastic flow at least 50 m thick, formed of several flow units, is exposed on the upper northwest flank, close to the caldera rim (site C). This flow appears to overlie the welded zone seen to the northwest, and probably belongs to the same group of deposits.

The younger group of pumice deposits is overlain by up to 2 m of airfall pumice layers on the north and northwest flanks of Witori. These layers are not present on Pago volcano inside Witori caldera, and are less than 1 m thick on the coast northeast of Witori (site D), but they thicken westwards and are over 8 m thick on the caldera rim of Busui volcano (Fig. 2). Charcoal fragments from buried soils within the airfall sequence on the lower western and northern flanks of Witori and on the caldera rim of Busui have been dated at 830 ± 110, 320 ± 170 and 1990 ± 90 years B.P. respectively (Blake & McDougall, 1973; Bleeker & Parfitt, 1974). Most of the pumice forming the layers may be derived from volcanoes several kilometres to the northwest, such as Garbuna and Dakatau (Fig. 1) or from Buru volcano to the south (Fig. 2), but some may have been deposited during the emission of the cumulodome in Witori caldera.

Pyroclastic flow deposits

The flow deposits are composed predominantly of dacitic pumice, but they also contain some angular fragments of andesite probably derived from lavas in the core of Witori, and some on the east flank contain fragments of eutaxitic rhyolitic obsidian which may have come from the densely welded zone of a flow deposit no longer exposed.

Most of the flow deposits consist of several highly friable non-welded flow units, some less than 1 m thick, which are indicated by abrupt vertical changes in average grain size. Each flow unit generally appears unsorted and is characterised by abundant very fine pumice fragments. The largest constituent pumice and other fragments are generally angular and less than 5 cm across. However, large lava blocks are present locally, mainly in deposits on the upper flanks, and tend to be concentrated near the base of a flow unit.

The non-welded deposits form vertical cliffs along the valleys on the flanks, and are eroded by lateral corrasion — the cliffs are undercut by streams during floods, and landslips result. Pumice from the landslips is carried downstream, but large lava blocks remain, more or less in situ, on the valley floor.

Locally the non-welded flows have cross-bedded basal layers consisting of coarse sandy to gravelly rounded fragments, with little or no fine pumiceous matrix. Such basal layers may be ground surge deposits of the type described by Sparks & Walker (1973).

The pumice deposits forming hills, ridges, and mounds within the caldera are chaotic and friable, like the non-welded deposits on the flanks. Some may be the products of pyroclastic flows, but some may have been laid down by lahars.

Flow deposits with moderately welded zones are present on the northwest and southwest flanks of Witori. One such zone is well exposed in the northwest in gorges along the Gavuvu River (Fig. 4) and its tributaries near Waisisi. It is at least 7 m thick in places, and passes upwards into non-welded pumice; its base is not exposed. This welded zone can be traced for over 8 km southeast of Waisisi, before it disappears beneath non-welded pumice deposits, but to the northwest it appears to grade laterally into non-welded pumice within 1 km of Waisisi, where the valley of the Gavuvu River abruptly widens out. A similar welded zone at least 3 m thick forms vertical cliffs along valley sides on the southwest flank of Witori (as at site A,

Fig. 4: Moderately welded zone of pyroclastic flow deposit on the east side of the Gavuvu River at Waisisi, northwest flank of Witori.

Fig. 5: Moderately welded zone of pyroclastic flow deposit containing blocks of pumice. Side of fluted gully south of Waisisi, northwest flank of Witori. Scale shown is 14 cm long.

Fig. 6: Volcano-karst. Small gully eroded in the top of a moderately welded pyroclastic flow deposit, 1.5 km southeast of Waisisi, northwest flank of Witori.

Fig. 3). The welded zones are probably restricted to the thickest flow deposits, in which temperatures remained high long enough for some welding to take place.

In marked contrast to the non-welded deposits, the welded zones are coherent rather than friable (Fig. 5), and consequently are less porous and permeable. They also tend to contain a higher proportion of large pumice fragments, many of which are distorted and partly moulded around other fragments. In addition, the welded zones show distinctive volcano-karst features (Fairbridge, 1968), such as fluted gullies (Figs 6, 7) and small sinkholes connecting with tunnels.

Airfall deposits

Highly friable, unconsolidated airfall pumice deposits are preserved on ridge crests and planezes on the flanks of Witori, overlying flow deposits, and are also present at various levels between successive flow deposits exposed in valley sides, as, for instance, at site A on the southwest flank. The individual fall deposits form layers generally less than 50 cm thick which mantle the pre-existing land surface, and they commonly occur in groups of two or more layers separated from one another by buried soils or flow deposits. They consist of pumice fragments ranging up to over 10 cm across, minor andesitic lava fragments generally less than 2 cm across, and very rare bomb-like fragments of vesicular lave. Some of the layers show internal stratification (cf. Sparks & Walker, 1973).

The thickest fall deposit on Witori covers the northeastern flank, where it represents the younger group of pumice deposits and overlies a soil or erosion surface developed on the older pumice deposits. This fall deposit is at least 3 m

Fig. 7: Volcano-karst. Narrow gully 4 m deep with fluted sides in a moderately welded pyroclastic flow deposit, 1 km southeast of Waisisi.

thick near the caldera rim and decreases to about 2 m thick on the lower flank. It was previously thought to be a flow deposit (Blake & Bleeker, 1970), but it is now known to form a uniform layer mantling the former irregular land surface, instead of filling in depressions, and it also lacks the very fine material characteristic of flow deposits. Some of the pumice fragments within it are more than 25 cm across, even on the coast 10 km from the caldera, and there is no apparent decrease in grain-size with increasing distance from the caldera. The largest fragments have pinkish interiors similar to those recorded by Williams (1942), who attributed this to atmospheric oxidation of iron-bearing gases held within large pumice fragments.

Grain-size analyses

Grain-size analyses of several samples from pyroclastic flow and fall deposits were carried out by B.G. Griffiths, Division of Land Use Research, CSIRO, so that these deposits could be compared with similar deposits elsewhere. Following Walker (1971), two parameters were chosen to quantify the main characteristics of the cumulative curve for each sample — the median diameter $Md\phi$ and the deviation $\sigma\phi$ of Inman (1952), where d is the diameter in millimetres and $\phi = \log_2 d$.

The results are plotted in Figure 8, which is based on Figure 2 of Walker (1971). Details of the samples are given in Johnson & Blake (unpublished BMR Record 1973/133). All the samples from fall deposits and all but one of the samples from flow deposits lie within the respective fields determined by Walker. Two of the five samples from pumice deposits within the caldera unexpectedly lie outside the field of flow deposits; the significance of this has not been determined.

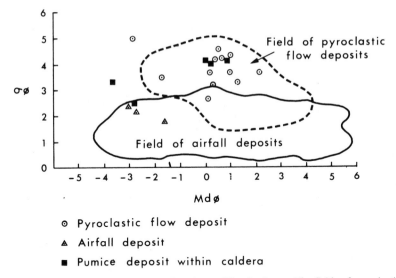

○ Pyroclastic flow deposit

▲ Airfall deposit

■ Pumice deposit within caldera

Fig. 8: Grain-size characteristics of pumice deposits on Witori volcano. The fields of pyroclastic flow
deposits and air fall deposits are taken from Walker (1971).

Petrography

Over 100 samples of pumice from flow and fall deposits on Witori have been
examined microscopically and two have been chemically analysed (Blake & Ewart,
1974). The samples contain phenocrysts of plagioclase, augite, orthopyroxene,
opaque minerals, and in some cases hornblende, quartz, and rare olivine, enclosed
in pumiceous glass, R.I. 1.498-1.518. Internal variations within a single deposit
were found to cover the range between different deposits, hence it was not possible
to distinguish individual deposits petrographically.

DISCUSSION

Pumiceous pyroclastic deposits similar to those forming most of Witori are
also present on four other volcanoes in the eastern part of the Bismarck volcanic
arc: Dakatau, on the northern end of Willaumez Peninsula (Lowder & Carmichael,
1970); Hargy and Lolobau (Johnson, unpublished information), respectively 70 km
east-northeast and 100 km northeast of Witori; and Rabaul (Heming &
Carmichael, 1973; Heming, 1974), at the northeast end of New Britain (Fig. 1).*
Like Witori, these volcanoes have calderas containing younger cones, and the
pumice deposits on their flanks overlie older lava flows and associated agglomerate
and ash. On Hargy volcano both non-welded and moderately welded flow deposits
are present, as well as airfall deposits, and some of these contain fragments of
obsidian formed of highly deformed pumice, indicating intense welding. A flow
deposit showing partial welding is also present on Rabaul volcano. All four
volcanoes and Witori were originally high-angle cones formed of lava flows and
associated agglomerate and ash. However, following abrupt changes in type of
volcanic activity, from relatively quiet emission of lava flows to catastrophic
caldera-forming eruptions accompanied by the emission of pumice, they are now

*For a description of the caldera and associated volcaniclastic deposits of a volcano in the western part
of the Bismarck volcanic arc, see the paper on Long Island by Ball & Johnson (this volume). Ed.

broad low-angle cones with central calderas, and their flanks are formed mainly of pumiceous pyroclastics. In each case post-caldera eruptions have taken place inside the calderas, and prominent cones such as Pago have been formed.

The pumice on Witori and the other volcanoes was laid down as pyroclastic flow and fall deposits. These are the typical products of caldera-forming eruptions (see, for example, Williams, 1942, and Smith, 1960). The last such eruption of Witori probably took place about 2600 years ago, and the products of this eruption overlie a thick group of older pumice deposits which may have been erupted during one or more earlier phases of caldera formation. The complex succession of airfall deposits and flow units within the younger group of pumice deposits on Witori, as exposed at site A on the southwestern flank, indicates that the last caldera-forming eruption of this volcano consisted of several phases and was not a single catastrophic eruption. The original central cone probably subsided in stages during an eruption which may have lasted for several days or weeks, each stage being accompanied or perhaps preceded by the emission of vesiculated lava as pumice and separated by relatively short quiescent periods. Whether the eruptions took place from ring fractures around the edge of the caldera or from a central vent is not known.

ACKNOWLEDGEMENTS

I am grateful to G.P.L. Walker, B.S. Oversby, R.J. Bultitude, and C.D. Ollier for their helpful comments on the original manuscript.

REFERENCES

Blake, D.H., & Bleeker, P., 1970: Volcanoes of the Cape Hoskins area, New Britain, Territory of Papua and New Guinea. *Bull. volc., 34*, pp. 385-405.

Blake, D.H., & Ewart, A., 1974: Petrography and geochemistry of the Cape Hoskins volcanoes, New Britain, Papua New Guinea. *J. geol. Soc. Aust., 21*, pp. 319-331.

Blake, D.H. & McDougall, I., 1973: Ages of the Cape Hoskins volcanoes, New Britain, Papua New Guinea. *J. geol. Soc. Aust., 20*, pp. 199-204.

Bleeker, P., & Parfitt, R.L., 1974: Volcanic ash and its clay mineralogy at Cape Hoskins, New Britain, Papua New Guinea. *Geoderma, 11*, pp. 123-135.

Fairbridge, R.W., 1968: Volcano-karst: *in* Fairbridge, R.W. (Ed.), *The Encyclopedia of Geomorphology*, pp. 1205-1208. New York, Reinhold.

Heming, R.F., & Carmichael, I.S.E., 1973: High-temperature pumice flows from the Rabaul caldera, Papua New Guinea. *Contr. Miner. Petrol., 38*, pp. 1-20.

Heming, R.F., 1974: Geology and petrology of Rabaul caldera, Papua New Guinea. *Bull. Geol. Soc. Am., 85*, pp. 1253-1264.

Inman, D.L., 1952: Measures for describing the size distribution of sediments. *J. sediment. Petrol., 22*, pp. 125-145.

Lowder, G.G., & Carmichael, I.S.E., 1970: The volcanoes and caldera of Talasea, New Britain: geology and petrology. *Bull. geol. Soc. Am., 81*, pp. 17-38.

Smith, R.L., 1960: Ash flows. *Bull. geol. Soc. Am., 71*, pp. 795-842.

Sparks, R.S.J., & Walker, G.P.L., 1973: The ground surge deposit: a third type of pyroclastic rock. *Nature phys. Sci., 241, no. 107*, pp. 62-64.

Walker, G.P.L., 1971: Grain-size characteristics of pyroclastic deposits. *J. Geol., 79*, pp. 696-714.

Williams, H., 1942: The geology of Crater Lake National Park, Oregon, with a reconnaissance of the Cascade Range southward to Mount Shasta. *Carnegie Inst. Washington Publ. 540*.

1941-42 ERUPTION OF TAVURVUR VOLCANO, RABAUL, PAPUA NEW GUINEA

N. H. FISHER

68 National Circuit, Deakin, Canberra, A.C.T. 2600

ABSTRACT

The eruption of Tavurvur between 6 June 1941 and March 1942 was characterised by a long build-up period which began with the violent explosive outburst of Tavurvur that accompanied the Vulcan eruption of 1937. Pre-eruption phenomena included: a general but intermittent increase in size of the gas and vapour cloud given off from the central crater; increases in the contents of sulphur dioxide, hydrogen chloride and, later, other high-temperature gases; development of new active fumaroles in the 1878 crater and on the outer southeastern slopes of the volcano; and some minor explosive activity at the southwestern end of the main crater between February and May 1940. The temperatures of the gases given off by the fumaroles in the 1878 crater first rose above 100°C in November 1940, and had risen to nearly 400°C just before the eruption started on 6 June 1941. Pre-eruption seismic effects were too small to be identified on the Observatory seismograph (7.2 km away, with a magnification of 150:1), or on a similar instrument installed in April 1941 at an observation post 1 km from the crater. After the eruption began, individual outbursts were preceded and accompanied by distinctive tremors. From June to September 1941, eruptions were almost continuous, except for some minor breaks of a few days. After September, the breaks became longer, though individual eruptions were commonly violent. No detailed record is available of the eruptions after 23 January 1942. However, reports indicate that the main eruptive period terminated about March 1942, though there is evidence of small later eruptions, probably in October and December 1943.

INTRODUCTION

The eruption of Tavurvur, or Matupi, volcano that began on 6 June 1941 provides an interesting example of the phenomena that accompany the approach of an eruptive period, corresponding to a slow rise in the lava column in the conduit of the volcano. The build-up of this eruption really began with the 1937 Blanche Bay eruption (Fisher, 1939) where the main activity took place at Vulcan, 6 km distant, on the southwestern side of Blanche Bay. Tavurvur is situated on the eastern side of Matupi Harbour, a subsidiary harbour on the eastern side of Blanche Bay (Fig. 1).

In 1937, a violent eruption at Vulcan built up a cone of pumice 226 m high and about 2 km in diameter at the base. An eruption at Tavurvur began when the Vulcan eruption has been in progress for 21 hours, and consisted of a violent steam explosion that blasted out parts of the super-structure of the volcano along a southwest-northeast fissure. Before 1937, Tavurvur had been in a quiet solfataric stage with water vapour and other gases, probably mostly carbon dioxide and hydrogen sulphide, gently escaping from several vents on and around the volcano. After 1937, Tavurvur did not return to this quiet solfataric condition, but continued to produce a large vapour cloud, which issued from fairly vigorous vents within the blasted-out main central crater and commonly containing obvious sulphur dioxide.

Sources of information for this paper, in addition to the author's own recollections of direct observations, include: weekly vulcanological reports for the period 1937 to 1942 prepared by the author, L. E. Clout, C. L. Knight, and L. C. Noakes, and published in the *Rabaul Times;* daily reports on the 1941 eruption from 6 June to 23 July by C. L. Knight; an incomplete unpublished description of the eruption by C. L. Knight written about 1946; and aerial photographs taken during and after the Second World War. Pre-eruption seismograph records were lost in the war, but the results were recorded in the weekly vulcanological reports.

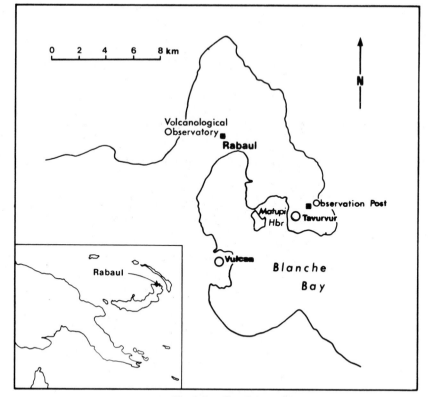

Fig. 1: Locality plan.

PRE-ERUPTION PHENOMENA

Temperature

After the 1937 eruption a number of observation points were established at which temperatures were recorded weekly; these temperatures were normally 100°C or less. In the later part of August 1940, however, a new area of solfataric activity was noticed on the northern side of the 1878 crater, where only one or two patches of hot ground had previously existed, about 20 m from the top of the steep southern wall of the 1937 crater (Fig. 2). A temperature observation point was immediately established here, but the temperature of the gases stayed at 100°C until early November, when 110.8°C was recorded. The temperature remained about 110°C until 8 December, when a spectacular rise up to 224°C took place over a 10-day period (Fig. 3). By 7 January 1941 the temperature had reached 260°C, and it continued to rise steadily but spasmodically until by 20 March the temperature had passed 300°C. From 15 April, the temperature rose more rapidly and by about 10 May had reached 380°C. The last temperature recorded, three days before the eruption started, was 392°C.

Only in the fumaroles within the 1878 crater were temperatures above 100°C recorded. The bottom of the 1937 crater (Fig. 2) was at that time extremely difficult to reach because of vertical walls exuding hot gases. When increased

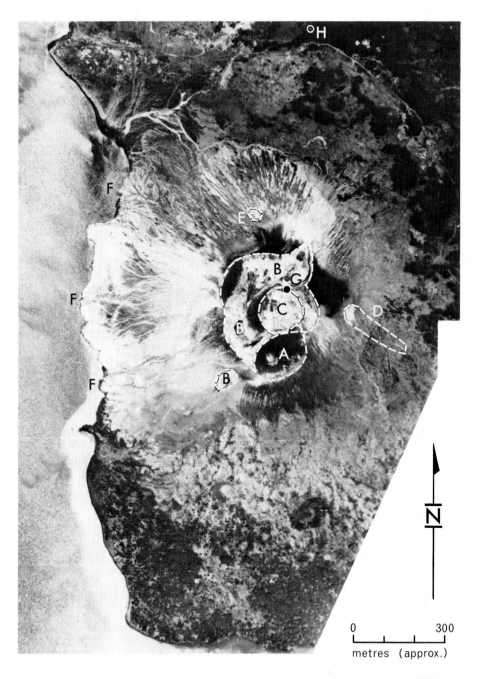

Fig. 2: Aerial photograph of Tavurvur taken 17 March 1943. A: 1878 crater. B: 1937 crater. C: 1941-42 crater. D: southeastern solfataric area 1941. E: solfataras on northern slope. F: solfataric areas along Matupi Harbour beach front. G: central lava plug. H: Matupi observation post.

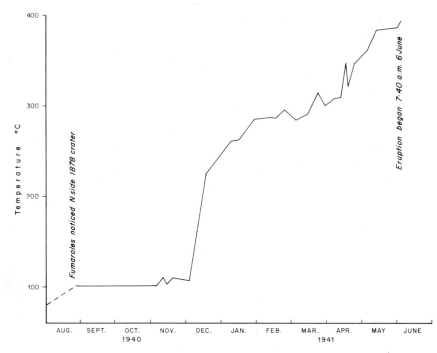

Fig. 3: Pre-eruption rise of temperature in the fumaroles in the 1878 crater.

activity in the 1878 crater was first noticed, a somewhat precarious descent was made with the aid of a rope, and temperatures of the gases in the crater were measured, but nothing above 100°C was recorded. The experience was not repeated, but no doubt the gases here, at a lower elevation and closer to the source of heat, also participated in the subsequent rise in temperature. Although there were signs of increased activity at other areas around the volcano, temperatures remained at 100°C, and there was no visible increase in activity in the solfataric areas along the Matupi Harbour waterfront.

Volume and distribution of gas emission.

After the 1937 eruption, emission of hot gases was confined to (1) the 1937 crater, mainly at the southwestern end and near the foot of the central lava plug (Fig. 2; see also Fisher 1939, pp. 36, 62), (2) quietly steaming vents in the blow-out craterlet on the southwestern flank of Tavurvur, (3) patches of hot ground on the southeastern outer slope of the volcano, (4) similar patches on the northern slope, and (5) the solfataras at the western foot of the volcano along the harbour front (Fig. 2). The last two areas were not affected by the build-up to the 1941 eruption. In the 1937 crater, minor changes in the relative strength and even in the location of the various gas vents took place from time to time. For about a year after the 1937 eruption no significant change was recorded in the total volume of gas emitted, although there was possibly a slight decrease in the first few months, with minor variations thereafter. In June 1938 it was observed that a slight increase had taken place in the previous two months, and again, in October 1938, an increase in gas pressure and volume was noted. Thereafter, up to the time the eruption began, increases were frequently noted, but there was no way of measuring these quantitatively.

Until August 1940 the only signs of activity in the 1878 crater were one or two patches of hot ground on the northern and northeastern inner slope mentioned above. In the latter part of August it was noticed that the new series of quite vigorous vents which had developed in this area were giving off a strong smell of sulphur dioxide. At the same time it was observed that in what had previously been just an area of hot ground on the outer southeastern slope of Tavurvur, a number of active fumaroles now existed, also emitting sulphur dioxide, but not as strongly as those in the crater area. Temperatures then and later did not exceed 100°C. Many vents developed along and near a gully running from this area off the southeastern slope of Tavurvur, and this became quite a large area of solfataric activity. New vents also appeared about this time (August-October 1940) on the southern and southwestern outer slopes, but these did not individually show much subsequent increase in activity, although white incrustations developed over a considerable area of the southern slope.

When, in November 1940, the temperatures of the gases in the 1878 crater fumaroles began to rise above 100°C, an increase in the vigour of the gas emission within the 1937 crater was noticeable, and the solfataric areas on the southeastern outer slope increased in size, extending onto the flat land at the foot of the mountain and giving off increasing amounts of sulphur dioxide, particularly from the higher vents. The active area in the 1878 crater continually increased in size, number of vents, vigour of gas emission, and amount of sulphur dioxide and other 'high temperature gases'. By 22 April 1941 the main fumarole, now at a temperature of about 350°C, was no longer accessible, but several others in the vicinity were at temperatures equally or nearly as high. At this time the production of gas within the 1937 crater was increasing and on 20 May the largest discharge was noted from the fumarole at the bottom of the central basalt plug, the subsequent eruption point. New vents were observed nearby on the southern wall of the 1937 crater. On 27 May, gas emission had further increased and was noted to be issuing with a 'roar like a train' and the steam cloud was visibly larger. This increase was maintained until the eruption began at 7.40am on 6 June.

Composition of gases

Before the 1937 eruption the only gas, apart from water vapour, that was evident in the solfataric discharges was hydrogen sulphide. When conditions settled down after the eruption, again only hydrogen sulphide was noticeable in the gases. In September and October 1938, gas with a pungent odour and irritating to the eyes — probably hydrogen chloride — was noticed, and this was recorded again the following March, when sulphur dioxide was noticed more and more frequently, and it increased intermittently in quantity. By January 1940, the smell of sulphur dioxide was so strong that it was rarely possible to detect any hydrogen sulphide that might have been present in the gas, though it was recorded in May and again, for the last time in the gases from the 1937 crater, in October 1940. Hydrogen chloride, on the other hand, was observed quite frequently.

No gas analyses of samples from the 1937 crater were made. Tests on the gas from the Matupi Harbour beach front solfataras indicated that the gas associated with the water vapour (the main consitituent) consisted of carbon dioxide and hydrogen sulphide in the proportion of at least 5 to 1. No apparent change occurred in the composition of the gas from these solfataras at any time.

The fumaroles in the 1878 crater were rich in sulphur dioxide from the time of their inception in August 1940. Hydrogen chloride was noticed for the first time in early December, and the sulphur dioxide content continued to increase both from these vents and from those on the outer southeastern slope. As the temperature rose, so did the quantity of sulphur dioxide, hydrogen chloride, and other acid gases, and they became so corrosive that it was very difficult to maintain the pyrometers that were used for taking the temperatures. Mercury thermometers reading to 360°C had been obtained but the limit of these was passed in early May 1941.

Seismic effects

In 1940, seismographs were installed at the Rabaul Observatory, on the north side of Blanche Bay 7.2 km northwest of Tavurvur (Fig. 1), and recordings were maintained fairly continuously until January 1942. The instruments were locally built smoked-paper-recording horizontal seismographs with a magnification about 150:1. In early April 1941, a similar installation was completed at an Observation Post on the caldera wall 1 km north of Tavurvur crater (Fig. 2). The sensitivity of this instrument was increased on 20 July.

Because of the low sensitivity of the seismographs and the gradual nature of the build-up towards the eruption, little significant information on seismic events was recorded before the eruption began. All identifiable events on the seismograms were earth tremors of tectonic origin. After the eruption began, tremors of volcanic origin preceded each eruptive outburst by at least one hour, commonly by several hours, and each explosion also registered its signature on the seismograms.

Tiltmeters at the Observatory and at the observation post, reading to about 1 second of arc, were not sufficiently sensitive to register any tilting that may have occurred at Tavurvur before the 1941 eruption, although a tilt aggregating about 30 seconds of arc was recorded just before a severe regional earthquake of 14 January 1941 (Fisher, 1944). An analysis of tide-gauge records at Rabaul compared with tide markers that were read regularly every week indicated that Tavurvur sank gradually about 7.5 cm relative to Rabaul over the two years prior to the eruption.

Minor explosive activity

Although the catchment area of the Tavurvur main crater was small, after the 1937 eruption a certain amount of siltation occurred in the bottom of the crater, mainly from material washed down from the northwestern wall. Also, depending on the season, a pool of water up to about 2 m deep generally occupied the southwestern end of the crater (one of the main centres of gas emission). In February and March 1940 (at the end of the wet season) a series of small steam explosions took place in this part of the crater, which were attributed to the clearing of vents that had been blocked by mud and water. In early February fountains of muddy water a few metres high were observed, and in early March quite substantial explosions occurred. On 3 March there were four such explosions which hurled mud and small stones up to about the height of the crater rim. These

Fig. 4: Tavurvur in eruption sometime during the southeast season (about April to November) of 1941. Photograph taken from Matupi Harbour looking east-northeast, and showing the South Daughter in right background. The southeasterly wind is blowing the eruption dust-cloud towards Rabaul.

explosions were followed by greater than usual emissions of water vapour and sulphur dioxide, and a geyser action developed in the muddy pool, with fountains up to 10-12 m high at 10-minute intervals. On 12 and 16 March even more severe explosions took place; mud, stones, and later dry dust were thrown out to a height of a hundred metres or more above the crater rim, depositing 15 to 20 cm of fine volcanic dust around the upper part of the crater. Hydrogen chloride was noted in the gases during the next few weeks, and on 17 and 18 May 1940 mild explosions similar to those on 3 March occurred, but these were the last recorded.

1941 ERUPTION

Although it provided at times spectacular pyrotechnic displays and, while the southeast season continued, was a great nuisance to the town of Rabaul because of the dust cloud swept across by the wind (Fig. 4), the 1941 eruption was not at any time of major proportions. There were no lava flows or anything resembling a nuée ardente; the largest individual explosion reached only a height of about 1200 m above sea-level (although dust was carried much higher by the wind and convection currents), and only a small crater was built up, in the northeastern part of the bottom of the 1937 crater (Fig. 2). The eruption point was near the base of the lava plug in the 1937 crater, and because the southern wall of the 1937 crater was practically vertical, it was possible, with care, to approach the southern side of the crater even when an eruption was in progress. The series of active fumaroles that occupied the 1878 crater before the eruption subsided completely soon after it began.

The general course of the eruption was an almost continuous period of eruptive activity for four months, with minor breaks, and a particularly active period during August and September 1941. After September the breaks between eruptive periods became longer, although some individual eruptions were more violent than any that had gone before. When the Japanese invasion force captured Rabaul on 23 January 1942 the volcano was in eruption, but it appears to have become quiescent about March the same year.

The first eruptive period lasted from 6 to 29 June and consisted of irregularly spaced explosions, of varying intensity, the largest of which occurred on 17 June when ejecta were hurled beyond the foot of the outer slopes of the volcano. During this period, activity was more or less continuous for the first 16 days and intermittent for the next eight. For the first two days of the eruption only dust and rocks were ejected, but on the evening of the second day red-hot rocks were thrown out. The initial eruption was followed by 'puffs' up to 60-120 m above the crater at five-minute intervals. A 10-hour lull occurred after the eruption had been in progress for 14 hours, and 10 hours later came a large explosion to 300 m with red-hot rocks. The eruption continued irregularly, with strong explosions on 15 and 17 June. The tremor accompanying the latter was recorded at both the Observatory and the observation post; it produced the largest dust cloud up to that time, and rocks, including bombs up to 1 m in diameter, were hurled beyond the foot of the outer slopes of the volcano. Similar eruptions occurred over the next few days, but activity declined on 21 and 22 June and ceased from 2 a.m. on 24 to 7.40 p.m. on 26 June. In the 10 hours before the eruption resumed, eight large and many small local tremors were recorded on the seismograph at the observation post. At 1.15 p.m. on 27 and 4.25 p.m. on 28 June very violent explosions took place, rising to 1200 m above sea level, with spectacular ejection of red-hot rocks. This eruptive period ceased at 8 pm. on 29 June. By this time a crater wall about 20 m high had been built up to the west of the vent, which was 6-10 m wide.

Activity resumed on 3 July at 9 a.m. and was fairly continuous but not severe until 11 July. From then until 17 July there was a marked reduction, and, after a renewed burst of activity on 16 and 17, the volcano lapsed into quiescence. On 27 July eruptions took place between 1.30 and 2.00 p.m., but activity was very mild until 9 August, increasing gradually over the next two days, and inaugurating the longest and most continuous period of activity in the whole eruption. Billowing dust clouds rose every 3 or 4 minutes, and red-hot bombs were thrown out. Particularly violent explosions occurred on 22 and 25 August, and on 27 August it was observed that the central section of the crater was occupied by solidified lava. Water vapour and gases escaped through small vents in the lava and gave rise to a continuous roaring sound, which ceased when the lava was removed by an eruption late in the afternoon of 27 August. The week of eruption ending on 27 August was easily the most violent since commencement of activity in June (Knight, vulcanological report for the week ending 27 August 1941).

A short break in activity occurred between 30 August and 1 September, and eruptions continued, but generally at reduced severity, until 7 October. This break lasted until 22 October, when powerful explosions occurred, hurling hot rocks beyond the foot of the slopes of the volcano — still hot enough to set fire to dry grass 1 km from the crater. Another lull followed from 27 October to 8 November, and another from 13 to 15 November, followed by strong activity until 25 November with powerful explosive outbursts, some of which, on 19 November, hurled bombs the farthest yet. After 25 November, the volcano was quiet except for short bursts of activity in the afternoon of 4 December and on 9 January 1942, until the week beginning 19 January, when another eruptive period began.

No record is available of details of the volcano's activity after the Japanese occupied Rabaul on 23 January. However, the main eruption probably came to an end about March 1942, although there are reports of some small later outbursts of activity. T. Kizawa (pers. comm., 1961) referred to an eruption in October 1943, and has provided copies of photographs of an eruption dated 23 December 1943. Comparison of aerial photographs taken during or before March 1943 with those taken in March 1944 and later show that in this period a small crater had developed at the northern end of the 1941-42 crater adjacent to the northern wall of the 1937 crater and the central basalt plug referred to earlier.

The whole course of the eruption is consistent with the slow rise of a small column of lava underneath the Tavurvur crater, activated in the first place by the major disturbance of the 1937 eruption, until the point of eruption balance was reached; thereafter explosive release of gases and fragmentation of the upper part of the lava column took place with the formation of some pumice and glass, followed by a gradual settling down after the main pressure was released, with longer and longer periods between eruptions. Some of these later eruptions, in late 1941 and early 1942, were particularly violent because of the energy required to reopen the vent.

PETROLOGY

Petrological information on the products of the eruptions of Tavurvur is very limited. No comprehensive study has been made and the precise localities of some of the specimens examined have not been recorded.

C. L. Knight (in Fisher, 1939, p.51) described specimens from the central basalt plug and from the ejecta of the 1878 eruption. Heming (1974) referred to the Tavurvur rocks and gave a chemical analysis of a glassy andesite bomb from Tavurvur. Miyake & Sugiura (1953) recorded chemical analyses of three samples from Tavurvur; these analyses, while very similar to each other, are markedly different from the one recorded by Heming (e.g. silica content 47.82 percent compared with Heming's 62.26 percent). In the absence of detailed information about the location and nature of the samples it is not possible to comment further on these differences, although it might be expected that the composition of the bombs thrown out by Tavurvur in 1941-42 would be similar in composition to the material of the 1878 eruption. Further study of the detailed petrology and chemistry of the Tavurvur lavas and ejecta is needed.

REFERENCES

Fisher, N.H., 1939: Geology and vulcanology of Blanche Bay and the surrounding area, New Britain. *Terr. New Guinea Geol. Bull.* 1.

Fisher, N.H., 1944: The Gazelle Peninsula, New Britain, earthquake of January 14, 1941. *Bull. seismol. Soc. Amer., 34.* (1), pp. 1-12.

Heming, R.F., 1974: Geology and petrology of Rabaul Caldera, Papua New Guinea. *Bull. geol. Soc. Amer., 85*, pp. 1253-1264.

Miyake, Y. & Sugiura, Y., 1953: On the chemical compositions of the volcanic eruptives in New Britain island, Pacific Ocean. *Proc. Seventh Pac. Sci. Congr.* 1949, Vol. II Geology pp. 361-363.

AERIAL THERMAL INFRARED SURVEY, RABAUL AREA, PAPUA NEW GUINEA, 1973

W.J. PERRY and I.H. CRICK

Bureau of Mineral Resources, P.O. Box 378, Canberra City, A.C.T. 2601

ABSTRACT

An airborne thermal infrared survey of the Rabaul caldera was carried out in May 1973 to try to detect previously unknown hot spots, both on land and within the waters of the harbour. All imagery was acquired before sunrise at wavelength intervals of 3.5-5.5 micrometres and 8-14 micrometres (μm). Both sets of imagery showed the location and extent of all known thermal areas, but no new areas were discovered. Because of its greater contrast, the longer wavelength displayed the details of the terrain better than the shorter wavelength record. An airborne radiometer provided general information on the temperatures of sea and land, its broad field of view (2 degrees) combined with its response time (50 milliseconds) resulting in averaging of the surface temperatures of the small thermal areas.

INTRODUCTION

The town of Rabaul, on the northeastern end of the island of New Britain, Papua New Guinea, is built on the shores of Simpson Harbour (Fig. 1), within a breached volcanic caldera.

The latitude is 4 degrees south, and the climate humid tropical. The northwest monsoon season generally begins in December and continues to March, followed by a period of relatively calm weather until the southeast season starts in June. The average annual rainfall is approximately 2000 mm.

Two important volcanic centres in the caldera have been active in recent times. Vulcan erupted in 1937 with simultaneous minor activity at Tavurvur, and Tavurvur erupted again between 1941 and 1943 (Fisher, this volume; Taylor, 1973). A volcanological observatory was established on the caldera rim in 1940 (Fisher, 1940). During the Second World War the observatory was destroyed, but it was re-established in 1950, and since then the surveillance instrumentation has been undergoing development and is now of a high standard (Taylor, 1973).* The temperatures of vapours from fumaroles and water from thermal springs in the vicinity of Tavurvur are recorded each week. The thermal infrared survey was undertaken to find out whether previously unknown hot spots could be detected, particularly springs that might issue from the sea-bed within Simpson Harbour and the shallower waters of Blanche Bay.

KNOWN THERMAL ACTIVITY

Thermal activity within the Rabaul caldera is confined to mild fumarolic emission in the craters and on the slopes of Tavurvur and Rabalankaia and to thermal waters which flow into the bay, mostly from areas near and below the high-tide mark along the shoreline of Matupi Harbour, the northern shoreline of Vulcan, and in Sulphur Creek (Fig. 2).

Fumarolic emission in Rabalankaia crater is confined mainly to two concentric rings open to the southeast which are probably related to ring fractures. The outer ring of fumaroles coincides with the inner slopes and base of the crater wall, whereas the smaller circle of fumaroles lies centrally on the flat floor of the

* See Myers (this volume). Ed.

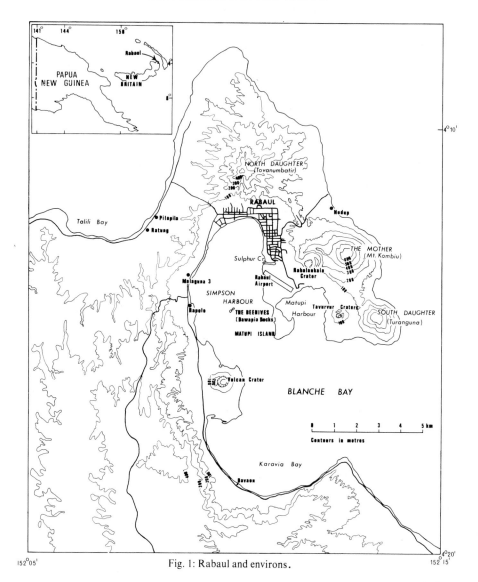

Fig. 1: Rabaul and environs.

crater. Heming (1973) suggested that the smaller circle of fumaroles may have
formed along a fault around a collapsed central plug. Two small thermal springs,
which flow intermittently, are located in a gully half-way up the western slope of
Tavurvur. Minor gas ebullition occurs in several thermal springs along the
northern shoreline of Matupi Harbour.

Over the past twenty years fumarolic temperatures have rarely risen much
above 100°C. Thermographs show little variation except for one small fumarole in
one of Tavurvur's craters which exhibits a regular rise and fall of as much as 37°C
over a twelve-hour period, a phenomenon which may be related to tidal rhythms.
Heavy rainfall can produce sharp falls in fumarolic temperatures, but they recover

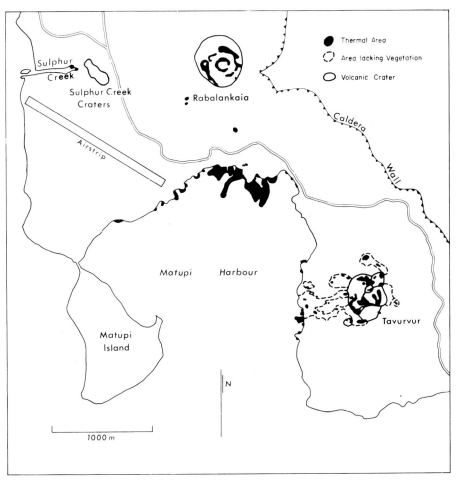

Fig 2: Thermal anomalies of Matupi Harbour from infrared imagery (8-14μm).

to original levels within several hours of cessation of rain. This factor did not influence infrared imaging of the fumarolic areas as no rain fell for at least 48 hours before imaging commenced or during the period of imaging. Temperatures of the fumaroles at the time of imaging ranged from 87°C to 99°C, most fumaroles recording a temperature of 99°C.

Shoreline thermal spring temperatures fluctuate widely depending on tidal conditions and on the amount of recent rainfall and consequent height of the water-table.

INFRARED SURVEY*

The area was surveyed with a Daedalus optical-mechanical infrared line scanner operating in both the 8-14 μm and the 3.5-5.5 μm intervals. The detectors

* For details the reader is referred to an unpublished account by Perry & Crick in Bureau of Mineral Resources Record 1974/97, "Aerial thermal infrared survey, Rabaul, Papua New Guinea, 1973."

had an instantaneous field of view of 2.5 milliradians, and a thermal resolution of 0.3°C (C. Ellyett, pers. comm.). Their output was recorded on magnetic tape and the data later converted to film records. From the survey altitude of 914 m (3000 ft) the detectors viewed a square of terrain 2.28 m (7.5 ft) on a side at any instant. Preferred flying time was pre-dawn to avoid the daily heating effect of the sun, so that thermal anomalies would show up most strongly against the background, but local restrictions limited flying to the period between first light and sunrise. The scanner was fitted with a roll stabilizer that provided electronic compensation for aircraft roll to the scanner imagery by reference to a gyroscope.

The aircraft, an Aero Commander 680F, also carried a Barnes PRT 5 radiometer with a field of view of 2 degrees which recorded surface temperatures with a specified accuracy of ± 0.5°C on a strip chart. During daylight hours the location of the radiometer trace could be determined by reference to photographs taken with a 35-mm tracking camera. Because flying took place after first light, navigation was by visual reference to ground features, and the coastlines in the area made this a relatively simple task. A Doppler navigation system did not provide useful information from most survey runs because they were partly over land and partly over calm water. During the period of the infrared survey (3-11 May 1973) 1:15,000 scale vertical aerial colour photographs of the caldera were obtained for comparison with the infrared imagery.

CORRELATIONS BETWEEN THERMAL AREAS AND INFRARED ANOMALIES

The grey tones of the imagery (Figs. 3 and 4) give an indication of relative temperature of the terrain. The hotter areas are light toned and the cooler areas relatively dark.

8-14 μ m band (Fig. 3)

The most noticeable features of the imagery are light and dark bands at right-angles to the flight-lines, which are caused by an instrumental effect due to amplification of the detector output practically to the maximum to extract as much information as possible on temperature variations of the water surface (C. Ellyett, pers. comm.). Neglecting these it can be seen that the sea is generally warmer than the land at this time of day.

Bright areas within the water, principally in Matupi Harbour, are caused by hot water from thermal springs. The main activity is at the head of Matupi Harbour, with other hot zones at the southeast end of the airstrip and west of Tavurvur; the water in Sulphur Creek is slightly warmer than the harbour water, and north of Vulcan Crater a small body of water isolated from the harbour by a narrow sand bar is also warmer than the harbour water.

On the land Rabalankaia is outlined by concentric bright rings (Fig. 5), which can be seen on the aerial photographs to correspond with areas of bare ground on the crater floor and walls. It is concluded that the bright returns on the imagery are produced by warm ground. Similarly, small areas of bare ground between Rabalankaia and Matupi Harbour show as hot spots on the imagery.

The bright, rather complicated pattern at Tavurvur corresponds in part to

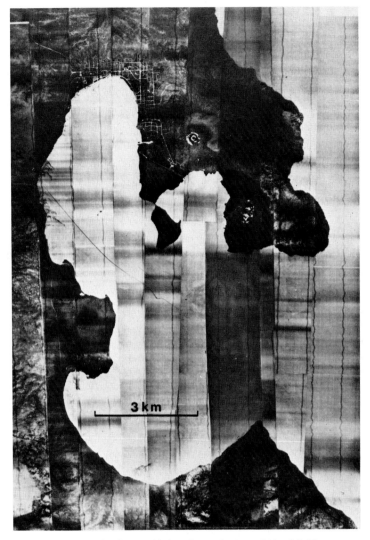

Fig 3: Mosaic of thermal infrared runs, 8-14 μm, Rabaul Caldera.

bare ground, but exceptions are dark toned bare areas that image cool, for example the northeast slope and the upper southwest flank. Bare areas on the lower southwest flank are warm in only a few places. Also imaging cool are the floors of the four individual craters within the main crater rim, and the floor of one small parasitic cone on the southwest flank of Tavurvur.

3.5-5.5 μm band (Fig. 4)

In general the contrast is less in this band than in the 8-14 μm interval, and as a result details of the terrain are less apparent. The obvious features of the imagery are the dark zones along one side of each of the strips, which were flown from south to north. Four runs not shown in the figure were flown round the coastline from

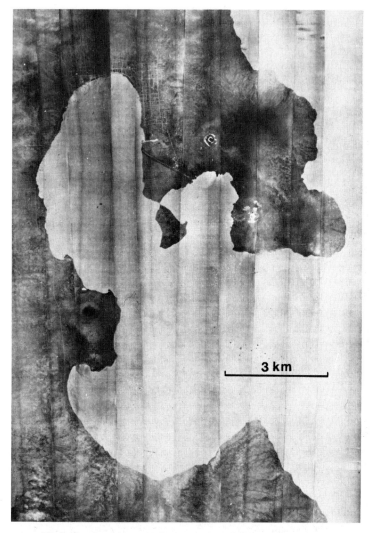

Fig 4: Mosaic of thermal infrared runs, 3.5-5.5μm, Rabaul Caldera.

west to east, and it was apparent that the dark zone remained on the starboard side of each strip irrespective of aircraft heading, therefore it is concluded that the dark zones are caused by some instrumental factor.

All the 'hot spots' visible on the long-wavelength record can be seen on the original 3.5-5.5 μm imagery, but on the latter the details of the background are less well shown because of the lack of contrast.

Radiometer

It is evident from a comparison of radiometer readings with the infrared

Fig. 5 (opposite): Infrared imagery (8.14 μm) of Tavurvur (upper) and Rabalankaia (lower) matched with radiometer traces. Both runs flown south to north.

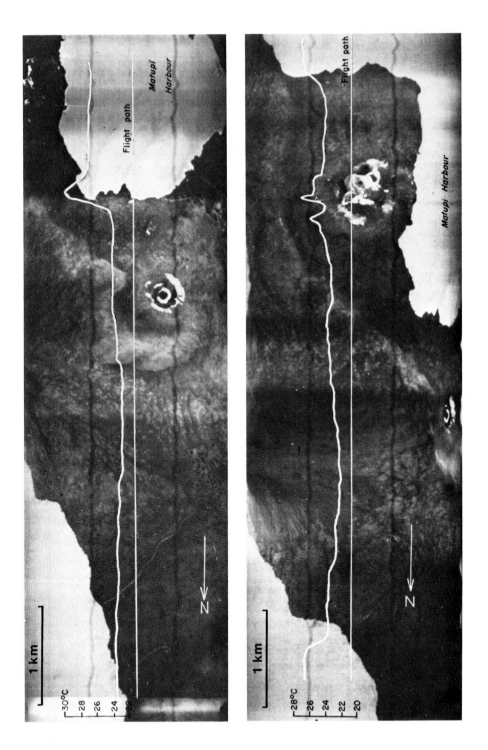

imagery of Tavurvur and Rabalankaia (Fig. 5), that the recorded temperature variations are less than one would expect from the tonal variations on the imagery. These differences are ascribed to the larger field of view of the radiometer (nearly 35 milliradians as compared to the scanner's 2.5 milliradians) and to its response time (50 milliseconds), and it is concluded that in surveys of this type the radiometer should have a field of view comparable with that of the scanner; alternatively, a calibrated thermal scanner should be used.

OTHER USES FOR AERIAL INFRARED TECHNIQUES IN PAPUA NEW GUINEA

Airborne infrared imaging systems have been used in the thermal surveillance of more than 23 volcanoes around the world (Friedman & Williams, 1968; Moxham, 1971). Use of such systems over volcanoes in Papua New Guinea could provide information on the location and extent of thermal areas on active and dormant volcanoes to facilitate ground surveillance studies, and in the case of calibrated systems could give an estimate of temperatures. Infrared imaging could prove valuable in monitoring rapid changes in thermal activity, particularly effusive eruptions. If a calibrated system were used, volcanic thermal energy yield and partition estimates could be determined in conjunction with volcanological ground observations.

Prediction of an impending eruption by means of airborne infrared imaging systems is possible but not yet demonstrated. Thermal activity increased before the eruption of Tavurvur in 1941 (Fisher, this volume) and at some other volcanoes around the world (Moxham, 1971). Regular flights would need to be maintained over the volcano in question to detect changes in thermal activity; a calibrated device would be preferable, and the results would be analysed in conjunction with other surveillance techniques.

A potential major use for infrared imaging in Papua New Guinea is in the mapping of possible sources of geothermal power. Several possible sites are known, and although thermal activity in these areas is usually indicated on normal aerial photography by the lack of vegetation, in some places, as for example along the Num River, Ambitle Island (off New Ireland; see Johnson et al., this volume) thermal activity occurs along stream and river courses and may be overlooked in the initial mapping if only normal aerial photography is used.

CONCLUSIONS

The thermal infrared scanner successfully mapped the location and extent of most of the known thermal areas with the exception of a few springs below high-water level, and it is concluded that there are no other major 'hot spots' within the surveyed area. Of the two infrared records, the 8-14 μm imagery is preferred for this sort of general volcanological work, because of its better contrast, although had only the 3.5-5.5 μm detector been available the mapping of thermal areas could still have been satisfactorily achieved.

The radiometer contributed little useful information to the study of the thermal areas; in the writers' opinion the use of a radiometer of this type in similar aerial surveys in future would be worthwhile only in situations in which the thermal areas are larger than the instantaneous field of view of the instrument at the operational altitude. The temperature of large areas of warm surface water produced by subaqueous thermal activity could probably be accurately measured, but that of small or patchy fumarolic areas commonly found on slopes and in craters of volcanoes probably could not.

ACKNOWLEDGEMENTS

Grateful acknowledgement for help received during the survey is made to the staffs of the Department of Civil Aviation and the Meteorological Office, Rabaul. This paper is published with the permission of the Director of the Bureau of Mineral Resources, Canberra.

REFERENCES

Fisher, N.H., 1940: Note on the Volcanological Observatory at Rabaul, *Bull. Volcanol. 6*, pp. 185-187.

Friedman, J.D. & Williams, R.S. Jr, 1968: Infrared sensing of active geologic processes, *Proc. Fifth Symposium on Remote Sensing of Environment*, 1968, University of Michigan, Ann Arbor, pp. 787-819.

Heming, R.F., 1973: Geology and petrology of Rabaul Caldera: an active volcano in New Britain, Papua New Guinea, *Ph.D. Dissertation, University of California, Berkeley (unpublished)*.

Moxham, R.M., 1971: Thermal surveillance of volcanoes. In *The surveillance and prediction of volcanic activity, a review of methods and techniques*, pp. 103-124 UNESCO, Paris, 1971.

Taylor, G.A.M., 1973: *In* Davies, H.L., Explanatory notes on the Gazelle Peninsula 1:250 000 Sheet, *Bur. Miner. Resour. Aust. Explan. Note SB/56-2*.

NATURE AND ORIGIN OF LATE CAINOZOIC VOLCANOES IN WESTERN PAPUA NEW GUINEA

D. E. MACKENZIE[1]

Department of Geology, University of Melbourne, Parkville, VIC 3052

ABSTRACT

Sixteen late Cainozoic centres in western Papua New Guinea rest on 25-30 km of Palaeozoic sialic crust and up to 10 km of post-Palaeozoic sedimentary and volcanic rocks. One rests on a 35 km-thick pile of eugeosynclinal sediments and volcanic rocks. Fourteen of the centres are deeply eroded stratovolcanoes, and the last a ring-like cluster of lava domes. Basaltic rocks are dominant in eleven centres, and these range from high-K, oversaturated (shoshonitic) types, through lower-K saturated types to low-K types just saturated to slightly undersaturated in silica. Other rock types in these centres, and the dominant rock types in the other six centres, are low-Si andesite and andesite with high to moderate K_2O contents. The magmas originated in the upper mantle low-velocity layer following plate collision in the late Oligocene to early Miocene and subsequent crustal warping and uplift continuing into the Pliocene. Complex chemical variations were produced by mantle inhomogeneity, various types and amounts of partial melting in rising diapirs at various levels, and various rates of ascent and degrees and types of crystal fractionation at the base of and within a mechanically inhomogeneous crust.

INTRODUCTION

The rocks discussed in this paper are the eruptive products of a widespread group of large Quaternary stratovolcanoes in central mainland Papua New Guinea (Fig. 1). These volcanoes first came to the attention of geologists in 1939 when L. C. Noakes, chief geologist of the Papuan Administration, visited Mounts Hagen, Giluwe, and Ialibu. Subsequently, Rickwood (1955) and the Australasian Petroleum Company (1961) mapped all of the volcanic centres in the area except Mount Yelia, which was recognised as volcanic in 1962 (Branch, 1967). In 1968, Bain et al. (1975) sampled Mounts Hagen, Giluwe, Ialibu, Suaru, and Karimui, and Crater Mountain, and Taylor (1971) visited Doma Peaks to investigate reported eruptive activity.

Following upon the work of several geologists (e.g. Dickinson & Hatherton, 1967), Jakeš & White (1969) related shoshonitic rocks from Mounts Hagen, Giluwe, and Ialibu to a deep (ca. 200 km) portion of a supposed Benioff zone. Johnson et al. (1971) and Mackenzie & Chappell (1972), discussing chemical analyses of samples collected by Bain et al. (1975), showed that the shoshonites were genetically related to calcalkaline andesites and that there was no direct evidence, in particular intermediate- and deep-focus seismic activity, to relate the volcanoes to a Benioff zone. Mackenzie & Chappell (1972) also pointed out that the volcanoes they described all rest on strongly buckled, but now stable, continental crust 30 to 40 km thick. They postulated partial fusion of eclogite detached from the base of downbuckled parts of the crust to explain the origin of the shoshonites.

New information on the volcanoes studied by Mackenzie & Chappell (1972), and on several other centres, reveals that basaltic rocks are dominant and range from silica-saturated or oversaturated high-K types to low-K types just saturated

[1]Present address: Bureau of Mineral Resources, P.O. Box 378, Canberra City, A.C.T., 2601.

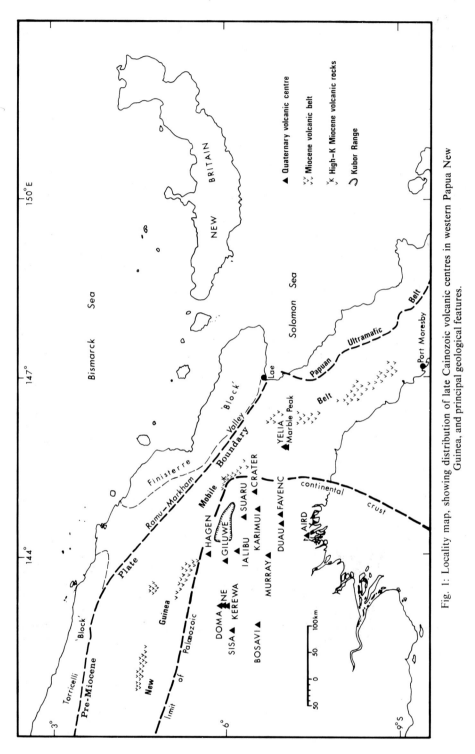

Fig. 1: Locality map, showing distribution of late Cainozoic volcanic centres in western Papua New Guinea, and principal geological features.

or slightly undersaturated in silica. In all but three of the centres, andesite overlies and grades into basalt.

Such a combination of features is difficult to explain in terms of conventional models of magma generation in island-arc or continental margin areas involving downgoing lithospheric slabs, particularly as seismic evidence is lacking.

GEOLOGY
Geological and tectonic setting

All the volcanoes except Mount Yelia rest on folded Mesozoic and Tertiary sedimentary rocks up to 10 km thick which, in turn, overlie 20-30 km (St John, 1970) of Palaeozoic continental crust. Part of this basement is exposed in the Kubor Range, the part expression of a 120 km-long anticline in which Permian granitic and older metamorphic rocks crop out at up to almost 4000 m above sea level (Fig. 1). Mount Yelia rests on the New Guinea Mobile Belt (cf. Dow et al., 1974), a thick, strongly folded and faulted sequence of Mesozoic and Tertiary rocks which includes a discontinuous 1000 km-long belt of Miocene volcanic rocks. The Miocene volcanics close to the northeastern corner of the Palaeozoic continental plate include high-K undersaturated rocks which are not found elsewhere in the belt. Beyond the mobile belt to the north and northeast are the old island-arc-like features which make up the coastal ranges: the Torricelli 'block' and the Finisterre 'block'.

Gravity and seismic studies (e.g. St John, 1970; Jenkins, 1974) showed that most of western Papua New Guinea has a crustal thickness of 30-35 km, and revealed deep fault zones beneath many of the Quaternary volcanoes.

Seismic data (Denham, 1969; Johnson et al., 1971; Johnson & Molnar, 1972; Curtis, 1973; Dent, 1975) show that present-day seismicity is confined to crustal and shallow (50 km) subcrustal depths beneath all of the volcanoes except Mount Yelia, which is underlain by a small, pronounced pocket of intermediate-depth (175-190 km) seismicity. It is significant that some of the most frequent seismic activity in the area has occurred beneath Mount Yelia, and until the recent past, in the Doma Peaks area (Dent, 1975); these are the only two volcanoes still showing signs of thermal activity. Branch (1967) recorded cold solfataras and gas seeps from Mount Yelia, and noted that local people talked of activity on the mountain. Taylor (1971) found warm and cold springs and solfataras on Doma Peaks and also commented on local legends of eruption such as those recorded by Glasse (1963).

Geological and geophysical evidence (e.g. as summarised by Jenkins, 1974) indicate that the crust beneath the volcanoes was strongly warped (up to 11 km of relief) during the late Miocene to early Pliocene, and uplifted (by up to 3 km) during the late Pliocene.

Morphology

The volcanoes, except for Mount Giluwe which is shield-like in form and Aird Hills which is a ring-like dome cluster, are all stratovolcanoes surrounded by extensive aprons. All are deeply eroded except Mount Yelia, which shows only slight effects of erosion; basement is exposed in the erosion calderas of Mounts Hagen, Karimui, and Murray. Erosion is commonly asymmetric owing to concentration of orographic rainfall on the sides facing the wet seasonal winds (Ollier & Mackenzie, 1974).

Age

Volcanism began in the early Pleistocene or perhaps late Pliocene, and has continued virtually until the present day in at least two centres. The oldest dates are 1.1 ± 0.4 m.y. for a plug near Mount Ialibu (Jenkins, 1974), and 0.85 ± 0.05 m.y. for a flow near Mount Kerewa (Williams et al., 1972). Three dates of around 0.21 m.y. for lavas from Mounts Hagen and Karimui were reported by Page & Johnson (1974).

INTRODUCTORY PETROLOGY

The system of nomenclature used here is the same as that of Mackenzie & Chappell (1972), except for a change in slope of the lines separating higher-K from lower-K rocks on the K_2O/SiO_2 diagram: the line separating shoshonitic from high-K calcalkaline rocks passes through 2 percent K_2O at 50 percent SiO_2 and 3 percent K_2O at 70 percent SiO_2. The line separating high-K calcalkaline from calcalkaline rocks has the same slope. This change brings the system more into conformity with nomenclature used by other geologists, such as Joplin (1968). The term 'transitional' applies to basalts with low K_2O contents, saturated or just saturated in silica, which are neither shoshonitic nor alkaline. Those rocks which contain modal or normative nepheline and have low K_2O contents are termed 'slightly alkaline'.

Most of the volcanoes are made up of olivine-bearing basalt, overlain and in places intercalated with, or grading into, basaltic (low-Si) andesite and amphibole-bearing andesite. The basalt occurs mainly as lava flows, with some interbedded ash, agglomerate, and lahar deposits; a larger proportion of the andesites is fragmental. The proportion of andesite to basalt appears to differ from centre to centre. Andesitic rocks make up about 20 percent of Mount Hagen, about 5 percent of Mount Giluwe, 10 percent of Suaru, 20-30 percent of Crater Mountain, less than 1 percent of Karimui, 5 percent of Mount Murray, 30-40 percent of Duau and Favenc, and about 30 percent of Mount Bosavi. However, because of their stratigraphic position and the more fragmental nature of the andesitic rocks, they may have been far more extensively eroded away than the basaltic rocks.

Doma Peaks and Mounts Ne, Kerewa, and Ialibu are made up largely of low-Si andesite and andesite. Olivine basalt occurs in Ialibu and Doma, but only in minor quantities. Mount Yelia is made up largely of hornblende-pyroxene andesite resting on minor olivine-bearing low-Si andesite. Aird Hills is made up of generally deeply weathered leucocratic andesite and dacite which are characterised by plagioclase phenocrysts up to 2 cm long.

PETROGRAPHY

The rocks from the central Papua New Guinea volcanoes may be divided into three general petrographic groups: basalt, basaltic (low-Si) andesite, and andesite and dacite.

The basalts of each volcanic centre or group of centres are petrographically distinct. Those of Mounts Hagen, Giluwe, and Karimui are similar to the shoshonites of Wyoming (Nicholls & Carmichael, 1969). They differ in being fresher, generally holocrystalline, and coarser-grained, and in commonly containing orthopyroxene instead of biotite. Most basalts of Mount Suaru and Crater Mountain contain less interstitial alkali feldspar, and resemble calcalkaline basalts more closely. Two basalt specimens from Suaru have more alkaline affinities: one contains modal nepheline and the other normative (but not modal) nepheline. The scarce basalts of Doma Peaks and Mount Ialibu are poor in olivine and rich in plagioclase phenocrysts. In the Doma basalt, plagioclase phenocrysts are up to 1.5 cm across, and olivine is largely confined to the groundmass. In basalts of Mounts Murray, Duau, and Favenc, plagioclase and olivine phenocrysts are generally less prominent, and orthopyroxene is rarer than in basalts of the more northerly centres. In Mount Bosavi, plagioclase phenocrysts are commonly absent, and nepheline appears in the groundmass of some basalts: they are distinctive rocks, both petrographically and chemically.

Sparsely scattered phenocrysts of largely to completely oxidised (to magnetite + clinopyroxene $\pm$ plagioclase) basaltic or kaersutitic hornblende occur in some basalts of all centres.

Compared with the basalts, basaltic andesites contain more orthopyroxene, especially as phenocrysts, and less or no olivine. Basaltic hornblende is abundant in some rocks but absent in others, notably those of Doma, Ne, and Kerewa and most low-Si andesites of Mounts Giluwe and Ialibu.

The andesites are typically hornblende-two-pyroxene andesites, some of which contain tridymite, cristobalite, or β-quartz, and some of which contain a trace of biotite. Dacites, containing more than about 10 percent modal free silica, are rare.

Plagioclase is the dominant phenocryst phase in almost all these rock types. Phenocrysts are large, complexly zoned and twinned, commonly show resorption features, and are commonly clumped together. These features are more pronounced in the larger phenocrysts of the low-Si andesites and andesites. Compositions of the phenocrysts (determined by electron microprobe) range from An_{45} to An_{84} in the core, and from An_{60} to An_{36} at the rim, with most in the range An_{50} to An_{65}. Orthoclase content ranges from 0.5 to 4.5 percent.

Clinopyroxene forms almost ubiquitous euhedral phenocrysts (1-5 mm) and equant groundmass grains. Compositions (determined by electron microprobe) are in the ranges Ca_{42-54}, Mg_{42-54}, $Fe_{2.2-11}$ (cores) and Ca_{34-51}, Mg_{40-51}, $Fe_{4.5-18}$ (rims) — that is, in the diopsidic augite, diopside, and salite fields. Within each rock, there is an irregular trend towards higher Fe and lower Ca with advancing crystallisation. Calcium and titanium contents of clinopyroxene phenocrysts in the basalts are lowest in rocks from Hagen and Giluwe, higher in Murray, Duau, and Favenc, and highest in Bosavi. They follow the variations in bulk rock composition. Al in sixfold co-ordination is low and variable, giving no indication of any high-pressure influence.

Olivine occurs generally as small (0.5-1.5 mm) phenocrysts, and commonly also as groundmass grains. Some phenocrysts are 7 mm or more across and are commonly irregular or composite. Probe analyses show that the smaller

D.E. MACKENZIE

Table 1. Major and trace element analyses of representative specimens from some of the western Papua New Guinea volcanoes.

	MT HAGEN						MT GILUWE			
	1	2	3	4	5	6	7	8	9	10
SiO_2	50.5	52.2	54.4	56.8	58.4	60.8	51.2	54.9	56.1	56.8
TiO_2	0.92	1.01	1.00	0.81	0.72	0.50	0.99	0.85	0.78	0.62
Al_2O_3	13.5	15.1	15.3	17.2	18.0	18.2	17.3	16.7	15.8	18.8
Fe_2O_3	1.53	1.45	2.65	2.65	2.50	2.25	2.80	2.15	2.20	4.78
FeO	6.65	6.65	5.05	3.95	3.65	2.65	5.25	4.85	4.15	1.37
MnO	0.14	0.15	0.14	0.12	0.12	0.11	0.14	0.12	0.12	0.13
MgO	12.4	8.20	6.60	5.15	3.45	2.10	6.50	5.55	6.45	2.85
CaO	7.85	8.80	7.65	5.60	5.75	5.45	8.75	7.50	6.35	6.40
Na_2O	2.65	2.85	3.30	3.85	3.85	4.05	3.15	3.50	3.40	4.15
K_2O	2.35	2.35	2.32	1.85	1.70	2.30	2.35	2.30	2.90	2.40
P_2O_5	0.58	0.59	0.45	0.23	0.29	0.31	0.58	0.55	0.47	0.79
H_2O+	0.64	0.81	0.74	1.60	1.40	0.95	0.92	0.64	0.80	0.59
CO_2	0.03	0.07	—	0.05	0.05	—	0.07	—	0.08	0.03
S ·	0.03	0.03	0.02	0.02	0.02	0.02	0.02	0.01	0.02	0.03
Total	99.77	100.26	99.60	99.88	99.93	99.69	100.00	99.62	99.62	99.75
Rb	74	69	65	41.5	48.5	64.0	65.5	82.5	108	82.0
Sr	560	615	980	705	755	935	770	815	710	945
Y	20.0	21.0	24.5	18.0	17.5	18.0	21.0	21.5	22.0	21.5
Zr	83	92	155	155	145	180	84	98	145	125
Nb	3.1	2.1	3.7	5.0	6.3	4.8	2.4	2.1	5.8	4.4
Pb	6.8	9.2	17.5	12.0	14.0	17.0	8.0	11.5	14.5	13.5
Ba	405	485	935	705	680	815	480	560	650	655
La	112	13.0	35.5	23.5	25.5	31.0	16.5	19.0	25.0	21.0
Ce	45	25	51.5	44.0	36.5	39.5	33.0	22.5	46.5	32.5
Nd	—	—	—	—	—	—	—	—	—	—
Cu	75	105	129	59	53	20	122	85	78	30
Co	78	63	64	32	42	19	44	35	37	20
Ni	340	115	75	75	25	11	81	70	130	10
Sc	25	28	24	18	17	9	25	22	19	10
V	210	230	205	150	125	69	260	185	145	69.3
Cr	608	365	244	222	68.0	26.5	150	163	274	0.9
Th	—	—	—	—	—	—	—	—	—	—
U	—	—	—	—	—	—	—	—	—	—
CIPW norms										
Q	—	—	2.00	6.97	11.2	13.4	—	2.43	3.17	7.87
C	—	—	—	—	0.04	—	—	—	—	—
Z	0.02	—	0.03	0.03	—	0.04	0.02	0.02	0.03	0.02
or	13.9	13.9	13.6	11.0	10.1	13.6	13.0	13.7	17.2	14.3
ab	22.4	24.1	27.9	32.6	32.6	34.3	26.7	29.6	28.8	35.1
an	18.0	21.5	20.2	24.3	26.8	24.8	26.6	23.1	19.3	25.6
ne	—	—	—	—	—	—	—	—	—	—
di	13.8	14.7	12.4	1.73	—	0.65	10.4	8.85	7.24	0.99
hy	4.60	11.6	16.0	15.8	12.2	7.27	9.2	15.4	17.3	6.64
ol	20.9	8.07	—	—	—	—	5.65	—	—	—
mt	2.33	2.22	3.95	3.95	3.68	3.28	4.17	3.21	3.25	2.94
cm	0.26	0.13	0.11	0.10	0.03	0.01	0.06	0.07	0.12	—
hm	—	—	—	—	—	—	—	—	—	2.77
il	1.75	1.92	1.90	1.54	1.37	0.95	1.88	1.61	1.48	1.18
ap	1.37	1.40	1.07	0.55	0.69	0.74	1.37	1.31	1.12	1.88
cc	0.07	0.16	—	0.11	0.11	—	0.16	—	0.18	0.07
Others	0.73	0.87	0.79	1.66	1.45	0.99	0.97	0.66	0.84	0.63

Standard X-ray fluorescence and wet chemical techniques were used. H_2O- was eliminated from all but an earlier batch of analyses (Mackenzie & Chappell, 1972) by heating powdered samples at 110°C for 12 hours and then storing in a desiccator.

Table 1 cont.

MT MURRAY			MT DUAU		MT FAVENC		MT BOSAVI		
11	12	13	14	15	16	17	18	19	20
49.9	52.3	58.5	50.3	59.7	52.2	62.7	47.1	52.5	60.2
1.34	1.48	1.01	1.20	0.72	1.25	0.53	1.28	1.01	0.68
14.7	16.8	17.1	15.3	17.4	17.4	17.6	14.1	16.2	17.3
1.85	2.45	3.55	2.85	3.25	3.10	2.00	3.45	2.95	2.95
7.85	6.55	2.90	6.05	2.45	5.95	2.45	6.55	5.40	2.70
0.17	0.16	0.16	0.17	0.14	0.18	0.12	0.18	0.18	0.12
8.55	5.20	2.25	8.75	2.60	4.95	1.90	11.7	6.75	2.10
9.60	7.95	6.40	9.30	6.30	8.60	5.15	9.90	8.20	6.00
2.70	3.25	3.65	2.50	3.65	3.20	3.90	2.90	3.55	4.05
1.75	2.05	2.70	1.75	2.05	1.65	2.20	0.95	1.80	2.50
0.69	0.70	0.50	0.70	0.47	0.65	0.36	0.42	0.50	0.34
0.75	0.94	0.90	0.92	1.05	0.62	0.93	0.97	0.66	0.72
—	—	0.03	—	0.10	0.05	0.01	0.02	—	0.02
0.04	0.03	0.04	0.04	0.06	0.02	0.03	0.04	0.02	0.04
99.89	99.86	99.68	99.89	99.94	99.82	99.88	99.57	99.72	99.72
38.0	43.0	81.0	65.0	59.5	39.2.	67.9	29.0	39.5	71.0
635	765	785	680	925	703	820	675	690	765
21.4	24.5	23.5	24	23	25	17	20.5	22.0	21.5
115	135	285	—	—	—	—	146	187	256
6.3	9.0	12	—	—	—	—	8.9	13.5	14.0
8.7	10	14	12	16	13.5	15.5	4.8	7.3	8.6
295	330	700	195	480	308	526	355	520	760
14	21.5	39.0	9.6	25	17	23	26	31.5	34.0
21	43	70.5	26	45	38	48	64	51.5	58.0
—	27	—	—	—	—	—	—	—	—
135	100	22	95	37	84	18	68	52	42
66	50	21	47	23	36	19	72	46	40
95	33	8.0	163	22	45	6	222	97	10
31	24	12	29	13	22	8	30	22	12
285	280	130	284	111	237	61	280	195	120
355	75	0	294	47	75	10	780	335	7.1
—	—	—	1	5	4	7	—	—	—
—	—	—	0.3	1.4	0.2	1.3	—	—	—
—	1.27	12.2	—	15.0	2.49	17.6	—	—	12.5
—	—	—	—	—	—	0.22	—	—	—
0.02	0.03	0.06	—	—	—	—	0.03	0.04	0.05
10.4	12.1	16.0	10.5	12.0	9.78	13.0	5.64	10.7	14.8
22.7	27.5	30.9	21.0	30.9	27.1	33.0	21.6	30.0	34.3
22.9	25.3	22.4	25.4	25.1	28.3	23.4	22.7	23.0	21.7
—	—	—	—	—	—	—	1.60	—	—
16.6	8.12	4.98	13.2	2.18	8.20	—	19.2	11.9	4.97
8.13	16.7	4.17	16.6	6.17	15.0	6.87	—	13.7	4.36
11.5	—	—	4.18	—	—	—	19.3	2.43	—
2.76	3.69	5.23	4.16	4.71	4.49	2.93	5.18	4.38	4.34
0.15	0.03	—	—	—	—	—	0.33	0.14	—
2.54	2.81	1.92	2.28	1.37	2.37	1.01	2.43	1.92	1.29
1.63	1.66	1.19	1.66	1.11	1.54	0.85	0.99	1.19	0.81
—	—	0.07	—	0.23	0.11	0.02	0.05	—	0.05
0.82	1.00	0.97	0.96	1.13	0.63	0.97	1.04	0.70	0.80

Trace elements in 56 samples were measured by direct-reading optical spectrograph. Additional data may be obtained from the author upon request.

cont. overleaf

Table 1 cont.

1. High-K olivine basalt ('shoshonite'), SE summit area, Mount Hagen.
2. High-K olivine basalt ('shoshonite'), Turuk River, 20 km SE of Hagen trig.
3. High-K olivine-hornblende-2 pyroxene low-Si andesite, SE central summit area.
4. Olivine-2 pyroxene-hornblende low-Si andesite, SE summit area.
5. Hornblende-2 pyroxene andesite, central summit area.
6. Tridymite-2 pyroxene-hornblende andesite, central summit area.

7. High-K olivine basalt ('shoshonite'), flow, N summit area, Mount Giluwe.
8. High-K olivine-2 pyroxene low-Si andesite, lava dome, upper NE slopes.
9. High-K (olivine-)2 pyroxene low-Si andesite, lava dome, upper SW slopes.
10. Oxidised hornblende-2 pyroxene andesite, dome, N summit area.

11. Olivine basalt, crater area, Mount Murray.
12. Olivine basalt, northern slopes.
13. Hornblende-2 pyroxene andesite, crater area.

14. Olivine basalt, SE slopes, Mount Duau.
15. Hornblende-2 pyroxene andesite, S slopes.

16. Olivine basalt, S slopes, Mount Favenc.
17. Hornblende-2 pyroxene andesite, SE slopes.

18. Near-saturated alkali-olivine basalt, crater area, Mount Bosavi.
19. Olivine basalt, W slopes.
20. Hornblende-2 pyroxene andesite, Turama River (from crater area?).

phenocrysts range in composition from Fo_{70} to Fo_{82-85} in the core and from Fo_{60} to Fo_{75-80} at the rim. Some larger phenocrysts in basalts from Mount Hagen range from $Fo_{90-92.5}$ in the core to Fo_{70-84} at the rim, and contain chromite inclusions. Similar phenocrysts, with reaction rims of opaque oxides $\pm$ amphibole, or of pyroxene, occur in andesite from one locality. Calculations of Mg^{2+}/Fe^{2+} partitioning between olivine and liquid based on methods of Roeder & Emslie (1970) and Cawthorn et al. (1973) show that within experimental error, most of the high-Mg olivine compositions are compatible with bulk-rock chemistry. However, the olivine from the andesite (analysis 4, Table 1) is clearly shown to be residual and out of equilibrium with the host rock.

Large, irregular, commonly composite crystals or crystal clumps of olivine with chromite or chrome-rich spinel inclusions occur in basalts from several other centres, notably Mounts Duau and Bosavi. Features such as deformation lamellae and marginal corrosion (sponge-like texture) were observed in some of these olivines. These features, together with the presence of chromite inclusions and the low CaO contents (zoning from 0.02-0.11 in the core to 0.14-0.25 at the rim) suggest crystallisation under high pressure (non-surface) conditions (Simkin & Smith, 1970; Stormer, 1973).

Orthopyroxene compositions (based on 32 analyses of rocks from Hagen, Giluwe, and Murray) are in the range $Wo_{2.4-5.5}$, En_{57-75}, Fs_{22-39}, zoning out in one extreme case of $Wo_{7.0}$, En_{66}, Fs_{27}. Orthopyroxenes in the low-Si andesites are generally more magnesian than in the andesites. Aluminium contents are low and show no high-pressure influence.

Opaque minerals in the basalts are dominated by abundant titanomagnetite

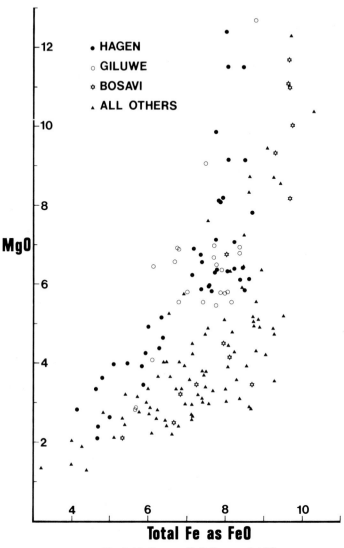

Fig. 2: MgO versus FeO diagram (wt %).

(magnetite-ulvöspinel solid solution) which contains between 10 and 21 weight percent TiO_2. Contents of MgO, Al_2O_3 and, most noticeably, Cr_2O_3 in the magnetites increase from Hagen and Giluwe to Murray, Duau, and Favenc, and then to Bosavi. Other opaque phases are the chromite inclusions in olivine (mentioned earlier), rare pyrite, and even rarer minute blebs of pyrrhotite or chalcopyrite, or both. *Chromite* in two Hagen basalts is high in Cr_2O_3 (46 to 53 percent), and contains 4 to 10 percent MgO and 4 to 11 percent Al_2O_3. Chromites in basalts from Duau, Murray, and Bosavi are lower in Cr_2O_3 (26 to 42 percent), with MgO 3 to 9.3 percent , Al_2O_3 10 to 17 percent, and 1 to 5 percent TiO_2. One analysed (by electron microprobe) grain in a Bosavi basalt is a chromian magnetite-ulvöspinel (TiO_2 20%, Al_2O_3 2.35%, Cr_2O_3 6%, MgO 2.0%). *Magnetite*

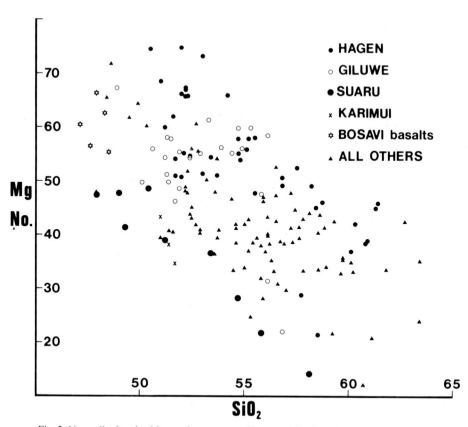

Fig. 3: Normalised molar Mg-number versus wt % SiO2. 'All others' includes Bosavi andesites.

(sensu lato) is also dominant in the low-Si andesites and andesites, but there is an increasing amount of exsolution of ilmenite as the host rocks approach dacitic compositions. The magnetites contain between 4 and 14 percent TiO_2, and always contain less MgO and more Al_2O_3 than the co-existing ilmenite.

CHEMISTRY

A total of 185 samples have been analysed for major and selected trace elements; representative analyses are presented in Table 1.

Major elements

Overall, the major element data show a broad diffuse spectrum of compositions. However, on closer examination, each volcanic centre is seen to have its own chemical characteristics and trends. Some of these are illustrated in Figures 2 to 7.

The plot of MgO against total-Fe (Fig. 2) shows the overall lack of iron enrichment, the low total Fe and high Mg contents of the Hagen and Giluwe rocks relative to the other rocks, and the clear separation of the basalts and andesites of

Bosavi. These, and the following important features, are shown by the Mg-number versus silica plot (Fig. 3)*:

1. The high Mg-numbers of most Hagen and Giluwe rocks, in particular a group of Hagen basaltic rocks, relative to the other rocks.
2. A concentration of analyses of Hagen and Giluwe rocks in the ranges SiO_2 51 to 56 percent and Mg-number 50 to 60.
3. The distinct group of transitional (to alkaline) Bosavi basalts; and the group of Suaru and Karimui rocks with low Mg-numbers.
4. Two Hagen and two Giluwe high-K andesites with very low Mg-numbers: these are high-level intrusive and late-stage cumulodome rocks respectively.

The total-alkalis versus silica plot (Fig. 4) shows the distinct trends, or grouping of analyses, of rocks from most centres (Doma-Ne, Kerewa, and Favenc produce scattered plots), and, again, the distinct 'cluster' of Bosavi basalts. It also shows the relation between total-rock and groundmass compositions for some of the Hagen and Murray rocks. In the case of Mount Hagen, there are two distinct populations, one basaltic (or shoshonitic) the other high-K andesitic, each with a distinct fractionation trend. In the case of Mount Murray, the total-rock/groundmass tie lines are roughly tangential to the total-rock trend, probably reflecting simple crystal fractionation.

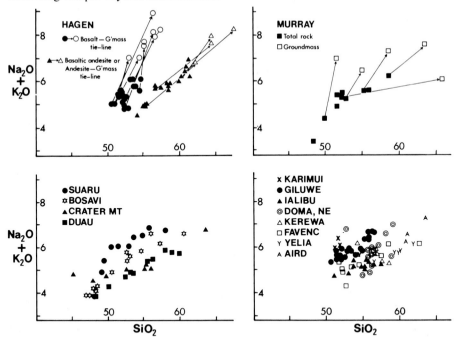

Fig. 4: $Na_2O + K_2O$ versus SiO_2 diagram (wt %) Groundmass compositions (open symbols in top diagrams) were obtained using a non-dispersive solid-state detector on a JEOL microprobe and manually scanning a defocussed beam.

* Mg-number is calculated as molar MgO divided by molar FeO normalised to 80 percent of total molar Fe as FeO. This eliminates the effect of late-stage oxidation and normalises FeO/Fe_2O_3 ratio to a value proposed (Nicholls & Whitford, this volume) as the average value for the mantle.

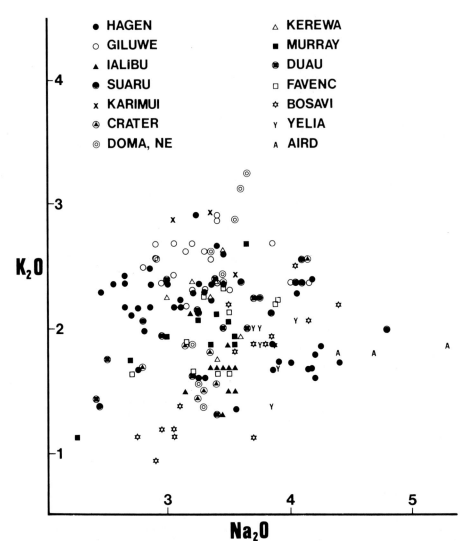

Fig. 5: K2O versus Na2O diagram (wt %).

The K2O versus Na2O and K2O versus SiO2 diagrams (Figs 5, 6) show the high K/Na ratios and K2O contents of the Hagen, Giluwe, Suaru, and Karimui rocks relative to those from the other centres. There is also a tendency in the more basic rocks for K/Na and K2O to decrease with increasing distance from the edge of the Palaeozoic continental plate: values are slightly lower in Murray, Duau, and Favenc than in Hagen, Giluwe, Suaru, and Karimui, and lower still in the Bosavi rocks. The diagrams also show the separation of the basalts and the andesites from Hagen and Bosavi, and the diffuse overlapping fields of the more acid rocks of the other centres.

The TiO2 versus SiO2 plot (Fig 7) shows a broad, roughly linear trend of low

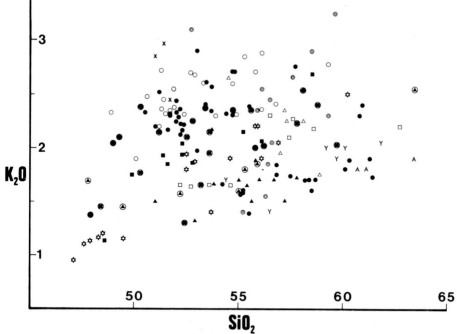

Fig. 6: K2O versus SiO2 diagram (wt %). Symbols as for Fig. 5.

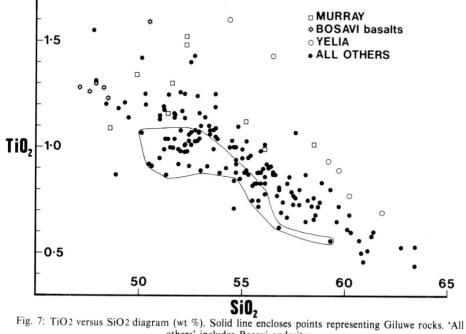

Fig. 7: TiO2 versus SiO2 diagram (wt %). Solid line encloses points representing Giluwe rocks. 'All others' includes Bosavi andesites.

TiO2 values decreasing with increasing SiO2. The rocks of Mounts Murray, Yelia, and Giluwe can be distinguished from the remainder on the basis of higher or lower TiO2. The Bosavi basalts form a distinctive group at the basic end of the trend.

Other variations in major-element chemistry (not illustrated) are:

1. High variable Al2O3 contents, reflecting mainly plagioclase phenocryst content. There is a strong negative correlation between Al2O3 and SiO2 in Hagen and Giluwe rocks in the 51-55 and 50-56 percent silica ranges respectively. In the Hagen rocks, this cuts across a strong overall positive correlation.

2. CaO content shows a strong linear negative correlation with SiO2 , and a narrow range of values for any one value of SiO2 . There is a cross-trend in the same ranges of Hagen and Giluwe rocks that show the Al2O3 cross-trend.

3. Na2O contents increase from 2.5-3.5 percent in the basalts to about 4.5 percent in the andesites in contrast with potassium, which increases only slightly or, in some centres, decreases from basalt to andesite.

4. Moderately high P2O5 contents (0.3-0.7 percent) relative to calcalkaline rocks from other areas (cf. Taylor, 1969).

Trace elements

Content of the 'incompatible' trace elements such as Rb, Sr, and Ba is generally high and correlates with K2O content; this appears to be typical of 'shoshonitic' volcanic rocks described by numerous geologists (e.g. Nicholls & Carmichael, 1969). Rb is highest (55 to 90 ppm) in the high-K olivine basalts and low-Si andesites (shoshonites) of Hagen, and in the samples from Giluwe, Karimui, and Doma. The high-K low-Si andesites and high-K andesites from Mount Hagen are generally significantly lower in Rb than the shoshonitic rocks. Sr shows no systematic variation from centre to centre. Rocks from Suaru, Karimui, and Aird Hills are high in Sr (over 900 ppm) and those from Yelia low (490-580 ppm) relative to their respective silica contents when compared with rocks from the other centres. Barium is lower (250-400 ppm at 53% SiO2) in Murray, Duau, and Favenc, and in the Bosavi basalts than in the other centres (420-820 ppm at 53% SiO2); it is particularly high in rocks from Suaru and Doma (590-920 ppm). Pb, La, and Ce are also high relative to calcalkaline rocks from other regions (e.g. Jakeš & White, 1972), but like Rb, Sr, and Ba, are 'normal' for potassic (shoshonitic) volcanic rocks. Y, Zr, Nb, Pb, La, and Ce are higher in rocks from Doma and Kerewa than in those from any of the other centres.

Contents of V, Ni, and Cr, particularly in some of the basaltic rocks of Hagen, Crater Mountain, Bosavi, Murray, and Duau, are very high relative to other shoshonitic and calcalkaline volcanic rocks (e.g. Gill, 1970; Jakeš & White, 1972). This is at least in part attributable to the high ferromagnesian phenocryst content of these rocks, in particular chromite-bearing olivine.

K/Rb ratio ranges from 225 to 660, with values for most rocks between 225 and 400; the most common values lie in the ranges 225-295 and 325-385. Similar ratios are reported in, for example, high-K andesites from Mount Bagana on Bougainville (Taylor et al., 1969a) and from Peru (Dupuy & Lefèvre, 1974), in the New Zealand andesites (Taylor & White, 1966), and in the Aeolian Islands trachybasalts and trachyandesites (Keller, 1974).

Strontium isotope ratios range from 0.7037 to 0.7059 (Page & Johnson, 1974; R.W. Page, written communication, 1975), within the range for 'modern depleted mantle' (Gill & Compston, 1973). Each volcanic centre has a characteristic narrow range of Sr isotope ratios, with lowest values in Hagen, Giluwe, Suaru, Karimui, and Crater Mountain, higher values in Murray and Doma, and highest values in Bosavi.

DISCUSSION

The data presented above show the central Papua New Guinea volcanoes to be a scattered group, each centre with its own petrological 'stamp', and together covering a broad range of compositions. There are, however, regional chemical differences: shoshonitic and related rocks occur near the northeastern corner of the Australian Palaeozoic continental plate, and just-saturated and slightly undersaturated basaltic and related rocks farther to the south and southwest.

Some rocks from the volcanoes have many features in common with island-arc and continental margin volcanic rocks, such as some of those in the Andes (e.g. Lefèvre, 1973), Fiji (Gill, 1970), and the Mediterranean area (e.g. Keller, 1974), which are generally believed to be related to processes in or above Benioff zones. However, the central Papua New Guinea volcanoes, with only one exception, are underlain by relatively stable old continental crust beneath which there is no intermediate- to deep-level seismicity attributable to a currently active Benioff zone. The one exception is Mount Yelia, which as pointed out previously, is unlike any of the other centres either petrologically or in its geological and tectonic setting. It is situated almost directly above a concentrated pocket of intermediate-depth (175-190 km) seismicity which is probably connected with some form of underthrusting.

Following collision with the Finisterre and Torricelli "blocks" in the late Oligocene – early Miocene (cf. Johnson, this volume) crustal buckling and shortening, culminating in extensive Middle Miocene igneous activity (Page & McDougall, 1970) took place. This was followed by rapid uplift in the Pliocene, immediately preceding initiation of volcanism in the late Pliocene to early Quaternary. Geological evidence (Jenkins, 1974; BMR, 1974; Bain et al., 1975) shows that deformation and uplift were most intense in the Kubor Range area (Fig. 1) and decreased to the west, south, and southeast.

The volumetric abundance of basic lavas, commonly with high Mg-numbers, the presence in some of these of possible high-pressure, Mg-rich chromite-bearing olivine, high Cr, Ni, and V contents (in particular — Taylor et al., 1969b), and low Sr isotope ratios all indicate an ultimate source of the magmas in the mantle. It is therefore speculatively suggested that the Miocene-Pliocene buckling and uplift triggered off partial melting and diapiric rise in the low-velocity layer (cf. Green, 1972, 1973) and that the compositions of the melts derived from these diapirs reflect chemical inhomogeneities in the low velocity layer perhaps caused by migration of incompatible elements into the zone(s) of greatest uplift and disturbance. Further variations in magma composition, such as high-K basalt to high-K low-Si andesite, were probably produced by variations in the amount and depth of partial melting in the diapirs. These variations could in turn have been at least partly controlled by regional variations in the intensity of Miocene-Pliocene lithospheric disturbances. Olivine, pyroxene (probably including or perhaps exclusively orthopyroxene), and possibly also garnet may have been residual phases. These minerals could account for much of the major-element variation and the enrichment in volatile trace elements (cf. Green, 1972; Ringwood, 1974; Nicholls, 1974), but could not produce the enrichment in K over Na. The compositions of the magmas that finally reached the surface reflect the parent compositions, and the amount and type of subsequent fractionation. Fractionation may also be partly controlled by the degree of deformation of the crust, the less deformed and fractured areas affording easier and quicker access to the surface and hence less opportunity for fractionation.

The type and amount of fractionation vary widely from centre to centre, and it is not proposed to deal with this problem in detail: it will be the subject of part of a later, more comprehensive publication. In Mount Hagen, for example, basalts and andesites cannot be related by simple low-pressure fractionation. The main chemical variations (increased Si, Al, and Na, and decreased K, Mg, Fe, Ca, and Ti) from basalt to andesite can best be explained by fractionation at depth involving a relatively potassium-rich sodium-poor amphibole, and olivine, which may be liquidus phases in the basaltic compositions at or near the base of the crust (cf. Nicholls, 1974). There is some evidence, in the form of cross-trends perpendicular to the overall trends of increasing Al and decreasing Ca with increasing Si, of limited plagioclase accumulation in the Hagen rocks. Mount Giluwe is petrologically similar to Mount Hagen, but the volume of andesite is very small. However, if Giluwe and Ialibu are genetically related, as is considered possible, the Giluwe-Ialibu pair is a close parallel to the basalt-andesite association of Hagen. If Ialibu is not related to Giluwe, it is then in the same category as Doma, Ne, and Kerewa and also Aird, where the parent basaltic magma does not appear to have reached the surface and only fractionated magmas were erupted. Crater Mountain, which seems to be out of place among the centres near the Kubor Anticline, and Mounts Murray, Duau, and Favenc all show chemical variation trends compatible with fractionation of a low-K saturated basaltic parent involving residual olivine and clinopyroxene. Mount Bosavi appears to have erupted two separate and distinct batches of magma, one a basic low-K basalt on the borderline between under- and oversaturated, and the other andesitic.

CONCLUSION

The central mainland Papua New Guinea volcanoes are an example of the value of an integrated study of regional geology and tectonics, as well as volcanic petrology. Their study has shown that although a group of volcanoes is situated in an area in which subduction has occurred, they should not automatically be regarded as the direct products of subduction. Not only each group of volcanoes, but each volcano should be treated on its own merits.

ACKNOWLEDGEMENTS

I am greatly indebted to R. W. Johnson, and also to I.P. Sweet and D. S. Hutchison, without whose help in the field much of the data used in this paper could not have been obtained. I am also grateful to I. A. Nicholls for beneficial discussions, to R. Freeman of Australian National University for XRF analyses, and to R. W. Page for strontium isotope ratios. D. H. Blake and A. J. R. White kindly donated specimens from Mount Giluwe. Financial support was provided by an Australian Government postgraduate research award and an Australian Government Public Service Board scholarship. The paper is published with the permission of the Director, Bureau of Mineral Resources.

REFERENCES

Australasian Petroleum Company, 1961: Geological results of petroleum exploration in western Papua 1937-1961: *J. geol. Soc. Aust., 8,* pp. 1-133.

Bain, J.H.C., Mackenzie, D.E., & Ryburn, R.J., 1975: Geology of the Kubor Anticline, central highlands of Papua New Guinea. *Bur. Miner. Resour. Aust. Bull.* 155.

BMR, 1974: Geological map of Papua New Guinea, 1:1,000,000. *Bur. Miner. Resour. Aust.*

Branch, C.D., 1967: Volcanic activity at Mount Yelia, New Guinea; *in* Short papers from the Vulcanological Observatory, Rabaul, New Britain. *Bur. Miner. Resour. Aust. Rep. 107,* pp. 35-39.

Cawthorn, R.G., Ford, C.E., Biggar, G.M., Bravo, M.S., & Clarke, D.B., 1973: Determination of the liquid composition in experimental samples: discrepancies between microprobe analysis and other methods. *Earth Planet. Sci. Letters, 21,* pp. 1-5.

Curtis, J.W., 1973: Plate tectonics and the Papua New Guinea — Solomon Islands region. *J. geol. Soc. Aust, 20,* pp. 21-36.

Denham, D., 1969: Distribution of earthquakes in the New Guinea — Solomon Islands region. *J. geophys. Res., 74,* pp. 4290-4299.

Dent, V.F., 1975: Felt earthquakes from the Southern Highlands, Western, and Gulf Districts of Papua New Guinea, 1953-1973. *Papua New Guinea Geol. Surv. Rep. 74/31* (unpubl.).

Dickinson, W.R., & Hatherton, T., 1967: Andesitic volcanism and seismicity around the Pacific. *Science, 157,* pp. 201-203.

Dow, D.B., Smit, J.A.J., Bain, J.H.C., & Ryburn, R.J., 1974: Geology of the south Sepik region. *Bur. Miner. Resour. Aust. Bull.* 133.

Dupuy, C., & Lefèvre, C., 1974: Fractionnement des éléments en trace Li, Rb, Ba, Sr dans les séries andésitiques et shoshonitiques du Pérou. Comparaison avec d'autres zones orogéniques. *Contr. Mineral. Petrol., 46,* pp. 147-157.

Gill, J.B., 1970: Geochemistry of Viti Levu, Fiji, and its evolution as an island arc. *Contr. Mineral. Petrol., 27,* pp. 179-203.

Gill., J.B., & Compston, W., 1973: Strontium isotopes in island arc volcanic rocks. *in* P.J. Coleman (Ed.): *The Western Pacific: island arcs, marginal seas, geochemistry.* Univ. West. Aust. Press.

Glasse, R.M., 1963: Bingi at Tari. *J. Polynes. Soc. 72,* p. 270.

Green, D.H., 1972: Magmatic activity as the major process in the chemical evolution of the Earth's crust and mantle. *Tectonophysics, 13,* pp. 47-71.

Green, D.H., 1973: Experimental melting studies on a model upper mantle composition at high pressure under water-saturated and water-undersaturated conditions. *Earth Planet. Sci. Letters, 19,* pp. 37-53.

Jakeš, p., & White, A.J.R., 1969: Structure of the Melanesian arcs and correlation with distribution of magma types. *Tectonophysics, 8,* pp. 223-236.

Jakeš, P., & White, A.J.R., 1972: Major and trace element abundances in volcanic rocks of orogenic areas. *Bull. geol. Soc. Amer., 83,* pp. 29-40.

Jenkins, D.A.L., 1974: Detachment tectonics in western Papua New Guinea. *Bull. geol. Soc. Amer., 85,* pp. 533-548.

Johnson, R.W., Mackenzie, D.E., & Smith, I.E., 1971: Seismicity and late Cenozoic volcanism in parts of Papua-New Guinea. *Tectonophysics, 12,* pp. 15-22.

Johnson, T., & Molnar, P., 1972: Focal mechanisms and plate tectonics of the southwest Pacific. *J. geophys. Res., 77,* pp. 5000-5032.

Joplin, G.A., 1968: The shoshonite association: a review. *J. geol. Soc. Aust., 15,* pp. 275-294.

Keller, J., 1974: Petrology of some volcanic rock series of the Aeolian arc, southern Tyrrhenian Sea: calc-alkaline and shoshonitic associations. *Contr. Mineral. Petrol., 46,* pp. 29-47.

Lefèvre, C., 1973: Les charactères magmatiques du volcanisme plio-quaternaire des Andes dans le Sud du Pérou. *Contr. Mineral. Petrol., 41,* 259-271.

Mackenzie, D.E., 1975: Volcanic and plate tectonic evolution of central Papua New Guinea. *Bull. Aust. Assoc. Expl. Geophys., 6,* pp. 66-68.

Mackenzie, D.E. & Chappell, B.W., 1972: Shoshonitic and calc-alkaline lavas from the highlands of Papua New Guinea. *Contr. Mineral. Petrol., 35,* pp. 50-62.

Nicholls, I.A., 1974: Liquids in equilibrium with peridotitic mineral assemblages at high water pressures. *Contr. Mineral. Petrol., 45,* pp. 289-316.

Nicholls, J., & Carmichael, I.S.E., 1969: A commentary on the absarokite-shoshonite-banakite series of Wyoming, U.S.A. *Schweize Mineral. Petrog. Mitt., 49,* pp. 47-64.

Ollier, C.D., & Mackenzie, D.E., 1974: Subaerial erosion of volcanic cones in the tropics. *J. trop. Geog., 39,* pp. 63-71.

Page, R.W., 1975: The geochronology of igneous and metamorphic rocks in the New Guinea region. *Bur. Miner. Resour. Aust. Bull.* 162 (in press).

Page, R.W. & Johnson, R.W., 1974: Strontium isotope ratios of Quaternary volcanic rocks from Papua New Guinea. *Lithos, 7,* pp. 91-100.

Page, R.W., & McDougall, I., 1970: Potassium-argon dating of the Tertiary f1-2 stage in New Guinea and its bearing on the geological time-scale. *Amer. J. Sci., 269,* pp. 321-342.

Ringwood, A.E., 1974: Petrological evolution of island arc systems. *J. geol. Soc. London, 130,* pp. 183-204.

Roeder, P.L., & Emslie, R.F., 1970: Olivine-liquid equilibrium. *Contr. Mineral. Petrol., 29,* pp. 275-289.

Simkin, T., & Smith, J.V., 1970: Minor-element distribution in olivine. *J. Geol., 78,* pp. 304-325.

St John, V.P., 1970: The gravity field and structure of Papua and New Guinea. *APEA J. 10,* pp. 41-55.

Stormer, J.C., 1973: Calcium zoning in olivine and its relationship to silica activity and pressure. *Geochim. Cosmochim. Acta, 37,* pp. 1815-1821.

Taylor, G.A.M., 1971: An investigation of volcanic activity at Doma Peaks. *Bur. Miner. Resour. Aust. Rec. 1971/137* (unpubl.).

Taylor, S.R., 1969: Trace element chemistry of andesites and associated calc-alkaline rocks. *in* A.R. McBirney (Ed): *Proceedings of the Andesite Conference,* Dept. Geol. Miner. Resour. Oregon Bull., 65, pp. 43-64.

Taylor. S.R., Capp, A.C., Graham, A.L., & Blake, D.H., 1969a: Trace element abundances in andesites. II: Saipan, Bougainville and Fiji. *Contr. Mineral. Petrol., 23,* pp. 1-26.

Taylor, S.R., Kaye, M., White, A.J.R., Duncan, A.R., & Ewart, A., 1969b: Genetic significance of Co, Cr, Ni, Sc, and V in andesites. *Geochim. Cosmochim, Acta., 33,* pp. 275-286.

Williams, P.W., McDougall, I., & Powell, J.M., 1972: Aspects of the Quaternary geology of the Tari-Koroba area, Papua. *J. geol. Soc. Aust., 18,* pp. 333-347.

LATE QUATERNARY TEPHRAS AROUND MOUNT HAGEN AND MOUNT GILUWE, PAPUA NEW GUINEA

C.F. PAIN and R.J. BLONG

Department of Geography, University of Papua New Guinea, Box 4820, University Post Office, Papua New Guinea

School of Earth Sciences, Macquarie University, Sydney, N.S.W. 2113

ABSTRACT

Nine tephra formations of Late Quaternary age in the area surrounding Mount Hagen and Mount Giluwe, Papua New Guinea, were mapped using marker beds, including palaeosols and distinctive tephra beds. The nine tephras are named and their characteristics and known distribution presented. Other undifferentiated tephra deposits are noted. One tephra formation, Bune Tephra, erupted from south of Mount Giluwe. Most of the others appear to have erupted from sources on the Mount Hagen Range. The youngest, Tomba Tephra, is more than 30,000 years old, and on the basis of isopachs can probably be attributed to a source within the central caldera complex of the Mount Hagen volcano. The antiquity of the tephras suggests that major volcanic activity on Mount Hagen and Mount Giluwe ceased more than 30,000 years ago.

INTRODUCTION

Deposits of volcanic ash, or tephra beds, were recognised and reported by the first geologists to work on cover beds in the Papua New Guinea highlands (e.g., Perry, 1965; Bik, 1967). More recently, we have identified numerous tephras in the course of investigations into the Quaternary histories of the Kaugel Valley (CFP) and the Wahgi Valley (RJB). In an attempt to correlate events in the two valleys, the value of the tephras as marker beds was recognised. Accordingly, joint fieldwork was undertaken to identify and map tephras occurring in the general area between and surrounding the two valleys.

Tephra is a convenient term that may be used to refer to all airborne pyroclastic material including both tephra-fall and tephra-flow deposits (Thorarinsson, 1974). The term is used here to refer to all unconsolidated and airborne pyroclastic deposits irrespective of the size of the material, which ranges up to block size in rare cases, and irrespective of degree of weathering.

Tephras in the study area fall into two groups. One group consists of thin tephra beds (less than about 0.1 m) occurring within peat throughout the area; these tephras, derived from outside the study area, will be described elsewhere by RJB. The other group consists of tephras sufficiently thick to retain their identity within the cover beds around Mount Hagen and Mount Giluwe. This report provides a summary of the upper part of the column of 'thick' tephras, and considers characteristics, correlations, and chronology of the tephras. Possible sources are also considered, but our main aim is to present an account of the tephras that will be useful to workers wishing to use them as marker beds in geomorphology and related sciences (see Pullar, 1973). For this reason our descriptions are based on field data, since the usefulness of marker beds is considerably lessened if laboratory analyses are necessary for identification.

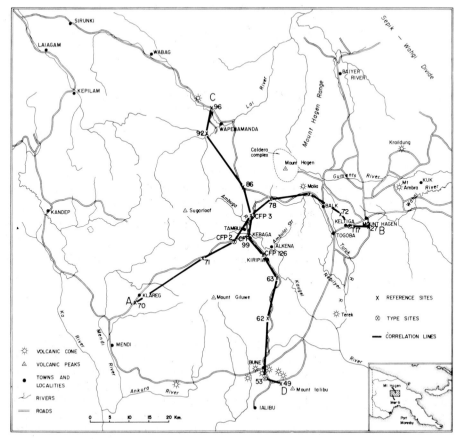

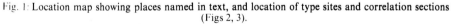

Fig. 1: Location map showing places named in text, and location of type sites and correlation sections (Figs 2, 3).

AREA AND MAPPING PROCEDURES

Mapping was carried out in the area surrounding Mount Hagen and Mount Giluwe and west to Laiagam and Kandep, south to Ialibu, and east to include the upper part of the Wahgi Valley (Fig. 1). Within this area the most detailed preliminary mapping was carried out in the Kaugel Valley (Pain, 1973). From this 'core' area we have undertaken mapping within the broader area shown in Figure 1.

The procedures adopted during mapping are similar to those used in New Zealand by Vucetich & Pullar (1964, 1969). Within the tephra column, distinctive marker beds, which may be tephra units or palaeosols, are first identified, and then interbedded units may be identified by their stratigraphic position. Ideally each tephra unit should consist of a basal part composed of little-weathered material and an upper part where soil formation took place before the following tephra was deposited leaving a buried palaeosol. In the study area, however, most of the tephras are weathered throughout their thickness, and post-burial change appears to have destroyed most of the buried palaeosols that presumably occurred between many of the units in the tephra column.

Table 1. Summary of tephra names, type site locations and distribution

Tephra name and symbol	Name derivation (Fig. 1)	Type site location (Fig. 1)	Distribution and source
Tomba Tephra (tm)	Tomba Village (Ramu, SB55-5/AP6956)	CFP 2 on Tambul Mendi Rd, 8 km west of Tambul (Wabag, SB54-8/ZU2245)	Widespread, found over whole of study area. Erupted from Mount Hagen.
Bune Tephra (bn)	Bune locality (Karamui, SB55-9/AP7918)	53, 2.5 km west of Bune (Lake Kutubu, SB54-12/ZU3118	Surrounds southern half of Mount Giluwe. Probably erupted from a source on the southern side of Giluwe.
Kiripia Tephra (ki)	Kiripia Catholic Mission (Wabag, SB54-8/ZU3140)	CFP2	Kaugel Valley and western and southern slopes of Mount Hagen. Source may be on Mount Hagen.
Kebaga Tephra (kb)	Kebaga Village (Wabag, SB54-8/ZU2945)	CFP2	Kaugel Valley, northern slopes of Mount Giluwe, and western and southern slopes of Mount Hagen. Source may be on Mount Hagen.
Balk Tephra (ba)	Balk locality (Ramu, SB55-5/AP8253)	T17, at Keltiga, 6 km west of Mount Hagen town, on the Highlands Highway. (Ramu, SB55-5/AP8749)	West and south of Mount Hagen township. May be found to correlate with Kiripia or Kebaga Tephras. Source unknown.
Ambulai Tephra (am)	Ambulai Stream (Ramu, SB55-5/AP6945)	T17	Kaugel Valley, northern slopes of Mount Giluwe, and southern and southeastern slopes of Mount Hagen. May have erupted from Mount Hagen.
Wanabuga Tephra (wn)	Wanabuga Stream (Wabag, SB54-8, SE part of sheet)	T17	Kaugel Valley, northern slopes of Mount Giluwe, southern and southeastern slopes of Mount Hagen. May have erupted from Mount Hagen.
Turuk Tephra (tk)	Turuk River (Ramu, SB55-5, SW part of sheet)	T17	A few exposures in the Kaugel Valley and southeast of Mount Hagen. Source unknown.
Togoba Tephra (tg)	Togoba settlement (Ramu, SB55-5/AP8447)	T17	A few exposures in the Kaugel Valley and southeast of Mount Hagen. Source unknown.

undifferentiated tephras

Note: Grid references are from the 1:250,000 topographic map series. An additonal reference for T17 is: 1:50,000 Mount Hagen (Special) sheet, AP873 495.

The nature of tephra deposition is such that tephra units can exhibit considerable lateral variation. This variation manifests itself both in initial depositional characteristics following the eruption, and in subsequent weathering changes. Generally, although there are exceptions, tephras decrease in thickness, grainsize, and number of shower beds away from the source. Differential weathering potentially follows from these decreases, and from the variety of weathering environments into which the tephra may be deposited. These variations are clearly discussed by Vucetich & Pullar (1964).

Problems arising from the variations in tephra characteristics can be overcome in the field if sufficient exposures are available. Sites examined should be sufficiently close together to allow recognition of additional tephra units as they enter the column, and loss of units as they lens out. In practice this meant studying as many sites as possible in systematic traverses across the study area (the 'hand-over-hand' mapping of Pullar, 1967). In this way variations in field characteristics could be recognised as they occurred. This method of working is necessary because the same tephra unit may exhibit quite different field characteristics at different points of its range.

In general, mapping was moderately successful, especially in the Kaugel Valley and on the western and southern slopes of Mount Hagen. However, uncertainties arise from a lack of identifiable marker beds in some localities, and long distances between suitable exposures in some parts of the study area. Because of a wide range of weathering and erosion environments, marker beds that are distinct in some places cannot be recognised in others. Moreover, lack of exposures means that coverage of the area is somewhat patchy. Our investigations were limited mainly to road cutting and stream bank exposures, there being few other kinds of exposure in the study area.

Nine tephra formations are named and described in this paper (Table 1). Some of these formations have only local distribution, but most were recognised over much of the study area. As there are few deep exposures, and thickness measurements on deeper tephras are lacking or do not show regular variation, only the uppermost tephra unit is shown on an isopach map.

Older tephras underlie those described here but these are exposed only rarely, and in no case could individual units be correlated between two sites. Many of these older tephras have only local distribution. Other deposits underlying and interbedded with the nine tephras described here include lahar deposits, and occasional agglomerates and colluvial deposits.

DESCRIPTIONS OF TEPHRA FORMATIONS

Locations of places from which names are derived, together with type site locations, are given in Table 1, and shown in Figure 1.

Undifferentiated tephras

At two sites, T17 at Keltiga and CFP 2 near Tambul, undifferentiated tephras are exposed at the base of the tephra column (Fig. 2). At Tambul these tephras rest on lavas from Mount Giluwe, while at Keltiga they rest on a palaeosol formed on a lahar deposit. At each site individual beds may be distinguished, but no correlations are possible on the available data.

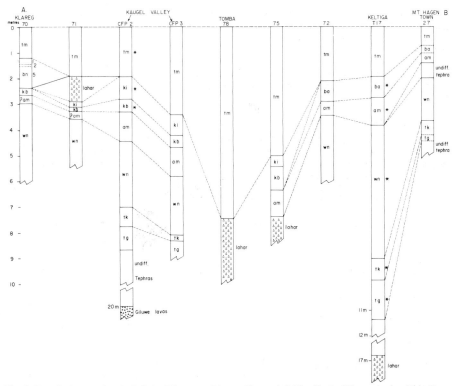

Fig. 2: Correlation section A-B from Klareg to Mount Hagen (cf. Fig. 1). At Klareg, beds within Bune Tephra are numbered. Asterisks beside tephra formations mark location of type sites for those tephras. For tephra symbols refer to Table 1. Note the decrease in thickness of Tomba Tephra both east and west of Tomba (site 78).

Togoba Tephra (tg)

At Keltiga (T17), the type site, Togoba Tephra is highly weathered and consists of 1.5 m of massive yellow-brown to dark yellow-brown clay with numerous grey clay inclusions. It has a firm consistency, and becomes sticky when worked. In the Kaugel Valley this unit is similar in appearance, but is more compacted and there is a slightly coarser layer at its base. This basal layer may be a shower bed, significantly coarser than the rest of the tephra when it was deposited but now weathered to a similar texture.

Togoba Tephra rests on older tephra beds in both the Kaugel Valley and on the southeastern slopes of Mount Hagen. The exposure at site 27, south of Hagen, is much thinner than elsewhere (Fig. 2), suggesting that it may thicken westwards from there. So far, however, it has only been seen in a few deep exposures, insufficient in number to enable any estimates of distribution or source.

Turuk Tephra (tk)

At T17 Turuk Tephra consists of 0.86 m of reddish brown firm massive clay. It contains a few grey clay inclusions, and many dark manganese-rich veins. In the Kaugel Valley this tephra unit is dark yellow-brown without the grey clay inclusions, but containing black manganiferous inclusions, both as nodules and in channels, similar to those at T17. The main criteria for its identification are the manganese-rich inclusions and its stratigraphic position immediately underlying Wanabuga Tephra.

The lower boundary of Turuk Tephra is diffuse at T17, and also irregular at CFP 2. It is generally darker than the underlying Togoba Tephra and may be recognised on this criterion. At T17 the contact between reddish brown Turuk Tephra and the markedly contrasting grey sandy Wanabuga Tephra is sharp, mammillated, and highlighted by the strong development of hard black and white mottled nodules. In other exposures the sharp upper boundary with its distinct colour change persists but is not so marked.

Turuk Tephra is exposed at a few sites in the Kaugel Valley and in the area southeast of Mount Hagen. There are no consistent changes in its thickness, which might indicate its sources, over the area studied.

Wanabuga Tephra (wn)

Wanabuga Tephra at Keltiga (T17) is very distinctive. It is 5.20 m thick and consists of four beds. Each is massive and sandy (their dominant characteristics) and generally grey, with inclusions of friable yellowish clay, and white to pink halloysite and manganese nodules. The clay inclusions are concentrated along cracks which extend downwards from the upper boundaries of the sandy beds, while the halloysite and manganese nodules are concentrated along these boundaries (Fig. 3). These characteristics make the Wanabuga Tephra formation a very distinctive unit at T17.

Elsewhere Wanabuga Tephra is also distinctive. At site 27, east of Keltiga and south of Mount Hagen town (Fig. 1), it is 1.70 m thick and comprises two beds. The upper bed retains the grey sandy features of the tephra at T17, and protrudes from the road cutting as a distinct band. The lower bed is more friable and is similar to the central beds at T17. At CFP 2, in the Kaugel Valley, Wanabuga Tephra appears to be more stongly weathered than at Keltiga, and has a finer texture and an orange-brown colour. However, it may be recognised by inclusions of grey silty sand similar to the deposits at Keltiga and site 27. These inclusions, which have the appearance of 'corestone areas' of little weathered tephra, range from a few centimetres to more than a metre in diameter. Farther west on the northern slopes of Mount Giluwe at sites 70 and 71, Wanabuga Tephra retains its grey sandy nature, the inclusions of friable clay, and the halloysite and manganese nodules which are associated with the contacts between the different beds.

Because of its distinctive character and sharp contact with the underlying Turuk Tephra, Wanabuga Tephra was used as a marker bed in mapping tephra formations around Mounts Hagen and Giluwe. Its upper boundary is marked by a blocky palaeosol which shows clearly as a dark band at Keltiga (Fig. 3) and on the northern slopes of Giluwe. At CFP 2, where the road cutting is older and has weathered, the palaeosol also has a fretted appearance where individual soil blocks have fallen out. The contact between Wanabuga Tephra and overlying tephra beds (generally Ambulai Tephra) is sharp and easily recognised by the contrast between the sandy grey Wanabuga Tephra and the yellowish clay of the overlying material.

Wanabuga Tephra has been found on the northern slopes of Mount Giluwe, in the Kaugel Valley, and on the southern and southeastern slopes of Mount Hagen, as far east as Mount Hagen town. Its distribution and thickness (Fig. 2) indicate that it may have erupted from the southern part of the Mount Hagen Range.

Ambulai Tephra (am)

At the type site (T17) Ambulai Tephra consists of 1.10 m of yellow-brown very friable silty clay with abundant pinkish-brown veins of silty clay material infilling channels and cracks. The unit has a prismatic structure which breaks down to medium to fine (5-20 mm) blocks in the upper part. Within the Kaugel Valley the formation is still very friable but contains fine white clay inclusions instead of the pinkish-brown silty clay, and exhibits a coarse basal layer about 0.1 m thick. These characteristics persist to the most westerly of the exposures examined. East of the type site the formation thins out rapidly and becomes less distinct. However, it may still be recognised at site 27 (Fig. 2). In most places its very friable nature and its position immediately overlying Wanabuga Tephra allow it to be readily identified.

The lower boundary of Ambulai Tephra, where it rests on Wanabuga Tephra, has been described above; it is quite distinct. The upper boundary of the formation is marked by a blocky palaeosol. At Keltiga this palaeosol is weakly developed (Fig. 3), but in the Kaugel Valley the palaeosol is distinct and is the uppermost of the only two palaeosols readily identified in the tephra column.

Fig. 3: Tephras exposed at Keltiga (site T17). Note inclusions in Wanabuga Tephra (wn), its sharp lower contact with Turuk Tephra (tk), and the blocky palaeosol underlying Ambulai Tephra (am). tm = Tomba Tephra. ba = Balk Tephra.

Ambulai Tephra is found on the northern slopes of Mount Giluwe, in the Kaugel Valley, and on the southern and southeastern slopes of Mount Hagen as far east as Mount Hagen town (Figs 2 and 4). Like Wanabuga Tephra, it may have erupted from Mount Hagen.

Balk Tephra (ba)

At T17, the type site, 0.8 m of bright yellow-brown clay with a coarse blocky structure is exposed; the clay is firm in the section but on working becomes very friable. Balk Tephra may be recognised easily by these characteristics over its restricted distribution. It rests on the palaeosol formed on Ambulai Tephra and so may also be identified from its stratigraphic position. The upper boundary, with

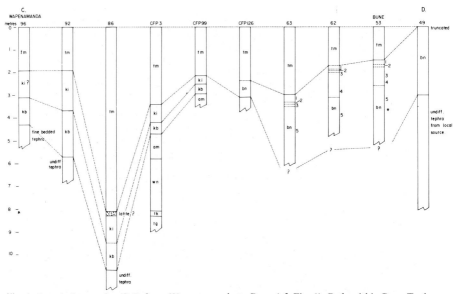

Fig. 4: Correlation section C-D from Wapenamanda to Bune (cf. Fig. 1). Beds within Bune Tephra are numbered. The asterisk indicates the type site for Bune Tephra (site 53). For tephra symbols refer to Table 1. Note the decrease in thickness of Tomba Tephra both north and south of site 86.

Tomba Tephra, another distinctive marker bed, may also be used to identify Balk Tephra. Both upper and lower boundaries are distinct to sharp.

Balk Tephra is locally distributed south and west of Mount Hagen town. As it occurs in the same stratigraphic position as Kebaga and Kiripia Tephras, it may on further work be found to correlate with either or both of these units. At present, however, it is regarded as a distinct unit. Its source is unknown.

Kebaga Tephra (kb)

At the type site (CFP 2) Kebaga Tephra consists of 0.5 m of light brown massive sandy clay, sticky when wet, with abundant fine black manganese-rich veins and a discontinuous sandy basal layer. At sites 96 and 92 (Fig. 4) the unit consists of light brown sandy tephra with numerous graded shower beds.

Kebaga Tephra may be recognised by its light brown colour which distinguishes it from the darker tephras which bracket it in the tephra column, and from its stratigraphic position as the second unit up from Wanabuga Tephra where the latter unit is exposed.

Kebaga Tephra rests conformably on Ambulai Tephra in the Kaugel Valley and on the southern slopes of Mount Hagen. On the western slopes of Mount Hagen it rests on undifferentiated tephras; here the contact may be conformable, or it may be an erosional disconformity. Its upper boundary is almost everywhere conformable with Kiripia Tephra, which is missing only at the far western edges of the mapped distribution of Kiripia Tephra. Near Mendi, Kebaga Tephra is overlain conformably by Bune Tephra (Figs 2 and 4).

Kebaga Tephra occurs on the northern slopes of Mount Giluwe, in the Kaugel Valley, and on the western and southern slopes of Mount Hagen. The thickest exposures (up to 2.50 m) are on the western slopes of Mount Hagen, suggesting a source in that vicinity.

Kiripia Tephra (ki)

At the type site (CFP 2) Kiripia Tephra consists of 0.9 m of massive yellow brown, sticky to slightly friable, sandy clay (designated KAU 7 in Pain, 1973). On the western slopes of Mount Hagen it is composed of up to 2 m of yellow-brown sandy tephra with numerous shower beds. The main internal criterion for its recognition is the abundance of white clay nodules throughout the unit. However, it is most readily recognised by its stratigraphic position immediately underlying the easily recognisable Tomba and Bune Tephras.

Kiripia Tephra has conformable contacts with both underlying and overlying units. At all exposures examined it is underlain by Kebaga Tephra, and at all but one site is overlain by Tomba Tephra. At site 71 (Fig. 2) it is overlain conformably by a lahar deposit which is in turn overlain by Tomba Tephra. Its stratigraphic position is thus clear.

This tephra is found on the western and southern slopes of Mount Hagen and in the Kaugel Valley. It is thickest near Mount Hagen and thins rapidly west from the Kaugel Valley (Figs. 2 and 4) and is therefore thought to have erupted from somewhere on the Mount Hagen Range.

Bune Tephra (bn)

The following is a description of Bune Tephra at the type site:

		.145 cm	Tomba Tephra
	Bed 1	30 cm	greenish-grey loamy sand
Bune	Bed 2	5 cm	reddish clay
Tephra	Bed 3	51 cm	yellow-brown and bluish-grey graded shower beds
	Bed 4	16 cm	grey olive fine sandy and clayey beds
	Bed 5	240 cm+	grey olive to white sandy clay massive tephra

base of road cutting

At other sections where the base of Bune Tephra is exposed, bed 5 in the above description is seen to be the lowermost. Beds 2, 3, and 4 are distinctive and form useful marker beds for correlation of Bune Tephra away from the type locality. This is particularly useful where Bune Tephra and Tomba Tephra are both present, since they have similar characteristics. Bed 2 is more persistent than the others and is the best marker bed; it is present in nearly every exposure of Bune Tephra examined. Thus the presence of one or more of beds 2, 3, and 4 is the main criterion for the recognition of Bune Tephra.

Where the lower boundary of Bune Tephra is exposed it rests on undifferentiated deposits including tephra and lahar materials; some contacts are conformable while others are disconformable. The upper contact with Tomba Tephra is conformable in all areas. Where these tephras are similar the upper boundary of Bune Tephra is defined as the contact between bed 1 and the coarse sandy basal layer of the overlying Tomba Tephra.

Bune Tephra surrounds the southern half of Mount Giluwe, from site 70 north of Mendi to site 63 south of Kiripia (Figs 1, 2, 4, 5). It has also been tentatively recognised at other sites in the Kaugel Valley (e.g. CFP 126, Figs 1 & 4). Because of this distribution, and because the thickest exposures and coarsest beds are found on the southern slopes of Mount Giluwe, it is thought to have erupted from a source in that area.

Tomba Tephra (tm)

At the type site (CFP 2) Tomba Tephra is 2.0 m thick and consists of two beds, both yellow-brown loamy sand, which contain numerous sand grains giving the unit a porphyritic appearance. The upper bed has a coarse sandy basal layer. Elsewhere the main criteria for recognition are: (1) the grey olive to light yellow-brown colour; (2) the sandy clay to loamy sand texture and porphyritic appearance; (3) the presence of two or more beds with coarse basal layers; and (4) the presence of distinctive clay channel-filling veins that are the result of post-depositional pedological alteration. The sand fraction is dominated by feldspars; heavy minerals are 90% green hornblende with minor amounts of biotite, zircon, and apatite. Allophane in the matrix and gibbsite in the channel fillings and nodules are the main clay minerals (Parfitt, 1975).

The above characteristics describe Tomba Tephra adequately over most of its range. As noted the number of beds is greater where the formation is thickest; 7 members occur at Tomba (site 78) and 8 at site 86, north of Tomba (Figs 1, 4). The basal layers of various beds range from lapilli to blocks at different sites. At T17, where Tomba Tephra is farther away from source, it consists of 0.90 m of light

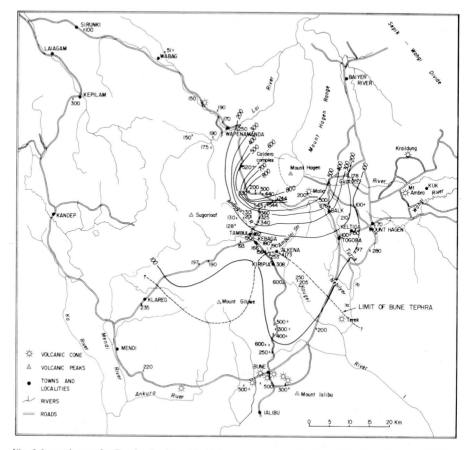

Fig. 5: Isopach map for Tomba Tephra. All thicknesses are in centimetres. Note the influence of Mount Giluwe on the shape of the isopachs, and the northern boundary of positively identified Bune Tephra. Thicknesses south of this boundary are for Bune and Tomba Tephras combined.

brown to yellow-brown sandy clay containing inclusions of greyish sandy material with a porphyritic appearance. It appears to be more strongly weathered at this site than elsewhere.

Tomba Tephra conformably overlies older tephra beds at many sites; where it does it exhibits mantle bedding. On valley floors (e.g. Kaugel Valley) it disconformably overlies lake beds or colluvial and alluvial deposits. In most places the base is defined by marked changes in colour and texture. Bune Tephra is distinguished from it by the presence of the reddish clay bed 2. In most exposures Tomba Tephra is the uppermost formation and occurs at the present landsurface. On fans, footslopes, and the valley floor of the Kaugel Valley it may be overlain by the Kaugel Diamicton (Pain, 1975). Elsewhere in similar situations it may be overlain by other colluvial deposits.

Tomba Tephra has been mapped over more than 3000 km² surrounding Mount Giluwe and west and south of Mount Hagen. It has also been tentatively recognised in the Margarima, Nembi, Wage, and Kandep valleys to the west, and also at Kuk, east of Mount Hagen. It is thickest near Tomba, and thins out with distance from this locality (Fig. 5). The number of beds is also greatest, and the unit coarsest, near Tomba (see Pain, 1973). These features, together with the isopachs (Fig. 5), indicate a source on the southern part of Mount Hagen, the most likely area being the caldera complex discussed by Mackenzie (1973).

DISTRIBUTION AND CORRELATION

The distributions of the various tephra formations are not known in their entirety, largely because of the lack of suitable exposures over large parts of the study area. However, other factors also play a part. East of Mount Hagen town and south of Togoba deep exposures of tephra were examined, but none of the tephras could be correlated with those described in this paper. More detailed field examination of sites close together may lead to such correlations, but additional work is required on trace element compositions of the kind carried out by Westgate and his colleagues in Canada (e.g. Smith & Westgate, 1969) and Kohn (1970) in New Zealand.

Detailed investigations of trace element compositions of Bune and Tomba Tephras may also help resolve the problem of distinguishing them in the areas where they overlap. From the field evidence it is clear that they are very similar in appearance, and that Bune Tephra appears to underlie Tomba Tephra; at least the distinctive marker beds within the former formation lie below the base of Tomba Tephra. However, part of the upper member of Bune Tephra may be interbedded with Tomba Tephra, and for tephrochronological purposes they may be regarded as being coeval.

Examination of the isopach map (Fig. 5) indicates that distributions of Tomba and Bune Tephras are influenced by the presence of Mount Giluwe. A lobe of Tomba Tephra extends down the Kaugel Valley and around the eastern side of Mount Giluwe towards Bune. The few sites examined in the northwest of the study area suggest that the major lobe of Tomba Tephra passed north of Mount Giluwe. Tomba Tephra has not been identified between Bune and Mendi on the south and southwest of Giluwe, suggesting that the mountain exercised a 'shadow' effect on the deposition of the tephra. Similarly, Bune Tephra does not occur on the northern slopes of Mount Giluwe (Fig. 5).

SOURCES

Only Tomba Tephra can definitely be assigned a source area, on the basis of the isopach map (Fig. 5) and an increasing size of material and number of beds. Bune Tephra was almost certainly erupted from a source south of the summit area of Mount Giluwe, while Wanabuga, Ambulai, Kebaga, and Kiripia Tephras probably came from sources on the Mount Hagen Range. Togoba, Turuk, and Balk Tephras give no indication of their likely sources.

Eruptive sources that were probably active during the period when the tephras were deposited include: the three major volcanic centres of Mounts Hagen, Giluwe, and Ialibu, as well as the Sugarloaf Plateau, and minor centres (in terms of size of cone) at Malia, on the southern slopes of Mount Hagen; Terek, near the confluence of the Kaugel and Nebilyer Rivers; Kraildung, north of Mount Hagen town; one in the Wabag-Wapenamanda Valley; and cones along a fault which runs across the southern footslopes of Mount Giluwe (Fig. 1).

The minor cones all have local tephra deposits associated with them; these are left undifferentiated in this report. The presence of these local deposits and the small size of the cones does not preclude the possibility that some of them produced tephras with a regional distribution. So far, however, no such tephras have been distinguished. The major tephras with regional distribution seem to have erupted from major volcanic centres, which in practice means Mount Hagen, since so far only one tephra can be attributed to Giluwe and none to Ialibu. Malia, a small parasitic cone on the southeastern slopes of Mount Hagen, seems likely as a source for some tephra deposits, but no definite formation can be attributed to this source.*

CHRONOLOGY OF ERUPTIVE EVENTS

At present there are few reliable absolute dates on Quaternary volcanic materials in the highlands apart from ^{14}C dates for young tephras found in swamps in the Gumants Valley (Blong, in prep.). Five ^{14}C dates relating to Tomba Tephra are reported in Pain (1973);

ANU	806	28,410	± 1500	yr B.P.
ANU	753	34,530	± 2360	yr B.P.
ANU	194	31,470	± 900	yr B.P.
ANU	752	31,900	± 1300	yr B.P.
ANU	254	34,000	± 1500	yr B.P.

When considered with a range of two standard deviations these dates cannot be separated statistically and thus an age of between 30,000 and 31,000 years B.P. can be assumed on this evidence for Tomba Tephra. However, a recently collected sample has given a dated of greater than 50,000 years B.P. for cellulose from wood collected at Kuk. This, together with doubts about the accuracy of the original dates because of contamination of samples with younger material before collection, suggests that Tomba Tephra may well be older than 50,000 years B.P.

The dominance of Mount Hagen in terms of tephra production appears at first sight to be surprising, since it is generally though to be the oldest of the Quaternary volcanoes in the highlands area on the basis of degree of dissection (see, e.g. Bain et al., 1970). However, the few available radiometric dates do not support this age relationship. Page & Johnson (1974) report K-Ar ages of 204,000 and 218,000 ± 10,000 years B.P. for specimens from Mount Hagen, whereas the only date so far available from Mount Giluwe is about 300,000 B.P. (Dr E. Löffler, pers. comm.).

Tephrochronological evidence presented here indicates that the last paroxysmal eruptions of tephra in the area were from Mount Hagen (Tomba Tephra) and Mount Giluwe (Bune Tephra), and that sources on the two stratovolcanoes were in eruption at approximately the same time prior to 50,000 years ago.

CONCLUSIONS

Of the more than 15 tephra formations indentified throughout the study area, only nine are sufficiently well exposed to allow them to be characterised and named. Of these nine, Tomba Tephra is the youngest, with an age greater than 30,000 years B.P. and probably greater than 50,000 years B.P. This formation was produced by Mount Hagen, as were at least four of the other named tephras. Of the

*For a distribution map of Late Cainozoic eruptive centres in central Papua New Guinea but outside of this area of study, see Fig. 1 in Mackenzie (this volume). Ed.

remaining four tephra formations, three come from unknown sources and the fourth, Bune Tephra, comes from the southern side of Mount Giluwe.

The tephra formations discussed in this paper may be identified on the basis of various kinds of criteria. Internal characteristics are important in some cases, particularly with regard to the important marker beds, Wanabuga Tephra, various beds of Bune Tephra, and Tomba Tephra. Other tephra formations are more easily recognised by their stratigraphic position in relation to distinctive marker beds.

The ages of the named tephras, ranging from greater than 30,000 years B.P. to perhaps several hundred thousand years B.P., indicate the considerable antiquity of the last stages of volcanism of Mount Hagen and Mount Giluwe. This paper provides a basis for further work on tephrostratigraphy and tephrochronology in the area; when undertaken, such work will help to elucidate the Quaternary geomorphic history of the Papua New Guinea highlands.

ACKNOWLEDGEMENTS

Field work was carried out while CFP was in receipt of an Australian National University Scholarship; this support is gratefully acknowledged. Financial support for RJB was provided by the Wahgi Project (J. Golson, ANU), Macquarie University, and the Myer Foundation. E. Löffler and D.E. Mackenzie read the manuscript and made many helpful comments.

REFERENCES

Bain, J.H.C., Mackenzie, D.E. & Ryburn, R.J., 1970: Geology of the Kubor Anticline — central highlands of New Guinea. *Bur. Miner. Resour. Aust. Rec.* 1970/79. (unpubl.).

Bik, M.J.J., 1967: Structural geomorphology and morphoclimate zonation in the central highlands, Australian New Guinea, *in* J.N. Jennings & J.A. Mabbutt (Eds.) *Landform Studies from Australia to New Guinea,* Canberra, A.N.U. Press, pp. 26-47.

Kohn, B.P., 1970: Identification of New Zealand tephra layers by emission spectrographic analysis of their titanomagnetities. *Lithos, 3,* pp. 361-8.

Mackenzie, D.E., 1973: Quaternary volcanics of the central and southern highlands of Papua New Guinea. *Bur. Miner. Resour. Aust. Rec.* 1973/89 (unpubl.).

Page, R.W. & Johnson, R.W., 1974: Sr isotopes of Quaternary volcanic rocks from Papua New Guinea. *Lithos, 7,* pp. 91-100.

Pain, C.F., 1973: The Later Quaternary geomorphic history of the Kaugel Valley, Papua New Guinea. *Unpublished PhD Thesis, Australian National University, Canberra,* 226 pp.

Pain, C.F., 1975: The Kaugel Diamicton — a Late Quaternary mudflow deposit in the Kaugel Valley, Papua New Guinea. *Z. Geomorph. 19,* pp. 430-442.

Parfitt, R.L. (1975): Clay minerals in recent volcanic ash soils from Papua New Guinea. *Quat. Studies N.Z. Roy. Soc. Bull. 13,* pp. 241-245.

Perry, R.A., 1965: Outline of the geology and geomorphology of the Wabag-Tari area. *in* R.A. Perry et al. *General report on lands of the Wabag-Tari area, Territory of Papua and New Guinea, 1960-1961. CSIRO Land Res. Ser., 15,* pp. 70-84.

Pullar, W.A., 1967: Volcanic ash beds in the Waikato district. *Earth Sci. J., 1,* pp. 17-30.

Pullar, W.A., 1973: Tephra marker beds in the soil and their application in related sciences. *Geoderma, 10,* pp. 161-168.

Smith, D.G.W. & Westgate, J.A., 1969: Electron probe technique for characterising pyroclastic deposits. *Earth Planet. Sci. Lett., 5,* pp. 313-319.

Thorarinsson, S., 1974: The terms *tephra* and *tephrochronology. in* J.A. Westgate & CM. Gold, (Eds.): World Bibliography and Index of Quaternary Tephrochronology, INQUA, UNESCO, University of Alberta Press, pp. xvii-xviii.

Vucetich, C.G. & Pullar, W.A., 1964: Stratigraphy of Holocene ash in the Rotorua and Gisborne districts. *N.Z. Geol. Surv. Bull. 73,* pp. 43-88.

Vucetich, C.G. & Pullar, W.A., 1969: Stratigraphy and chronology of late Pleistocene volcanic ash beds in central North Island, New Zealand, *N.Z.J. Geol. Geophys., 12,* pp. 784-837.

MADILOGO, A LATE QUATERNARY VOLCANO NEAR PORT MORESBY, PAPUA NEW GUINEA

D.H. BLAKE[1]

Division of Land Use Research, C.S.I:R.O., P.O. Box 1666, Canberra City, A.C.T. 2601

ABSTRACT

A small extinct volcano, Madilogo, was discovered in 1969 on the southwest side of the Owen Stanley Range, 50 km northeast of Port Moresby. It consists mainly of a well preserved, forest-covered cone about 250 m high with a breached summit crater and a lava flow to the west. The vegetation on the volcano, together with the uneroded form of the cone, a thin and impersistent soil cover on the lava flow, and the fresh unweathered nature of the exposed lava, indicate that the volcano is Quaternary, and it may have been formed less than 1000 years ago. The lava contains phenocrysts of olivine, augite, biotite, and minor apatite, together with cognate xenoliths, set in a very fine to glassy groundmass which, where crystalline, consists of potassium feldspar, clinopyroxene, and opaques. Chemically it has anomalously High K_2O, TiO_2, and K_2O/Na_2O values compared to lavas from shoshonitic volcanoes in eastern Papua, and its origin is uncertain.

INTRODUCTION

A small extinct central-vent volcano situated 50 km northeast of Port Moresby (Fig. 1), on the southwest side of the Owen Stanley Range, at latitude 9°12′S, longitude 147°34′E, is named Madilogo, after a small village on its lower western flank. The volcano is bounded to the west and south by the Naoro River, a tributary of the Brown River, and to the north and east by high mountain ridges (Fig. 2). It was recognised as a probable volcano by the author early in 1969 during an airphoto interpretation study prior to a land resources survey of mainland eastern Papua (Blake et al., 1973) carried out by the Division of Land Research (now Division of Land Use Research), CSIRO. The existence of the volcano was confirmed during field work in June 1969.

GEOLOGICAL SETTING

Madilogo volcano lies in eastern Papua, the geology of which has been summarised recently by Davies & Smith (1971) and Blake (1973). It is situated well away from the large stratovolcanoes of eastern Papua — Lamington, Hydrographers, Victory, and Trafalgar — and also from the volcanoes of the Managalase Plateau (Fig. 1), and it is one of the few Quaternary volcanic centres known in the Owen Stanley Range, southwest of the Papuan Ultramafic Belt. The mountain ridges on the north side of the volcano are formed predominantly of the Owen Stanley Metamorphics, which here consist of Mesozoic schist, slate, and phyllite. These are metasediments which belong mainly to the greenschist facies of regional metamorphism. The high ridges to the south of the volcano are probably formed partly of similar metamorphic rocks, but they also include some basaltic igneous rocks.

FORM OF THE VOLCANO

Madilogo volcano covers about 3 km², and consists mainly of a well preserved symmetrical cone with a summit crater breached to the west. The cone rises about

[1] Present address: Bureau of Mineral Resources, P.O. Box 378, Canberra City, A.C.T. 2601

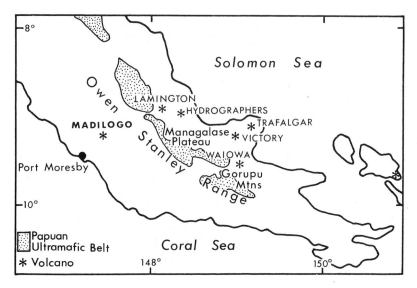

Fig. 1: Locality map.

250 m from the flood plain of the Naoro River, which is approximately 600 m above sea level. Its flanks have steep slopes (up to 30⁰), and on airphotos appear to be quite smooth. On the west side of the cone, below the breach in the summit crater, a lava flow extends for 1 km away from the cone. The flow has an irregular surface, caused in part by transverse arcuate ridges and furrows, and has a general slope of less than 5⁰ down to the west. The flow appears to have suffered little, if any, erosion, but surface features are mainly masked by forest. The volcano is also taken to include a ridge on the east side of the cone. This ridge descends steeply southwards and is separated from the cone by a steep-sided V-shaped ravine. It is possibly a remnant of an older part of the volcano which was largely destroyed during the last eruption. The cone and lava flow were probably formed during one eruptive period, possibly lasting less than a year (cf. Waiowa volcano, described by Baker, 1946).

EVIDENCE FOR THE AGE OF THE VOLCANO

On airphotos taken in September 1956, the forest covering Madilogo volcano is seen to have a tonal pattern slightly different from that of the forest on the nearby mountain ridges, and it includes several small patches with an overall smooth tone, indicating groups of trees belonging to a single species ('pure stands'), in contrast to groups made up of many species elsewhere. The pure stands may be pioneering on former garden sites, representing a stage in the reversion of garden back to forest. On the other hand they may be remnants of a stage in the succession from bare ground present immediately after the last volcanic eruption, to eventual mature mixed rain forest. If the latter is the case, the volcano was probably active within the last few hundred years. This estimate is based on the suggestion of Paijmans (1973), who has studied the vegetation on volcanoes in the Cape Hoskins area, New Britain, where the climate is generally similar to that at Madilogo and where mature mixed rain forest has become established within a period of 1500 years.

A young age for Madilogo volcano is supported by the well preserved form of the volcanic cone, and the relatively thin and impersistent soil cover present on the lava flow. The cone, from its shape and size, is probably formed of cinder and ash, which are readily eroded, yet on airphotos it appears to have suffered no erosion. However, small radial gullies of the type commonly developed on such cones (as, for example, on Pago volcano, New Britain, described by Blake & Bleeker, 1970) may be present but masked by the forest cover. The breach in the summit crater is considered to be a volcanic rather than an erosional feature: it was probably formed during the latter part of the last eruption, when part of the cone was displaced by lava flowing westward from the central vent.

Auger holes indicate that the soil on the lava flow is generally less than 0.5 m thick, and that it terminates abruptly downwards, overlying hard unweathered lava. The soil appears to be mainly a yellowish-brown fine sandy loam, and may be developed from volcanic ash deposited during the waning stages of the last eruption from Madilogo. The absence of a transition zone between the soil and underlying fresh rock shows that the lava was not the parent material for the soil. Numerous rocky outcrops are present where scoriaceous lava protrudes through the soil, as at Madilogo village. The exposed lava is fresh and unweathered, even where it

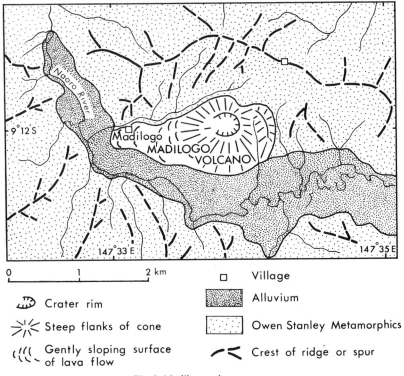

⌇ Crater rim	▨ Alluvium
⋗⃫⋖ Steep flanks of cone	⌑ Owen Stanley Metamorphics
⟨⟨⟨ Gently sloping surface of lava flow	⌒⃖ Crest of ridge or spur
⌑ Village	

Fig. 2: Madilogo volcano.

Table 1. Chemical analyses, CIPW norms, and modal analyses of lava from Madilogo

Specimen No.	45A	45B	45C
SiO2	51.7	52.4	48.5
TiO2	1.95	1.70	1.95
Al2O3	13.8	13.3	14.4
Fe2O3	2.65	3.55	2.90
FeO	4.55	3.70	4.65
MnO	0.12	0.11	0.12
MgO	7.90	8.45	8.55
CaO	6.90	7.05	6.75
Na2O	1.95	2.05	1.50
K2O	5.70	5.50	5.15
P2O5	1.10	1.00	1.25
H2O+	0.97	0.61	2.60
H2O-	0.39	0.17	1.26
CO2	0.1	0.05	0.05
Total S as SO3	0.14	0.13	0.10
Total	99.9	99.7	99.7
K2O/Na2O	2.9	2.7	3.4

CIPW Norms

		45A	45B	45C
	or	34.22	32.89	31.77
	ab	16.76	17.55	13.25
	an	12.27	10.98	18.12
	wo	6.08	7.43	3.34
di	en	4.68	6.15	2.59
	fs	0.75	0.37	0.38
hy	en	6.33	8.01	9.47
	fs	1.02	0.48	1.40
ol	fo	6.28	5.00	7.12
	fa	1.12	0.33	1.16
	mt	3.90	5.21	4.39
	il	3.76	3.27	3.87
	ap	2.65	2.40	3.09
	cc	0.23	—	0.12

Chemical analyses by Australian Mineral Development Laboratories, Adelaide.

Modal Analyses

Phenocrysts

	45A	45B	45C
Olivine	6	5	9
Biotite	2	2	1
Augite	5	3	3
Groundmass (vesicle free)	87	90	87

consists largely of volcanic glass, which is highly susceptible to chemical weathering in the humid tropics.

The above evidence indicates that the volcano is Quaternary, and may have been formed less than 1000 years ago.

PETROLOGY OF THE MADILOGO LAVA

Three samples of grey vesicular lava collected from exposures between Madilogo village and the base of the cone have been studied. Modal and chemical analyses and CIPW norms are given in Table 1. The samples contain phenocrysts of olivine, augite, biotite, and minor apatite, which show a complete range in size up to 2 mm, and also porphyritic and vesicular cognate xenoliths. The phenocrysts and xenoliths are set in a vesicular groundmass which in two of the samples (45A, 45C) consists of minute potassium feldspar laths, clinopyroxene rods, and opaque granules, and in the other sample (45B) is glass containing needle-like crystallites of clinopyroxene. The phenocrysts in the cognate xenoliths are similar to those in the host rock.

In thin section the olivine phenocrysts are seen to contain small equant and mainly opaque inclusions. The augite phenocrysts are very pale greenish and show slightly anomalous birefringence colours. The biotite phenocrysts are strongly pleochroic from almost colourless to pinkish brown, and in the two crystalline samples they have reaction rims formed by three sets of minute opaque plates, possibly of ilmenite, intersecting one another at about 60°; these plates appear to be perpendicular to the cleavage of the biotite.

The chemical analyses and CIPW norms show that the lava erupted from Madilogo volcano is silica-saturated (it contains normative olivine and hypersthene), is unusually rich in K_2O, P_2O_5, and TiO_2 (i.e. the oxides of incompatible elements), and has exceptionally high K_2O/Na_2O values. The high P_2O_5, TiO_2, and K_2O/Na_2O and relatively low Al_2O_3 contents show that the lava is chemically unlike other Late Cainozoic potassium-rich volcanic rocks in eastern Papua, such as those of the large calcalkaline stratovolcanoes (Jakeš & Smith, 1970), and the shoshonitic volcanoes of the Managalase Plateau (Ruxton, 1966), Waiowa (Baker, 1946; Blake, unpublished data), and other small centres now known within the Owen Stanley Range (P. E. Pieters, pers. comm., 1974).

Jakeš & White (1969) suggested that the chemical compositions of the potassium-rich lavas of eastern Papua are related to the positions of the volcanoes relative to the depth of a postulated underlying southerly-dipping Benioff zone. However, there has been relatively little recorded seismic activity in the area, and there is no geophysical evidence for a present-day Benioff zone beneath the volcanoes (Johnson et al., 1971). Even if it is accepted that the volcanoes of eastern Papua are correlated with some form of subduction, the Madilogo lava, because of its high K_2O/Na_2O and TiO_2 values, still does not fit in with any of the tectonic models proposed for Papua New Guinea, and for the present it is regarded as a petrological anomaly (R.W. Johnson, pers. comm. 1975).

REFERENCES

Baker, G., 1946: Preliminary note on volcanic eruptions in the Goropu Mountains, southeastern Papua, during the period December, 1943, to August, 1944, *J. Geol., 54,* pp. 19-31.

Blake, D.H., 1973: Geology of eastern Papua. *CSIRO Aust. Land Res. Ser. 32,* pp. 62-71.

Blake, D.H., & Bleeker, P., 1970: Volcanoes of the Cape Hoskins area, New Britain, Territory of Papua and New Guinea. *Bull. volc., 34,* pp. 385-405.

Blake, D.H., Paijmans, K., McAlpine, J.R., & Saunders, J.C., 1973: Land-form types and vegetation of eastern Papua. *CSIRO Aust. Land Res. Ser. 32.*

Davies, H.L., & Smith, I.E., 1971: Geology of eastern Papua. *Geol. Soc. Amer. Bull, 82,* pp. 3299-3312.

Jakeš, P., & Smith, I.E., 1970: High potassium calc-alkaline rocks from Cape Nelson, eastern Papua. *Contr. Miner. Petrol., 28,* pp. 259-271.

Jakeš, P., & White, A.J.R., 1969: Structure of the Melanesian arcs and correlation with distribution of magma types. *Tectonophysics, 8,* pp. 223-236.

Johnson, R.W., Mackenzie, D.E., & Smith, I.E., 1971: Seismicity and late Cenozoic volcanism in parts of Papua-New Guinea. *Tectonophysics, 12,* pp. 15-22.

Paijmans, K., 1973: Plant succession on Pago and Witori volcanoes, New Britain. *Pacific Sci., 27,* pp. 260-268.

Ruxton, B.P., 1966: A late Pleistocene to Recent rhyodacite-trachybasalt-basaltic latite volcanic association in northeast Papua. *Bull. volc., 29.* pp. 347-374.

CRUSTAL STRUCTURE UNDER THE MOUNT LAMINGTON REGION OF PAPUA NEW GUINEA

D.M. FINLAYSON, B.J. DRUMMOND, C.D.N. COLLINS, and J.B. CONNELLY

Bureau of Mineral Resources, P.O. Box 378, Canberra City, A.C.T. 2601

ABSTRACT

The Mount Lamington stratovolcano in eastern Papua is underlain by the ophiolite suite of rocks which make up the Papuan Ultramafic Belt and is situated near a north-south magnetic anomaly offset which marks the western limit of current volcanic activity on the Papuan peninsula. The Moho in the region shallows from depths of about 21 km under the northeast Papuan coast to 8 km under Mount Lamington. The crustal layers are interpreted as having P-wave velocities of 2.8, 3.7, 5.66, and 6.86 km/s, which are similar to those found for oceanic layers 1, 2, and 3. However, the total crustal thickness along the coast is about twice that of normal oceanic crust. The crustal layers dip towards the Solomon Sea at angles between 13^0 and 19^0, which are higher than the 9^0 dip determined from magnetic models but much less than the 25^0-60^0 dips used in previous gravity modelling. The Moho depth along the southwest Papuan coast is about 26 km, decreasing towards the Coral Sea. There is evidence for a low-velocity zone under the Mount Lamington region at depths between 35 and 50 km. Deeper lithospheric processes are responsible for the calcalkaline volcanism of Mount Lamington and may be related to the minor earthquake activity at shallow and intermediate depth, possibly associated with a fossil Benioff zone.

INTRODUCTION

Mount Lamington (1585 m) is one of a number of large Quaternary strato-volcanoes which lie northeast of the Owen Stanley Range in eastern Papua (Fig. 1). Until 1951 it had been regarded as dormant because there was no history or legend of activity, but in January of that year a catastrophic Peléan-type eruption occurred which caused considerable loss of life and property damage (Taylor, 1958). A seismic monitoring system was subsequently set up on the northern slopes of the mountain to detect earth tremors. Data are telemetered to a recording station in Popondetta for immediate inspection.

In 1973, the Mount Lamington station was one of the recording stations operating during the East Papua Crustal Survey, a regional seismic survey designed to investigate some of the major structural features of the Papuan peninsula (Finlayson, in prep.). Preliminary interpretations of crustal structure in the Mount Lamington region have been derived from data recorded at stations in that region. These interpretations, together with those of unpublished aeromagnetic work (CGG, 1969, 1971, 1973), enable basic structures under Mount Lamington to be outlined.

REGIONAL GEOLOGY

The geology and structure of eastern Papua (Fig. 1) have been discussed by Davies (1971), Davies & Smith (1971), Milsom (1973), and St John (1970). The Mount Lamington volcanic region lies on the Solomon Sea side of the Papuan Ultramafic Belt. This Belt is an ophiolite suite of rocks regarded as the surface exposure of a dipping slab of Jurassic oceanic crust and mantle, which has been obducted on to sialic crust during an episode of crustal convergence in the late Eocene (Davies, 1971; Dewey & Bird, 1970; Coleman, 1971; Moores, 1973). During this episode, the Owen Stanley Metamorphics were formed from Cretaceous sediments deposited at the northeastern margin of continental

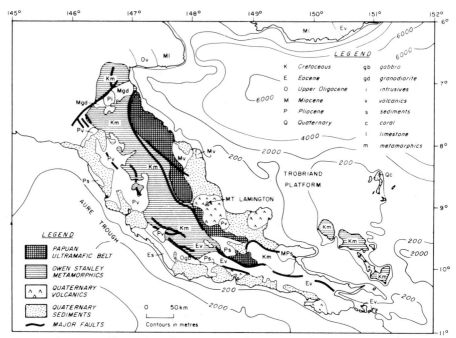

Fig. 1: Simplified geological map of eastern Papua.

Australia and rifted from Australia during the opening of the Coral Sea prior to separation of Australia from Antarctica in the early Eocene. The metamorphics now form the mountainous spine of the Papuan peninsula. Milsom (1973) has discussed the various proposed methods of emplacement of the ophiolite suite and concluded that there is a possibility that it may represent crust of a marginal basin type rather than a true oceanic type.

After emplacement of the Papuan Ultramafic Belt, volcanism took place in the late Miocene and Pliocene on the Papuan peninsula and offshore islands and continued through the Quaternary to the present time (Johnson et al., 1973; Smith, 1973; Taylor, 1958). It is difficult to relate the present episode of volcanism to the Tertiary tectonic episode which resulted in the emplacement of the ophiolite rocks. However, Mount Lamington and the other active volcanoes lie on the northeast margin of the Indian-Australian lithospheric plate where it abuts on a series of contrasting crustal types in a region of crustal convergence between the Indian-Australian and Pacific plates. The tectonic development of this northeast margin has continued since the separation of Australia from Antarctica in an episodic manner.

This development is illustrated in recent times by the pattern of earthquake activity (Denham, 1969) and volcanic activity (Johnson et al., 1973) in the region. The shallow and intermediate seismicity of the Papuan peninsula is minor compared with the major earthquake zones to the north and east but it is significant, and some form of left-lateral strike-slip movement in the general direction of the Owen Stanley Fault is essential to the tectonic synthesis of the

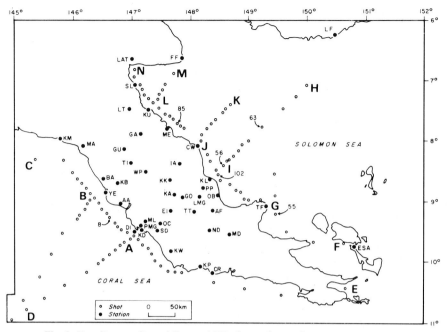

Fig. 2: East Papuan Crustal Survey 1973, shot and recording station locations.

Papua New Guinea region as a whole (Johnson & Molnar, 1972; Krause, 1973; Luyendyk et al., 1973; Milsom, 1970). Ripper (in prep.) has computed focal mechanism solutions for two earthquakes off the tip of the Papuan peninsula, one of which indicates strike-slip movement parallel to the geological strike and the other rifting.

SURVEY OPERATIONS AND DATA

The operational details of the East Papua Crustal Survey have been described by Finlayson (in prep.). Briefly, 111 shots were fired at sea (5 were small test shots) and 42 recording stations were positioned as shown in Figure 2. The stations were operated by the following institutions: Bureau of Mineral Resources (24), Australian National University (9), University of Queensland (3), Warrnambool Institute of Advanced Education and Preston Institute of Technology (3), PNG Geological Survey (3) and the University of Hawaii (1). The shots were usually either 1 tonne or 180 kg and were fired at approximately 100 m depth; a water replacement velocity of 4.0 km/s was used for shot-point corrections. Most of the recording stations were equipped with high-gain, slow-speed, automatic tape recorders but the permanent stations in the region (denoted by a 3-letter mnemonic in Fig. 2) were equipped with either photographic or smoked-paper drum recorders. Mount Lamington station (LMG) was of the latter type and was successful in recording 101 of the 111 shots, a performance bettered by only one other station (Mount Lawes, ML).

The regional gravity coverage of the Papuan peninsula was also completed during the 1973 survey period by helicopter, and this enabled a composite picture

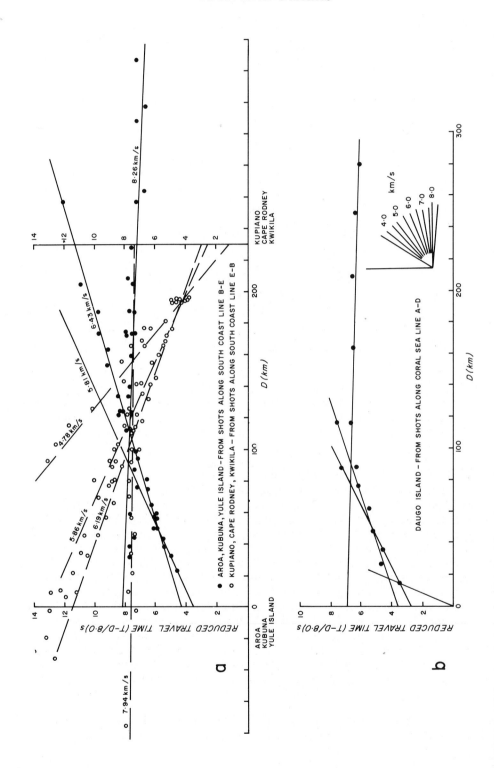

of the gravity field to be compiled using the BMR preliminary marine data (Tilbury, in prep.) and the land stations. Milsom (1973) has interpreted the regional gravity data straddling the Ultramafic Belt and constructed models to fit the observed values. The general gravity trends are consistent with the geological synthesis for the area, but the gravity models require a thicker crustal section. Much of the detail in the gravity models must be regarded as speculative at the survey station spacing used.

As described later in this paper, high-level aeromagnetic coverage of the Gulf of Papua, Papuan peninsula, and Trobriand Platform has also been completed (CGG, 1969, 1971, 1973), enabling depths to magnetic basement to be determined in these areas and some of the tectonic provinces to be delineated.

CRUSTAL STRUCTURE SOUTHWEST OF MOUNT LAMINGTON

Shots along line CBAE (Fig. 2) were recorded at several stations along the southwestern coast of the Papuan peninsula, giving approximately reversed profiles between Yule Island (YE) and Cape Rodney (CR). Unfortunately it was not possible to reverse the two lines of shots into the Coral Sea (lines AD and BD).

Representative time-distance plots from two groups of stations along the south coast are shown in Figure 3. Yule Island (YE), Kubuna (KB), and Aroa (AA) are grouped together, as are Cape Rodney (CR), Kupiano (KP), and Kwikila (KW). Also shown in Figure 3 is a time-distance plot of data from shots along line AD recorded at Daugo Island (DI). All travel-time data plots presented in this paper have been reduced by the shot-to-station distance divided by 8.0, which highlights changes in the apparent velocities near 8.0 km/s (parallel to the horizontal axis).

The data from the stations in Figure 3 and also from Mount Lawes (ML) were used to derive a simple three-layer crustal model along the southwest coast (Fig. 4). The uppermost layer has been assigned a velocity of 4.0 km/s. Survey design did not enable the recording of energy from this layer as first arrivals although it is sometimes present as later arrivals, as in Figure 3, where a refractor of 4.78 km/s is evident. Using 4.0 km/s as an average velocity, the uppermost layer appears to have an almost uniform thickness of 6-7 km along the southeast half of line CBAE, but thickens to approximately 10 km in the northwest. This coincides with the sediment distribution in the Aure Trough indicated by Mutter (1975) and CGG (1969).

Profiles both along the coast and in the Coral Sea indicate that velocities for the deeper crustal layers lie within the ranges 5.6-5.85 km/s and 6.0-6.45 km/s. Compilation of composite seismic record sections is necessary to define these velocities accurately because energy from the deeper layers commonly occurs as later arrivals. The velocities of 5.75 km/s and 6.3 km/s indicated in Figure 3 are considered representative of these refractors. From north to south the 5.75 km/s layer appears to thicken and the 6.3 km/s layer thins.

Fig. 3 (opposite): Seismic travel-time plots (reduced by shot-to-station distance/8.0) for stations along the southwest Papuan peninsula coast from (a) shots along the coast, and (b) shots out in the Coral Sea.

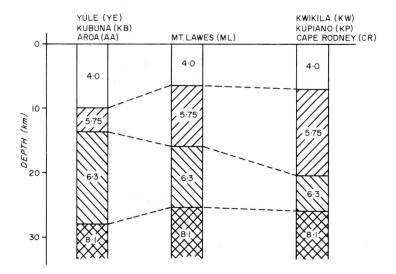

Fig. 4: Simplified crustal structure along the southwest Papuan peninsula coast (velocities in km/s).

Crustal information from the Coral Sea lines is available only near the coast, so it is necessary to interpret the Coral Sea structure by extrapolation using the data along the single-ended refraction lines AD and BD. An increase in the delay times of shots within 100 km of the coast implies an increase in the sediment thickness in the Aure Trough and its southeastern extension. The data are in good agreement with the model proposed by Mutter (1975) on the basis of gravity data, in which the sediments increase in thickness to about 7 km within the Trough. This model requires the mantle depth to decrease by about 6 km beneath the thickest part of the sedimentary pile to achieve isostatic equilibrium.

A mantle velocity of 8.1 km/s is found along the southwest coast, giving Moho depths under Cape Rodney (CR) and Mount Lawes (ML) of 25-26 km. The Moho dips to about 28 km below the northern coastal stations. An apparent Moho velocity of 8.2 km/s from shooting lines out into the Coral Sea suggests that the crust thins from the southwest Papuan peninsula coast towards the Coral Sea.

The crustal structure illustrated in Figure 4 is derived from simple plane layered models, but it is recognised that a more complex velocity/depth relation is likely and this is being investigated.

SEISMIC TRAVEL-TIME FEATURES THROUGH THE MOUNT LAMINGTON REGION

Two sets of shot and station data have been selected to illustrate the dominant features of seismic travel-times traversing the Mount Lamington region.

The first is illustrated by a seismic travel-time plot of recordings made at all stations from shots 56 and 8, on the northeast and southwest coasts of the Papuan peninsula respectively (Fig. 5).

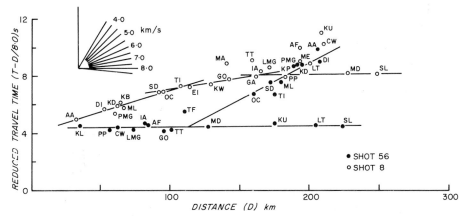

Fig. 5: Reduced seismic travel-times to all recording stations from shots 8 and 56.

The most pronounced feature of the plot is the asymmetry of the travel-times from the two shots. The arrivals from shot 8 on the southwest coast to all stations along azimuths ranging over 180⁰ indicate successive increases in apparent velocity as distance from the shot-point is increased. The departure of some travel-times from the approximately fitted lines detracts only slightly from the overall trend, and many of the departures are readily explained.

The travel-time plot of data recorded from shot 56 on the northeast coast, however, exhibits entirely different characteristics. Over an azimuth range of 180⁰, a high apparent velocity (8 km/s) is evident from distances less than 50 km out to distances of over 200 km. The departures from the approximately fitted line are small. The stations that indicate this high velocity at an intercept of approximately 4.5 s are all located northeast of the Owen Stanley Fault. The same stations that recorded shot 8 from the southwest coast indicate similar apparent velocities, but with intercepts greater than 8.0 s, demonstrating that a considerably longer ray path is required in traversing the peninsula from southwest to northeast.

A longer travel-time is also evident in the data from shot 56 recorded at stations southwest of the Owen Stanley Fault. It is not until the shooting distance is about 180 km, however, that the travel-times from the two shots are similar. The asymmetry is undoubtedly due to greatly differing crustal structure on the northeast and southwest coasts of the Papuan peninsula. A simplified interpretation of crustal structure for the southwest coast and Coral Sea is presented earlier in this paper, and an interpretation along the northeast coast is presented by Finlayson et al. (in prep.) and is summarised here.

There are distinct differences in the crustal structure between the areas where the Papuan Ultramafic Belt crops out along the northeast coast and the region of the Trobriand Platform. Between Salamaua (SL) and Morobe (ME), surface-layer P-wave velocities of 4.4 km/s are recorded and a prominent refractor with a velocity of 6.98 km/s is interpreted at depths between 2.5 and 6.0 km. The Moho is interpreted at 20 km depth with an upper mantle velocity of 7.96 km/s. Between Morobe (ME) and Tufi (TF) across the Trobriand Platform, upper crustal layers

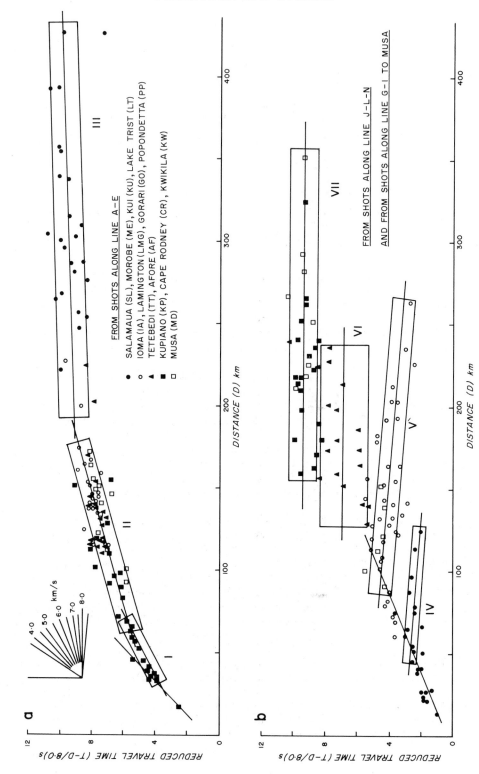

FROM SHOTS ALONG LINE A–E

SALAMAUA (SL), MOROBE (ME), KUI (KU), LAKE TRIST (LT)
IOMA (IA), LAMINGTON (LMG), GORARI (GO), POPONDETTA (PP)
TETEBEDI (TT), AFORE (AF)
KUPIANO (KP), CAPE RODNEY (CR), KWIKILA (KW)
MUSA (MD)

FROM SHOTS ALONG LINE J–L–N
AND FROM SHOTS ALONG LINE G–I TO MUSA

with velocities of 2.8, 3.7, and 5.66 km/s and total thickness between 10 and 15 km are interpreted, overlying lower crustal material with velocity 6.86 km/s. The Moho is interpreted as lying at depths between 20 and 23 km and the upper mantle velocity is 7.96 km/s. Computer modelling along the northeast coast indicates the existence of a low velocity zone at depths between 35 and 50 km.

The second set of data is illustrated in Figure 6(a and b), which shows two travel-time plots, one from shots along traverse A-E (Fig. 2) and the other from shots along traverse J-L-N (and traverse G-I to Musa), with common recording stations along the zone between the sets of shots. The time-distance plot in Figure 6(a) displays similar characteristics to that in Figure 5 from shot 8. The envelopes of the data sub-sets I, II, and III in Figure 6(a) are consistent with the seismic ray paths traversing a crust with velocities similar to those determined along the southwest coast (Fig. 4).

The data plotted in Figure 6(b) can be grouped into various sub-sets IV, V, VI, and VII, all with the apparent velocities near 8 km/s. Sub-set IV is from stations on the Papuan Ultramafic Belt and is clearly associated with high-velocity material relatively near the surface and steeply dipping refractors. Data sub-set V has a similar apparent velocity, but the seismic travel-times are influenced by the onshore structures of the Trobriand Platform's southern margin on which the recording stations Ioma (IA), Lamington (LMG), Gorari (GO), and Popondetta (PP) are situated. The arrivals recorded at Tetebedi (TT) and Afore (AF) are largely contained in sub-set VI and display a much less well defined trend than the other data sub-sets. Sub-set VI is similar to the transition data at distances between 150 and 180 km in Figure 5. The interesting point, however, is that the ray paths are all in the region northeast of the Owen Stanley Fault. Sub-set VII contains recordings from Musa (MD), which is also northeast of the Owen Stanley Fault, as well as data from KW, KP, and CR, southwest of the Fault, and no further offset in the time-distance plot is apparent on crossing the Fault.

Thus it appears that longer (and presumably deeper) ray paths from shots along traverse J-L-N are evident under the Mount Lamington region as recorded at stations southeast of Mount Lamington. This feature is not apparent in data recorded at Tufi (TF) from the same shots but may be inferred from recordings made at Salamaua (SL) and Kui (KU) from shots along line G-N (Finlayson et al., in prep.). It is also apparent from the data recorded at Musa (MD) from shots along traverse G-I that they fit into data set E in Figure 6(b). Thus the evidence for a deep 8 km/s layer described by Finlayson et al. (op. cit.) is substantiated and is certainly present under the Mount Lamington region.

An approximate model for the crustal structure under Mount Lamington can be determined by treating the line of stations Gorari (GO), Lamington (LMG), Popondetta (PP), Killerton (KL) and shots out from the coast along traverse I-H (Fig. 2) as a fan-shooting pattern. Morobe (ME) and shot 85 have been chosen as one pair of reference points, and in the opposite direction Tufi (TF) and shot 55 have been chosen as the other pair. The travel-times from these reference points are

Fig. 6 (opposite): Reduced seismic travel-times from shots along (a) shooting line A-E, and (b) shooting line J-L-N.

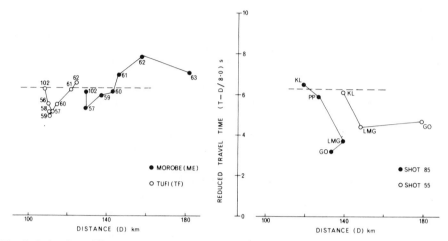

Fig. 7: Delay-time differences along a line of stations and shots normal to the northeast Papuan peninsula coast.

illustrated in Figure 7. The distances involved are all over 100 km, and the ray paths can be regarded as being from a refractor with a velocity of approximately 8.0 km/s (Finlayson et al., in prep.).

It is apparent from Figure 7 that Killerton (KL) and shot 102 (closest shot to KL) have approximately the same delay-times to the '8.0' refractor. The delay-times decrease as one moves along the line of shots out from Killerton (KL). At shots 57, 58, and 59 the delay times are a minimum and the 1.1 s delay-time difference from shot 102 can be attributed to thinning of the near-surface layer as the basement high is approached at the edge of the Trobriand Platform (CGG, 1971). The travel-times from shots in deeper water (60, 61, 62) indicate increasing delay-times, and these may be attributed to the structures on the submarine slope down from the Trobriand Platform to the floor of the Solomon Sea Basin (Finlayson et al., in prep.).

The right-hand side of Figure 7 illustrates the considerable decrease in the delay-times going from Killerton (KL) inland to Gorari (GO), the delay-time difference between Killerton (KL) and Lamington (LMG) being approximately 2.2 s. Aeromagnetic interpretation (CGG, 1971) indicates that sediment thickness decreases from approximately 3 km to zero along this line. The delay-time difference introduced by the near-surface layers can be taken as approximately the same as that in the zone extending offshore to the basement high referred to above, viz. 1.1 s. Interpretation of the aeromagnetic data later in this paper indicates that the basaltic province associated with the Papuan Ultramafic Belt extends under the Quaternary cover of the Trobriand Platform and the onshore areas northeast of the Owen Stanley Ranges, and that Mount Lamington is near its southwest boundary.

If the remaining 1.1 s delay-time difference between Killerton (KL) and Lamington (LMG) is attributed to a thinning of the basaltic and gabbroic layers and if the seismic velocity used for the volcanic pile is similar to that determined at Rabaul, i.e. 4.6 km/s (Finlayson & Cull, 1973), the structure under Mount

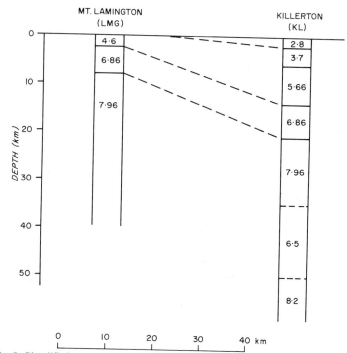

Fig. 8: Simplified crustal structure under Mount Lamington (velocities in km/s).

Lamington can be extrapolated from the crustal structure under Killerton (KL) as shown in Figure 8. The depth to the deep '8.0' refractor can be approximated from apparent velocities and intercepts of data sub-sets III and VII in Figure 6, but detailed analysis of the low velocity layer between the '8.0' refractors will be required in order to resolve the structure more accurately. Refractors may also dip considerably across the strike of the gross structures; if so, the refractor depths indicated would be offset from positions directly under Mount Lamington.

The first-arrival data from shot 56 to recording stations Killerton (KL), Popondetta (PP), Mount Lamington (LMG), and Gorari (GO) in Figure 5 have an apparent velocity of 8.4-8.5 km/s. This is the velocity to be expected if the 5.66/6.86 km/s interface dips at 13°-14° towards the Solomon Sea. However, seismic modelling indicates that the velocity/depth distribution may vary from the simple layered section indicated in Figure 8 to one of smoothly varying velocity increase with depth for the layers above the 6.86 km/s refractor. The apparent velocity indicates that the dip of this refractor is slightly less than the 16° obtained from Figure 8. This is much less than the 25°-60° dips used by Milsom (1973) in his gravity models for the dipping structures of the Papuan Ultramafic Belt.

Seismic first arrivals from shots along traverse I-H (Fig. 2) at distances greater than 100 km towards the same recording station array mentioned above (KL, PP, LMG, GO) give apparent velocities of 9.4-9.7 km/s. Modelling indicates that such velocities would be apparent from the 6.86/7.96 km/s interface at dip angles of 12°-15°, which are less than the 19° indicated in Figure 8. Further analysis will undoubtedly modify the simple dipping structures used in this paper.

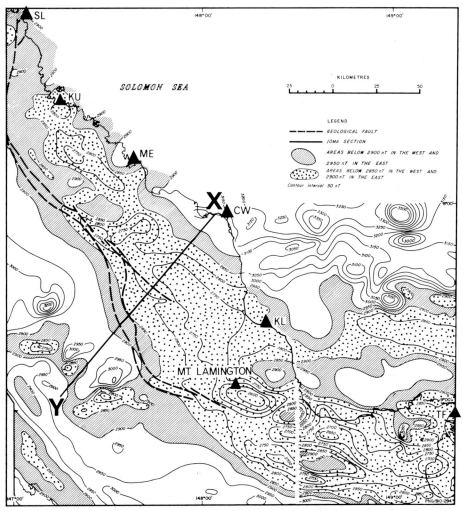

Fig. 9: Aeromagnetic anomaly map of the Mount Lamington region.

MAGNETIC EXPRESSION IN THE MOUNT LAMINGTON AREA

Aeromagnetic surveys covering the eastern Papuan peninsula, the Gulf of Papua and the Trobriand Platform have been undertaken by BMR (CGG 1969, 1971, 1973). Figure 9 shows a composite map of these surveys over the Papuan Ultramafic Belt in the Mount Lamington region and includes the major geological faults. The eastern half was flown at 2400 m and the western half at 4600 m. The regional gradient used to produce the contour map was that given by Parkinson & Curedale (1962) for the epoch 1957.5, but the contour values are not tied to any reference field. The dip, declination, and magnitude of the Earth's magnetic field in this region are approximately -31°, +6°E, and 42500 nT respectively (Finlayson, 1973).

The most obvious feature of the magnetic map is the prominent negative anomaly which is closely associated with the Papuan Ultramafic Belt particularly

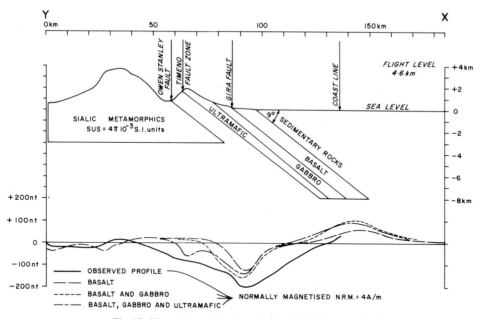

Fig. 10: Magnetic models along the line X-Y shown in Fig. 9.

from Kui (KU) to Mount Lamington (LMG). Immediately south Kui (KU) the negative anomaly is near the coast, and between Morobe (ME) and Cape Ward Hunt (CW) it is closer to the Owen Stanley Fault and shows rather irregular relief. South from Cape Ward Hunt it reflects the trends of the major faults indicated in Figure 9. The northernmost section of the Belt, which trends towards the coast near Salamaua (SL), is expressed by lack of relief in the magnetic contours.

In the region of Mount Lamington the anomaly is offset some 30 km to the south and its trend alters from southeast to east. The anomaly is very prominent in this area and the trend continues towards the east, but the relief of the anomaly is more irregular here than to the north of the offset. The magnetic expression of the Owen Stanley metamorphic zone southwest of the Papuan Ultramafic Belt is generally small except for a number of circular anomalies attributable to local intrusives.

In the northern part of the Belt the negative anomaly correlates well with two rock units mapped by Davies (1971) as the gabbroic and basaltic components of the ophiolite suite of rocks. Most of the areas mapped as being occupied by these two units have associated negative anomalies. Between Kui (KU) and Morobe (ME), however, the anomaly is rather irregular and does not correlate well with the large area of gabbro mapped in the region, an effect probably caused by the presence of numerous diorite intrusions in the area. The absence of magnetic expression in the most northerly part of the Belt reflects the absence of basalt and gabbro in this area and provides convincing evidence that the ultramafic component of the ophiolite suite has a very small magnetic effect.

Figure 10 shows three two-dimensional magnetic models along the profile XY (Fig. 9). The models show the effects of three layers with dip angles of about 9° and a natural remanent magnetism of 4 A/m in the direction of the Earth's present field. The terrain effects of the Owen Stanley Range are also included in the models. The three models illustrate the magnetic effects of the basalt, the basalt and gabbro, and the basalt, gabbro, and ultramafic rocks with layer thicknesses corresponding to the measured surface exposures.

Although none of the fits is good, that produced by the basalt and gabbro model is closest. The observed anomaly has a longer-wavelength component than the calculated anomalies, indicating that an accurate magnetic model of the Papuan Ultramafic Belt must differ from Figure 10. Detailed magnetic studies of oceanic crust (Talwani et al., 1971; Atwater & Mudie, 1973) indicate that the magnetic material may not be uniformly distributed within the oceanic layers 2 and 3, and the non-uniqueness of magnetic interpretation methods makes it difficult to construct detailed models based solely on magnetic observations.

The depth to the basalt layer at the coastline in Figure 10 is 6 km, which agrees with the depth to the 5.66 km/s layer indicated in Figure 8 but is 2 km deeper than that indicated by Finlayson et al. (in prep.) for the Cape Ward Hunt (CW) region. The dip of 9° is less than the dip of the 6.86 km/s refractor in the Mount Lamington region determined from seismic work described earlier in this paper and within the range of topographic dips (5°-15°) on the sea-floor towards the Solomon Sea basin (Finlayson et al., in prep.).

East of the magnetic anomaly offset mentioned earlier, outcrops of basalt and gabbro associated with the Ultramafic Belt are sparse and the main surface rock units are Quaternary volcanics and alluvium. However, the continuation of the regional negative magnetic anomaly suggests strongly that the basalt and gabbro units are present under the Quaternary cover. Their presence in this region has been postulated by Davies (1971) on geological evidence and by Milsom (1973) from the gravity modelling studies. The Quaternary volcanics have a relatively subdued magnetic expression. This may possibly be explained if they are composed of interlayered reversely and normally magnetised lava flows which would render them effectively non-magnetic. This effect was found in parts of Iceland (Piper, 1971).

The offset in the magnetic anomaly near Mount Lamington represents a boundary, possibly a north-south shear zone along which the Ultramafic Belt has been offset. Quaternary volcanics are prevalent east of the offset but absent to the west.

DISCUSSION

In eastern Papua the contemporary suites of calcalkaline and high-K calcalkaline lavas (Johnson et al., 1973) associated with recent volcanism cannot be associated with any currently active well defined Benioff zone in accordance with the model that Jakeš & White (1969) have proposed for island-arc areas. Such lavas are usually associated with magma generation at depths greater than 150 km. The lithospheric structures at such depths cannot be determined from the seismic and magnetic survey work interpreted in this paper. The simplified crustal structure proposed here indicates that the concept of a slice of crust and upper mantle obducted on to a less dense crust is substantially correct in the Mount Lamington region, but the development of the thicker crustal section (twice that of normal oceanic crust) along the northeast Papuan peninsula coast (Finlayson et al., in prep.) requires further investigation.

The crustal shortening required to generate the Papuan Ultramafic Belt structures in the Eocene-Oligocene (40 m.y. B.P. approximately) cannot be directly related to present-day volcanism in eastern Papua. However, the whole Melanesian region demonstrates episodic development since that time and is still a major interaction zone between the Australian and Pacific lithospheric plates. Thus it is possible that some latent geochemical separation has been maintained until the present to account for the current volcanic activity. A subduction episode of only 1-2 m.y. duration under east Papua would be sufficient to enable oceanic crust to descend to the 150 km depth which is generally considered necessary to generate calcalkaline volcanics. It may be speculated that this has occurred since the emplacement of the Papuan Ultramafic Belt. A few earthquakes and hypocentres in the depth range 70 to 300 km have been detected in the Mount Lamington area and also in the Ioma (IA) area, indicating that there is current lithospheric activity under east Papua.

ACKNOWLEDGEMENTS

The authors wish to acknowledge the contribution of the following institutions to the field work during the East Papua Crustal Survey: Research School of Earth Sciences, Australian National University; Department of Geology and Mineralogy, University of Queensland; Warrnambool Institute of Advanced Education; Preston Institute of Technology; Hawaii Institute of Geophysics, University of Hawaii; and Geological Survey of Papua New Guinea. The staff of the Papua New Guinea District Administration and many individuals rendered invaluable assistance with the logistic operations connected with the field work; the authors wish to thank them for their help towards the overall success of the survey. We also wish to pay tribute to the volcanological work of Tony Taylor who, with others, was instrumental in promoting and encouraging the deep seismic investigation of the Mount Lamington region of eastern Papua. This paper is published with permission of the Director of the Bureau of Mineral Resources, Geology & Geophysics.

REFERENCES

Atwater, T., & Mudie, J.D., 1973: Detailed near bottom geophysical study of the Gorda Rise. *J. geophys. Res.*, 78 *(35)*, pp. 8665.

Coleman, R.G., 1971: Plate tectonic emplacement of upper mantle peridotites along continental edges. *J. geophys. Res.*, 76 *(5)*, pp. 1213.

Compagnie General De Geophysique (CGG), 1969: Papuan Basin and Basic Belt aeromagnetic Survey, Territory of Papua and New Guinea 1967. *Bur. Miner. Resour. Aust. Rec.* 1969/58 (unpubl.).

Compagnie General De Geophysique (CGG), 1971: Eastern Papua aeromagnetic survey, Part 1; northeastern portion (mainly offshore) flown in 1969. *Bur. Miner. Resour. Aust. Rec.* 1971/67 (unpubl.).

Compagnie General De Geophysique (CGG), 1973: Eastern Papua aeromagnetic survey, Part 2; southwestern panel (onshore) flown in 1970-71. *Bur. Miner. Resour. Aust. Rec.* 1973/60 (unpubl.).

Davies, H.L, 1971: Peridotite-gabbro-basalt complex in eastern Papua: an overthrust of oceanic mantle and crust. *Bur. Miner. Resour. Aust. Bull.* 128.

Davies, H.L., & Smith, I.E., 1971: Geology of eastern Papua. *Bull. Geol. Soc. Am., 82,* pp. 3299.

Denham, D., 1969: Distribution of earthquakes in the New Guinea-Solomon Islands region. *J. geophys. Res., 74,* pp. 4290.

Dewey, J.F., & Bird, J.M., 1970: Mountain belts and the New Global Tectonics. *J. geophys. Res., 75 (14),* pp. 2625.

Finlayson, D.M., 1973: Isomagnetic maps of the Australian region for epoch 1970.0. *Bur. Miner. Resour. Aust. Rep. 159.*

Finlayson, D.M., & Cull, J.P., 1973: Time term analysis of New Britain-New Ireland island arc structures. *Geophys. J. Roy astr. Soc., 33,* pp. 265.

Finlayson, D.M. (ed.), in prep.: East Papua Crustal Survey 1973, operational report. *Bur. Miner. Resour. Aust. Rec.*

Finlayson, D.M., Muirhead, K.J., Webb, J.P., Gibson, G., Furumoto, A.S., Cooke, R.J.S., & Russell, A.J., in prep.: Seismic investigation of the Papuan Ultramafic Belt. *Geophysics. J. Roy. astr. Soc.*

Jakeš, P., & White, A.J.R., 1969: Structure of the Melanesian arcs and correlation with distribution of magma types. *Tectonophysics, 8,* pp. 223.

Johnson, R.W., Mackenzie, D.E., Smith, I.E., & Taylor, G.A.M., 1973: Distribution and petrology of late Cenozoic volcanoes in Papua New Guinea. In *The Western Pacific Island Arcs, Marginal Seas, Geochemistry* (ed. P.J. Coleman) *Perth University of Western Australia Press.*

Johnson, T., & Molnar, P., 1972: Focal mechanisms and plate tectonics of the southwest Pacific. *J. geophys. Res., 77,* pp. 5000.

Krause, D.C., 1973: Crustal plates of the Pacific and Solomon Seas. In *Oceanography of the South Pacific* (ed. R. Fraser) *New Zealand National Commission for UNESCO, Wellington.*

Luyendyk, B.P., Macdonald, K.C., & Bryan, W.B., 1973: Rifting history of the Woodlark Basin in the southwest Pacific. *Bull. Geol. Soc. Am., 84,* pp. 1125.

Milsom, J.S., 1970: Woodlark Basin, a minor center of sea-floor spreading in Melanesia. *J. geophys. Res., 75,* pp. 7335.

Milsom, J.S., 1973: Papuan Ultramafic Belt: gravity anomalies and the emplacement of ophiolites. *Geol. Soc. Am. Bull., 84,* pp. 2243.

Moores, E.M., 1973: Geotectonic significance of ultramafic rocks. *Earth Sci. Rev., 9,* pp. 241.

Mutter, J.C., 1975: A structural analysis of the Gulf of Papua and northwest Coral Sea region. *Bur. Miner. Resour. Aust. Rep. 179.*

Parkinson, W.D. & Curedale, R.G., 1962: Isomagnetic maps of eastern New Guinea for the epoch 1957.5. *Bur. Miner. Resour. Aust. Rep. 63.*

Piper, J.D.A., 1973: Interpretation of some magnetic anomalies over Iceland. *Tectonophysics, 16,* pp. 163.

Ripper, I.D., in prep.: Seismicity, earthquake focal mechanisms, volcanism and tectonics of the New Guinea Solomon Island region. *Bur. Miner. Resour. Aust. Rep.*

Smith, I.E., 1973: Late Cainozoic volcanism in the southeastern Papuan Islands. *Bur. Miner. Resour. Aust. Rec. 1973/67* (unpubl.).

Smith, I.E., 1975: Eastern Papua, evolution of volcanism on a plate boundary. *Aust. Soc. Explor. Geophys., 6,* pp. 68-69.

St John, V.P., 1970: The gravity field and structure of Papua and New Guinea. *Aust. Petrol. Explor. Ass. (APEA) Journal. 41.*

Talwani, M., Windisch, C.C., & Langseth, M.G. Jr, 1971: Reykjanes Ridge crest: a detailed geophysical study. *J. geophys. Res., 79 (2),* pp. 473.

Taylor, G.A.M., 1958: The 1951 eruption of Mount Lamington, Papua. *Bur. Miner. Resour. Aust. Bull. 38.*

Tilbury, L.A., in prep.: Geophysical results from the Gulf of Papua and Bismarck Sea. *Bur. Miner. Resour. Aust. Rec.*

PERALKALINE RHYOLITES FROM THE
D'ENTRECASTEAUX ISLANDS, PAPUA NEW GUINEA

I.E.M. SMITH

Department of Geology, Australian National University, P.O. Box 4, Canberra, A.C.T. 2600

ABSTRACT

Mildly peralkaline rhyolites (comendities) are the most abundant lava type in the Quaternary volcanic province centred on Dawson Strait in the D'Entrecasteaux Islands, southeastern Papua. The comendities are petrographically and chemically comparable to the comendites of oceanic islands but have compositions which trend toward the compositions of comendites from continental rift zones. Minor basaltic and intermediate rocks in the Dawson Strait area represent, with their associated comendites, a hypersthene-normative alkali basalt (transitional basalt) association, grading through benmoreite to oversaturated trachyte and peralkaline sodic rhyolite, which is comparable to the volcanic rock association found in extensional areas of the Earth's crust. The occurrence of such an association in south-eastern Papua is anomalous in the circum-Pacific context; however, recent work to the east of the D'Entrecasteaux Islands suggests that the tectonics of the area is dominated by crustal extension and rifting; the presence in the Dawson Strait area of a volcanic suite characteristic of a tensional tectonic regime provides further support for the hypothesis.

INTRODUCTION

Peralkaline silicic volcanics, unique among the volcanic products of west Melanesia, occur in the Dawson Strait area of the D'Entrecasteaux Islands, southeast Papua. These rocks, which appear to form part of a transitional basalt—peralkaline rhyolite series, overlap in time with Quaternary calcalkine volcanism in the western D'Entrecasteaux Islands and the east Papuan mainland. Mildly peralkaline rhyolites (comendities) are the characteristic siliceous rocks of oceanic islands (Baker, in press), and are also typically found on the continents in close association with major rift structures such as the East African rift (e.g. Bailey & Macdonald, 1970). Peralkaline rhyolites are rare in converging areas of the Earth's crust — for example, the island-arcs — and their presence in southeastern Papua has important implications for Quaternary tectonics in the area.

Published work on the comendites in the Dawson Strait area is limited to brief mention of their occurrence and chemical composition in work dealing with patterns of regional volcanism in Papua New Guinea (Morgan, 1966; Johnson et al., 1973). An account of their field occurrence is given in an unpublished BMR report (Smith, 1973).

Silicic volcanics form the southeastern part of Fergusson Island, the neighbouring island of Dobu, and a large part of Sanaroa Island; small outcrops are also found on Oiaobe Island (Fig. 1). The rocks are predominantly fragmental. Fine to coarse unconsolidated pumice ash predominates, but welded and non-welded ash-flow tuffs are common in some areas, notably the area to the west and southwest of Sebutuia Bay, on much of Sanaroa Island, and in basal coastal exposures on Dobu Island. Lava makes up less than five percent of the volcanic products in the area and typically occurs as small steep-sided lava domes or lobate lava flows, indicating that the lava was viscous at the time of eruption. Almost all of the lava is either glassy (obsidian) or crystalline comendite; it is not known

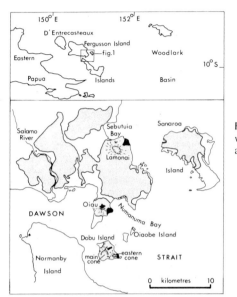

Fig. 1: Distribution of Quaternary peralkaline volcanism (shown stippled) in the Dawson Strait area, southeast Papua. Recent lava flows are shown in black.

whether blocks of trachyte which occur in the area southeast of Mount Lamonai are lava blocks from pyroclastic deposits or whether there are trachyte flows in the area.

Davies & Ives (1965) described an outcrop of basalt in the lower reaches of the Salamo River in southeastern Fergusson Island, but the relationship of this basalt to silicic volcanics in the area in unknown. Apart from this, basic and intermediate volcanics have been found only as inclusions in comendite lava flows and as lava blocks in ash deposits.

There are three recognisable volcanic centers in the Dawson Strait area, namely, Mount Lamonai and Mount Oiau (both on Fergusson Island) and Dobu Island. These cones are between 300 m and 500 m high; they are roughly aligned north-south and are spaced 10 and 6 km apart. These three centres display youthful volcanic landforms suggesting eruptive activity in the very recent past. G.A.M. Taylor (in Davies, 1973) suggested on the basis of devitrification rates that Mount Oiau was last active about 600 years ago.

Small hills in the area between Mounts Lamonai and Oiau probably represent plugs or crater rim remnants from an earlier period of activity. A near-circular bay 5 km southeast of Mount Lamonai has the appearance of a former volcanic centre, and arcuate steep cliffs and thermal activity in the northern part of Numanuma Bay suggest that there too is an extinct eruptive centre. Similarly the volcanics on northern Sanaroa Island show no development of volcanic landforms and also represent an earlier period of activity.

The lower age limit for volcanic activity in the Dawson Strait area is difficult to estimate. Because of the high rainfall, consequent high erosion rates, and abundant unconsolidated pyroclastics in the area, the absence of volcanic landforms on the older volcanics is not significant. Volcanic activity in the Dawson

Strait area is thought to have extended through Holocene times and possibly back into late Pleistocene times.

PETROGRAPHY AND CHEMISTRY

Comendite

The crystalline comendites are fine-grained, light grey to green-grey porphyritic rocks; the glassy comendites are typical porphyritic obsidians. The larger, thicker flows are predominantly crystalline but contain abundant glassy bands; smaller flows are mainly obsidian. The phenocrysts (typically less than 10 vol.%) are predominantly anorthoclase Or_{25-30} (0.5-2 mm long) with subordinate (less than 1 vol. %) clinopyroxene (sodic augite to sodic ferroaugite, less than 0.3 mm long), and either titaniferous magnetite or ilmenite; small crystals of fayalitic olivine, arfvedsonite, and aenigmatite are only rarely observed (phenocryst compositions determined by electron microprobe).

The crystalline lavas have a fine-grained groundmass composed of alkali feldspar, irregular patches of quartz, and minor clinopyroxene, brown-green to blue amphibole (arfvedsonite), iron-titanium oxides, and aenigmatite. The glassy lavas have a groundmass of colourless, or less commonly brown, isotropic glass containing microlites of alkali feldspar, clinopyroxene, iron-titanium oxides, amphibole, and aenigmatite. In some specimens this glass contains abundant flow-oriented ovoid bubbles.

Thirty-one comendite specimens from the Dawson Strait area have been analysed for major elements; six representative analyses are presented in Table 1. They show a greater range in chemical composition than that reported for comendities from both oceanic and continental environments by Bailey & Macdonald (1970). Silica ranges from 68 to 74 wt percent. Increase in silica is correlated with an increase in total iron, K_2O, and the ratio K_2O/Na_2O. The rocks are peralkaline with peralkalinity indices (molecular $(Na_2O + K_2O)/Al_2O_3$) within the range 1.0 to 1.4; there is an increase in peralkalinity with increase in silica content.

Trachyte

Two specimens of trachyte collected from the area northeast of Numanuma Bay are medium grey, porphyritic rocks containing phenocrysts of sodic oligoclase, augite, fayalitic olivine and titaniferous magnetite (phenocryst compositions determined by electron microprobe). The fine-grained groundmass is predominantly plagioclase with subordinate quartz, clinopyroxene, and amphibole. Chemical analyses of these two specimens are presented in Table 2 (33710, 33711). They are slightly oversaturated, high Na_2O/K_2O trachytes intermediate in composition between the comendites and the more basic rocks described in the following section. They probably represent a parental magma to the comendites.

Basaltic and intermediate rocks

The comendite flows at the eastern end of Dobu Island contain abundant rounded inclusions of dark grey porphyritic basaltic rock ranging from a few millimetres to tens of centimentres in diameter. These inclusions contain large (1-3 mm) phenocrysts of labradorite, clinopyroxene, and olivine in a fine-grained

Table 1. Major-element analyses, CIPW norms, and phenocryst abundances of six representative comendites from the Dawson Strait area.

No.	33714*	33720	33729	33731	33733	33741
SiO_2	69.03	69.65	70.68	70.73	70.18	72.43
TiO_2	0.33	0.42	0.28	0.26	0.29	0.25
Al_2O_3	14.64	14.03	14.09	12.97	12.52	10.50
Fe_2O_3	2.42	1.20	0.99	1.24	3.19	2.00
FeO	0.97	2.09	1.86	2.30	1.55	2.58
MnO	0.10	0.21	0.09	0.13	0.15	0.11
MgO	0.32	0.07	0.15	0.19	0.13	0.04
CaO	0.41	0.41	0.33	0.37	0.42	0.17
Na_2O	6.46	6.41	6.16	5.87	5.46	5.91
K_2O	4.84	4.80	4.94	4.95	4.59	4.35
P_2O_5	0.04	0.03	0.02	0.02	0.03	0.01
H_2O+	0.05	0.50	0.36	0.33	0.39	0.31
H_2O-	0.09	0.12	0.14	0.12	0.56	0.09
CO_2	0.15	0.04	0.11	0.05	0.08	0.07
Total	99.85	99.98	100.20	99.53	99.54	98.82
Peralkaline index	1.08	1.12	1.10	1.16	1.11	1.37
CIPW norms						
Q	13.26	15.62	17.10	19.85	21.20	28.54
or	28.60	28.36	29.19	29.25	27.12	25.71
ab	48.36	45.44	44.97	39.16	38.84	29.79
ac	5.55	3.47	2.86	3.59	6.48	5.79
ns	—	1.13	0.91	1.50	—	3.18
di	0.69	1.40	0.71	1.22	1.20	0.30
hy	1.46	2.97	3.12	3.87	1.56	4.47
mt	0.72	—	—	—	1.38	—
il	0.63	0.80	0.53	0.49	0.55	0.47
ap	0.09	0.07	0.05	0.05	0.07	0.02
cc	0.34	0.09	0.25	0.11	0.18	0.16
Modal phenocryst contents (vol. %)						
Anorthoclase	0.7	2.7	9.5	1.3	10.4	1.1
Clinopyroxene	+	0.3	0.5	+	+	0.1
Fe-Ti oxides	+	+	0.2	+	0.3	+
Olivine	—	+	+	—	—	—
Total	0.7	3.0	10.2	1.3	10.7	1.2

+denotes less than 0.1 vol. %

33714   Crystalline comendite, southeast of Mount Lamonai
33720   Glassy comendite, flow in summit crater, Mount Oiau
33729   Glassy comendite, northeast flank, Mount Lamonai
33731   Glassy comendite, west coast, Sanaroa Island.
33733   Crystalline comedite, Numanuma Bay.
33741   Glassy comendite, north of Numanuma Bay.

*Numbers refer to specimens held in the Department of Geology, A.N.U. collection. Analyses by mainly X-ray fluorescence methods; Na by flame photometry; FeO by wet chemistry.

Table 2. Major-element analyses, CIPW norms, and phenocryst abundances of basalts, benmoreites, and trachytes associated with comendites in the Dawson Strait area.

No.	33703*	33704	33705	33706	33707	33708	33709	33710	33711
SiO_2	47.63	48.91	50.02	50.16	52.00	54.22	54.93	63.70	64.25
TiO_2	1.12	1.17	2.01	2.58	1.98	1.62	2.30	0.89	0.81
Al_2O_3	20.61	20.77	17.07	16.09	16.65	15.65	15.30	16.33	16.49
Fe_2O_3	2.98	3.94	3.67	4.96	3.40	3.96	2.38	3.13	1.98
FeO	3.24	2.30	6.16	5.45	5.37	3.85	6.15	1.44	2.54
MnO	0.11	0.12	0.17	0.18	0.16	0.14	0.16	0.09	0.12
MgO	6.57	5.21	4.85	5.01	4.57	5.04	3.01	0.87	0.77
CaO	12.15	11.08	8.69	9.40	8.66	7.89	6.00	2.45	2.09
Na_2O	2.91	3.24	3.97	4.50	4.63	4.31	5.21	7.79	7.21
K_2O	0.33	0.74	1.03	0.79	1.19	1.14	1.49	2.79	2.49
P_2O_5	0.20	0.24	0.41	0.45	0.38	0.27	0.72	0.25	0.30
H_2O+	0.70	0.88	0.84	0.01	0.25	0.38	0.62	0.29	0.24
H_2O-	0.95	0.68	0.40	0.11	0.16	0.71	0.61	0.26	0.24
CO_2	0.22	0.24	0.16	0.18	0.12	0.17	0.89	0.03	0.20
Total	99.72	99.52	99.45	99.87	99.52	99.35	99.77	100.31	99.73

CIPW norms

Q	—	—	—	—	—	4.69	4.65	3.80	8.07
or	1.99	4.46	6.19	4.67	7.09	6.85	8.92	16.50	14.80
ab	24.97	27.95	33.61	37.85	39.49	36.74	44.52	65.94	61.24
an	43.03	40.73	26.43	21.54	21.30	20.48	14.23	1.36	5.36
di	12.43	9.61	11.20	16.78	15.17	13.04	4.66	4.68	1.77
wo	—	—	—	—	—	—	—	1.36	—
hy	3.19	7.36	10.50	2.72	3.44	8.04	11.21	—	2.98
ol	6.73	0.99	1.26	2.73	3.50	—	—	—	—
mt	4.40	4.48	5.41	7.20	4.97	5.83	3.50	2.36	2.89
il	2.17	2.27	3.88	4.91	3.79	3.13	4.43	1.69	1.55
hm	—	0.93	—	—	—	—	—	1.51	—
ap	0.48	0.58	0.99	1.07	0.91	0.65	1.73	0.59	0.71
cc	0.51	0.56	0.37	0.41	0.28	0.39	2.05	0.07	0.46

Modal phenocryst contents (vol %)

Plagioclase	26.3	31.1	+	6.3	6.5	4.7	2.5	16.8	11.0
Pyroxene	—	1.0	—	1.0	0.1	2.2	0.3	0.7	0.5
Olivine	12.6	3.1	—	0.4	—	2.2	+	0.1	+
Opaque	—	—	—	—	—	—	0.1	0.3	0.4
Total	38.9	35.2	< 0.1	7.7	6.6	9.1	2.9	17.9	11.9

+ denotes less than 0.1 vol. %

33703	Basalt inclusion in comendite flow, eastern Dobu Island.
33704	Basalt inclusion in comendite flow, eastern Dobu Island.
33705	Basalt boulder, west coast, Sanaroa Island.
33706	Basalt inclusion in comendite flow, eastern Dobu Island.
33707	Basalt inclusion in comendite flow, eastern Dobu Island.
33708	Benmoreite inclusion in comendite flow, eastern Dobu Island.
33709	Benmoreite, lava block, fragmental deposit northern coast Dobu Island.
33710	Trachyte, southeast from Mount Lamonai.
33711	Trachyte, southeast from Mount Lamonai.

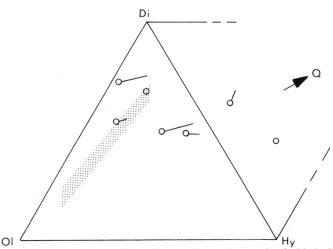

Fig. 2: Normative mineralogy of the basaltic and intermediate rocks associated with the Dawson Strait comendites plotted on a Di-Ol-Hy-Q diagram. Open circles represent the position of samples after re-calculation of Fe2O3 and FeO according to the method of Irvine & Baragar (1971); the end of the solid line extending to the right of the circles marks the position of the samples using Fe2O3 and FeO as measured. The position of the one atmosphere thermal divide separating tholeiitic from alkaline basalts (shown stippled) is taken from Bass (1971).

groundmass composed of plagioclase, clinopyroxene (in some specimens titaniferous), and abundant iron-titanium oxides. A similar specimen collected from a rounded boulder on the west coast of Sanaroa Island contains sparse small phenocrysts of labradorite in a fine-grained groundmass of plagioclase, clinopyroxene, altered olivine, and abundant iron-titanium oxides. The groundmass olivine in this rock typically forms larger grains than the other groundmass minerals, and textural relations show that it was an early crystallising phase. Orthopyroxene has not been recognised in any of these rocks and there is no evidence of reaction of olivine phenocrysts of form pyroxene.

Major-element analyses of five basaltic inclusions and lava blocks are presented in Table 2. The analyses show regular chemical variations which suggest that these analysed rocks are comagmatic. Thus with increase in silica there is an increase in Na_2O and K_2O and a decrease in Al_2O_3, total iron, MgO, and CaO. TiO_2 contents are variable and rather high. The inclusions are typically oxidised (Fe_2O_3/FeO, 0.6 to 1.7), but this may be due to their mode of occurrence rather than to secondary alteration.

Three of the basalts are of transitional character in the sense of Coombs (1963) and Bass (1971); the remaining two plot within the field of olivine tholeiites in the normative 'iron free basalt tetrahedron' (Fig. 2). The mildly alkaline character of all of the basalts is demonstrated on an alkalis-silica diagram (Fig. 3), where the analyses plot just within the Hawaiian alkaline field as defined by Macdonald & Katsura (1964). The basalts have a distinctly sodic character with Na_2O/K_2O ratios varying from 8.8 to 3.9. High Al_2O_3 values are correlated with abundant plagioclase phenocrysts (26 and 31 vol. %), which suggests that the chemical composition of these rocks has been modified by plagioclase accumulation; these two specimens are also notably lower in TiO_2 than the others.

Sparsely porphyritic lava blocks collected from basal pyroclastic deposits on the northern side of Dobu Island contain phenocrysts of calcic andesine, clinopyroxene, and rare olivine in a fine-grained groundmass of plagioclase, clinopyroxene, iron-titanium oxides, fine green-brown mesostasis, and minor secondary calcite. Plagioclase and clinopyroxene crystals, with or without olivine crystals, form clusters of possibly accumulative origin in some specimens. A similar rock type has also been found as inclusions in comendite flows at the eastern end of Dobu Island.

Analysis of two of these specimens (Table 2, 33708 and 33709) shows them to be intermediate in composition between the transitional basalts and the trachytes. Although higher in CaO and MgO and lower in alkalis than the rocks described as benmoreite by Tilley & Muir (1964), the intermediate character of these rocks is seen as justification for use of the term and they are accordingly classified as benmoreites.

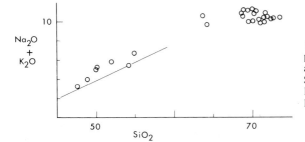

Fig. 3: Total alkalis against silica plot for the comendites and associated rocks from the Dawson Strait area. Solid line divides Hawaiin tholeitic lavas from Hawaiian alkaline lavas (Macdonald & Katsura, 1964).

ROCK ASSOCIATION

When plotted on conventional Harker-type diagrams (Fig. 4) the comendites, with their associated trachytes, benmoreites, and basalts, in general show the smooth variation curves to be expected of a comagmatic series. Minor discrepancies in this picture are the late-stage enrichment in iron in the more siliceous comendites and the kink in the Na_2O/SiO_2 curve, arising from high Na_2O contents in the analysed trachytes which probably result from a change in the nature of the early-crystallising mafic phases and in the composition of the fractionating feldspar (plagioclase to anorthoclase) respectively.

Clear evidence of plagioclase accumulation in at least two of the analysed rocks suggests that plagioclase fractionation may have been a major control on the evolution of the series. This suggestion can be tested by considering relationships within the quaternary system SiO_2-Al_2O_3-Na_2O-K_2O, and this is conveniently done by plotting the analyses on the ternary projection SiO_2-Al_2O_3-Na_2O+K_2O as used by Bailey & Macdonald (1969). In the ternary projection (Fig. 5) there is a well defined trend from the trachytes to the most siliceous comendites; a second trend at an angle to the first includes the basalts and benmoreites and the least siliceous comendites but not the trachytes. One important feature of this projection is that if feldspar is the fractionating phase the compositions of whole rock, derivative liquid and feldspar should be colinear. The relationships illustrated in Figure 5 indicate a trachyte-comendite series controlled by feldspar fractionation and a basalt-trachyte series which is controlled at least in part by other phases.

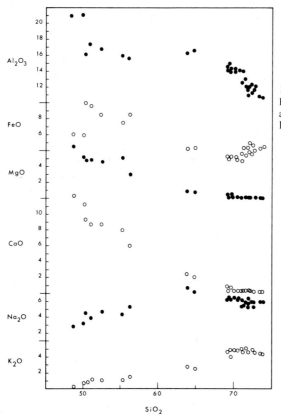

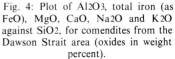

Fig. 4: Plot of Al2O3, total iron (as FeO), MgO, CaO, Na2O and K2O against SiO2, for comendites from the Dawson Strait area (oxides in weight percent).

Although a simple differentiation series cannot be clearly demonstrated, their close field relationship suggests that the basalts and benmoreites represent, with their associated rocks, a hypersthene-normative alkali basalt (transitional basalt) association, grading into oversaturated trachytes and peralkaline sodic rhyolites.

This transitional basalt-comendite association in eastern Papua is comparable to similar associations typically found in the late stage of evolution of ocean ridge islands (e.g. Easter Island, Bouvet Island, Kerguelen Island; Baker, in press), on some intra-plate oceanic islands (e.g. Ascension Island; Baker, in press), in continental rift systems (e.g. the East African rift; Bailey & Macdonald, 1970), and in some continental environments (e.g. western north America, Noble 1968; and eastern Australia, Abbott 1968, Bryan & Stevens 1973). This association has not been found elsewhere in western Melanesia although the St Andrew Strait area of northern Papua New Guinea (Johnson & Smith, 1974) has transitional basalts of a similar type to those found as inclusions in the Dawson Strait area and these are also associated with a predominance of rhyolitic rocks, some of which have peralkalinity indices close to one.

Bailey & Macdonald (1970) used the rather limited chemical data available in the literature to divide glassy comendites (obsidians) into two groups — oceanic and continental. Using their data, the continental comendites form a relatively

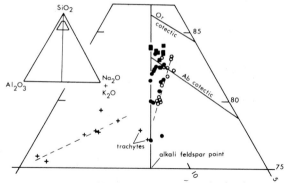

Fig. 5: Part of the ternary system SiO2-Al2O3-Na2O+K2O (molecular percentages) showing the relationships of the Dawson Strait glassy comendites described as either oceanic or continental by Bailey & Macdonald (1970). Crosses — basalts, benmoreites, and trachytes from Dawson Strait; open circles — glassy comendites from Dawson Strait; solid circles — 'oceanic' comendites; solid squares — 'continental' comendites. The positions of the Or and Ab coetectic are taken from Bailey & Macdonald (1970).

tight group which shows regular chemical trends, whereas the oceanic comendites are widely scattered. For example in the ternary projection SiO2-Al2O3-Na2O+K2O the continental comendities plot in a group within and along a quartz-feldspar cotectic zone defined by the field boundaries of Or-silica and Ab-silica taken from the systems K2O-Al2O3-SiO2 and Na2O-Al2O3-SiO2 at one atmosphere pressure (Schairer & Bowen, 1955, 1956). The oceanic comendites on the other hand plot as a scattered group trending from the feldspar point toward the quartz-feldspar cotectic zone. Bailey & Macdonald (1970) suggested that this difference is genetically significant, and that because of the close correlation with experimentally determined quartz-feldspar minima the continental comendites may have originated by partial melting within the continental crust, possibly in the presence of an alkali-bearing vapour. In contrast, the trend of the oceanic comendites from a simple trachytic composition (represented by the alkali feldspar point) toward the quartz-feldspar cotectic zone is consistent with derivation from a trachytic parent.

The scattered spread of Bailey & Macdonald's oceanic comendite field is defined largely by two analyses from Kerguelen Island and by two previously published analyses from the Dawson Strait area (Morgan, 1966). Apart from these the 'oceanic' comendites fall within a field which lies close to the Ab cotectic and overlaps the field of 'continental' comendites (Fig. 5). The comendites from the Dawson Strait area show a much greater variation in chemical composition than any of the comendite suites used by Bailey & Macdonald (1970) in defining their two types, and they appear to represent a more complete series than any previously described comendite suite. The trend shown by the Dawson Strait comendites in Figure 5 is that to be expected of a series fractionating feldspar (represented by the alkali feldspar point) and evolving toward a cotectic composition. On reaching the cotectic, further fractionation will be controlled by both feldspar and quartz, and the compositions of derivative liquids would lie along the cotectic zone. Although this argument does not preclude the possibility of quartz-feldspar equilibrium reactions between fractionated comenditic magma and sialic crust, the possibility that the difference between continental and oceanic comendites is nothing more than a difference in degree of fractionation seems equally acceptable.

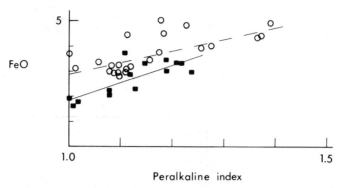

Fig. 6: Plot of total iron (expressed as FeO) against peralkaline index (molecular $(Na_2O+K_2O)/Al_2O_3$). Open circles — Dawson Strait comendites; solid squares — continental comendites of Bailey & Macdonald (1970). Least-squares regression lines for the continental (solid line) and Dawson Strait (dashed line) comendites are also shown.

Bailey & Macdonald (1970) suggested that a plot of total iron (calculated as FeO) against peralkaline index will distinguish between oceanic and continental comendites. In this plot (Fig. 6) most of the Dawson Strait comendites plot above the continental comendites. However, because both groups of points merge with one another, the conclusion is once again reached that there is no essential difference between these comendites and those labelled typically continental.

TECTONIC IMPLICATIONS

The transitional-basalt/comendite association is characteristically associated with extensional areas of the Earth's crust (mid-ocean ridges and continental rift systems) although they are also apparently associated with intra-plate mantle plumes (the Azores) and on the continents with block faulting (western north America) or rifted continental margins (Australia). The transitional basalt-comendite association does not form any part of the group of volcanic rocks characteristically associated with converging areas of the Earth's crust (the island-arcs).

Because it lies in the zone of interaction between the Indo-Australian and Pacific plates, the tectonic setting of eastern Papua has in the past been interpreted in terms of island-arc processes (eg. Karig, 1972). In fact, this is an oversimplified view of a complex plate boundary which has undergone dramatic polarity shifts within the past 50 m.y. (Smith, 1975). Calcalkaline volcanics, ranging in age from late Miocene to Recent, define a zone of island-arc type volcanism which passes through the D'Entrecasteaux Islands and to the west, on the Papuan mainland, includes the recently active volcanoes Waiowa, Mount Victory, and Mount Lamington (Johnson et al., 1973). The available evidence (Smith, 1973; unpublished data) indicates that this island-arc type volcanism becomes younger westward, and it is suggested that the tectonic regime (compressional) which gave rise to this volcanism has been progressively replaced by a tensional regime. Milsom (1970) and Luyendyk et al. (1973) provided evidence that the Woodlark basin which lies immediately to the east of the D'Entrecasteaux Islands has been opening by rifting for the past 3 m.y. The change in volcanic style from a calcalkaline suite, presumably related to compressional tectonics, to a transitional basalt-comendite association characteristically found in tensional areas of the Earth's crust provides further evidence for the hypothesis.

ACKNOWLEDGEMENTS

The author gratefully acknowledges the useful comments of R. Macdonald and R. Varne on a preliminary draft of the paper.

REFERENCES

Abbott, M.J., 1968: Petrology of the Nandewar Volcano, N.S.W. Australia. *Contr. Mineral. and Petrol., 20*, pp. 115-134.

Bailey, D. K., & Macdonald, R., 1969: Alkali-feldspar fractionation trends and the derivation of peralkaline liquids. *Am. J. Sci.,* 267, pp. 242-248.

Bailey, D.K., & Macdonald, R., 1970: Petrochemical variations among mildly peralkaline (comendite) obsidians from the oceans and continents. *Contr. Mineral. and Petrol., 28*, pp. 340-351.

Baker, P.E., in press: Peralkaline acid volcanic rocks of oceanic islands. *Bull. Volcanol.* Special issue on silicic peralkaline volcanic rocks.

Bass, M.N., 1972: Occurrence of transitional abyssal basalt, *Lithos, 5*, pp. 57-67.

Bryan, W.B., & Stevens, N.C., 1973: Holocrystalline pantellerite from Mt Ngun-Ngun, Glass House Mountains, Queensland, Australia. *Am. J. Sci. 273*, pp. 947-957.

Coombs, D.S., 1963: Trends and affinities of basaltic magmas and pyroxenes as illustrated on the diopside-olivine-silica diagram. *Miner. Soc. Am. Spec. Pap. 1*, pp. 227-250.

Davies, H.L., 1973: Fergusson Island, Papua New Guinea — 1:250,000 Geological series. *Bur. Miner. Resour. Aust. explan. Notes., SC/56-5.*

Davies, H.L., & Ives, D.J., 1965: The geology of Fergusson and Goodenough Islands, Papua. *Bur. Miner. Resour. Aust. Rep. 82.*

Irvine, T.M., & Baragar, W.R.A., 1971: A guide to the chemical classification of the common volcanic rocks. *Can. J. Earth Sci., 8*, pp. 523-548.

Johnson, R.W., Mackenzie, D.E., Smith, I.E., & Taylor, G.A.M., 1973: Distribution and petrology of late Cenozoic volcanoes in Papua New Guinea. *in* Coleman, P.J. (Ed.), *The Western Pacific: island arcs, marginal seas, geochemistry.* Univ. West Aust. Press, pp. 523-533.

Johnson, R.W., & Smith, I.E., 1974: Volcanoes and rocks of St Andrew Strait, Papua New Guinea. *J. Geol. Soc. Aust., 21*, pp. 333-352.

Karig, D.E., 1972: Remnant arcs. *Bull. geol. Soc. Amer., 83*, pp. 1057-1068.

Luyendyk, B.P., Macdonald, K.C., & Bryan, W.B., 1973: Rifting history of the Woodlark Basin in the south Pacific. *Bull. geol. Soc. Am., 84*, pp. 1125-1134.

Macdonald, G.A. & Katsura, T., 1964: Chemical composition of Hawaiian lavas. *J. Petrol., 5*, pp. 82-133.

Milsom, J.S., 1970: Woodlark Basin, a minor center of sea-floor spreading in Melanesia. *J. geophys. Research, 75*, pp. 7335-7339.

Morgan, W.R., 1966: A note on the petrology of some lava types from east New Guinea. *J. geol. Soc. Aust. 13*, pp. 583-591.

Noble, D.C., 1968: Systematic variation of major elements in comendite and pantellerite glasses. *Earth and Planet. Sci. Letters, 4*, pp. 167-172.

Schairer, J.F., & Bowen, N.L., 1955: The system K2O-Al2O3-SiO2. *Am.J.Sci., 253*, pp. 681-746.

Schairer, J.F., & Bowen, N.L., 1956: The system Na2O-Al2O3-SiO2. *Am.J.Sci. 254*, pp. 129-195.

Smith, I.E., 1973: Late Cainozoic volcanism in the southeast Papual islands. *Bur. Miner Resour. Aust. Red.* 1973/67 (unpubl.).

Smith, I.E., 1975: Eastern Papua — evolution of volcanism on a plate boundary. *Aust. Assoc. Explor. Geophys. J., 6*, pp. 68-69.

Tilley, C.E. & Muir, I.D., 1964: Intermediate members of the oceanic basalt-trachyte association. *Geol. Foren. For., 85*, pp. 434-443.

SUMMARY OF THE 1953-57 ERUPTION OF TULUMAN VOLCANO, PAPUA NEW GUINEA

M.A. REYNOLDS and J.G. BEST

Amax Petroleum (Australia) Inc., P.O. Box 7092, Perth, W.A., 6000.

Kajuara Mining Corporation Pty Ltd, P.O. Box 728, North Sydney, N.S.W., 2000.

ABSTRACT

Tuluman volcano, just south of Lou Island in the Admiralty Group, Papua New Guinea, began erupting in June 1953. Initial activity was submarine, but later the volcano gradually built up to sea level. Eruptions migrated between several centres which were numbered in order of their initial activity. The vents that reached the surface subsequently developed explosive phases, the most violent of which occurred in February-March 1955. After a series of submarine effusive, surface explosive, and later subaerial effusive phases, the eruption finished in January 1957. The eruption produced a small island, Tuluman Island, and a rocky islet to the northwest, composed of lava and pyroclastic deposits. Rock types were silica-rich and included highly vesiculated rhyolite and pumice, commonly with bands of black obsidian. Although some of the explosive outbursts were strong, the overall pattern was mild when compared with eruptions elsewhere in Papua New Guinea. No warning was given of the initial eruption and no earth tremors were felt. Damage to nearby islands was negligible.

INTRODUCTION

Tuluman volcano is a dominantly submarine volcano about 1 km south of the southwestern tip of Lou Island, a member of the Admiralty Group of Islands in the northern Bismarck Sea, Papua New Guinea (Fig. 1). Manus Island is the largest and best-known of the Admiralty Group; its main centre and airport is at Lorengau, 45 km north-northwest of Tuluman.

Tuluman and Lou Island are on the northwestern side of St Andrew Strait. The southeastern side is made up of a line of islands from Baluan in the south through Mok, Pam, and the St Andrew Islands, to the Fedarb Islands in the northeast. Except for the St Andrew Islands, which are composed of coral, all the islands adjoining the Strait are volcanic, and it is likely that the St Andrew Islands also have volcanic rock basements (Johnson & Smith, 1974).

Lou is the largest of the islands in St Andrew Strait and is formed by a chain of overlapping Quaternary volcanic cones. Admiralty Chart Aus. 054 (printed March 1944) shows a shoal in the area to the southwest of Lou near where Tuluman volcano now exists, and this could be the centre of the activity witnessed in 1883 by Miklouho-Maclay (1885). However, no islands or shallow sea bottom were indicated at Tuluman at the time of the first eruption in 1953. Neither did recorded seismicity indicate any local build-up of tectonic stress, or the formation or emplacement of magma in the crust. This seems unusual after a period of quiescence of some 70 years.

Submarine activity began at Tuluman on 27 June 1953. No earth tremors were felt on nearby islands either before or during the eruption, although the people of Pam Mandian Island reported small sea waves which caused minor damage to some canoes after the initial explosion on 27 June. The waves may have been triggered by tremors, or by a small submarine explosion when lava first breached the sea bottom.

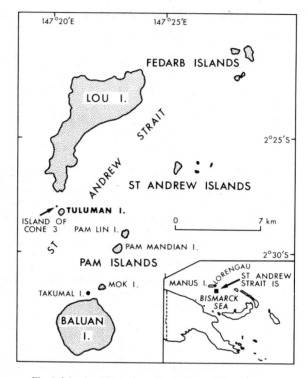

Fig. 1: Islands of St Andrew Strait, Papua New Guinea.

The eruption, more accurately described as a series of eruptions of different intensities, contined until 28 January 1957. Reynolds & Best (1957, in prep.) identified eight principal eruptive centres and referred to them as cones 1 to 6, 7A, and 7B, in order of initial activity. When last visited and mapped in 1971, Tuluman volcano was represented by two islands: the largest, Tuluman Island, was about 0.5 km in diameter and up to 23 m above sea level; the second was a small rocky islet, 400 m to the northwest, which has been called 'cone-3 island' (Johnson & Smith, 1974).

The products of the 1953-57 eruption were rhyolitic. Their petrology was discussed by Johnson & Smith (1974), who suggested a derivation from the melting of crust (about 25 km thick below St Andrew Strait). The rubidium/strontium ratios of the Tuluman lavas are exceptionally high compared with those of volcanic rocks from other sources in Papua New Guinea (Page & Johnson, 1974).

The 1953-57 eruption was a complex series of submarine and subaerial, effusive and explosive events. A detailed chronology of these events has been made by Reynolds & Best (in prep.) as a supplement to their early report (1957) which dealt with events only up to August 1955. This paper presents a summary of the main events and phases of activity; it identifies three main types of activity, and briefly describes the general pattern of the eruption.

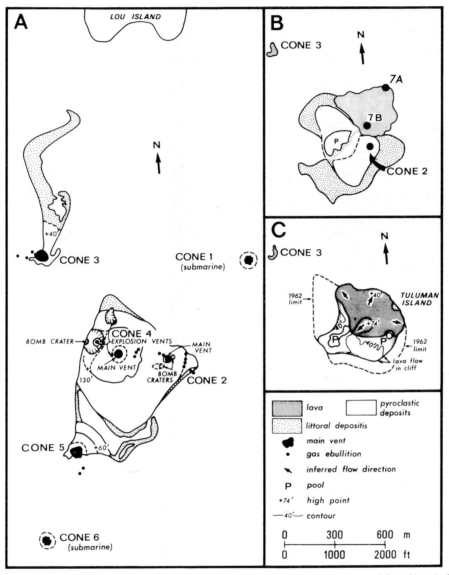

Fig. 2: Comparison of Tuluman islands and cones, 1955-1971. A: March 1955 (adapted from Plate 4 of Reynolds & Best, 1957). B: February 1957 (prepared from photographs taken during aerial inspection; position of cones doubtful). C: From plane table survey by J. Herlihy and F.E. Decker in June 1962 (broken lines), and modified by R.W. Johnson and R.A. Davies by mapping in July 1971.

OUTLINE OF EVENTS

27 June—6 July 1953. The first activity observed was the voluminous emission of water vapour at the location designated 'cone 1' (Fig. 2). The vapour was generated by the explosion, or fragmentation, of masses of lava which has flaked off submarine lava flows, or which were derived directly from the vent, and floated to the sea surface.

Fig. 3: Activity at cone 3, 14 July 1954, with small horseshoe island of cone 2 (temporarily dormant) in foreground. Southern tip of Lou Island at right. Aerial view from east.

14 November 1953—18 February 1954. A similar type of submarine activity took place from another location, 'cone 2' about 700 m southwest to cone 1. Aerial concussion effects were felt as tremors in buildings on Baluan Island to the south.

9-27 July 1954. Renewed submarine activity began at cone 2 and continued until a dome of lava built up to sea level on 11 July. Activity then declined at cone 2, but intensified at a new submarine centre 900 m northwest. This centre built up cone 3 which reached sea level on 13 July. On 14 July the first major explosive phase began at cone 3 (Fig. 3) but soon declined, followed by vapour emission. Lava flow resumed at cone 2 on 21 July from a small vent at sea level and moved outwards in a series of slow-moving concentric waves. Activity finished on 27 July leaving a prominent island at cone 3, consisting of a lava base overlain by a mound of pyroclastic material about 12 m high, and three small rocky shoals or islets up to 3 m high at cone 2.

20 October—6 November 1954. Submarine activity took place near cone 1.

10-15 February 1955. People travelling between Manus and Baluan Islands in January and early February noticed that a slight, but definite, emergence of the shoals at the summit of cone 2 to about 8 m above sea level had occurred. Strong explosions began at cone 2 on 10 February, but by 11 February it became clear that the major activity was from a centre west of, but adjacent to, cone 2. This new centre was called cone 4. Both cones built up a single island. In addition to the periodic violent explosive outbursts, with ejection of pyroclasts and release of large volumes of gas and vapour, jets of black dust and gas were emitted from the main vents with loud roars from time to time.

11 February—11 March 1955. Submarine activity at another centre, cone 5, southwest of cone 4, started on 11 February. It ceased on 13th, but had recommenced by 16th, and another island appeared on 23 February. A series of explosive and effusive phases followed, culminating with violent explosions on 9

Fig. 4: Activity at cone 7, 28 November 1956, with island of cones 2 and 4 in foreground and small remnant island of cone 3 on the left. Densely wooded southern tip of Lou Island at back. Aerial view from south.

March that rocked buildings on Baluan Island to the south, and deposited more than 2 cm of dust on the Pam Islands. The islands of cone 5 and cones 2 and 4 were joined by a low saddle of pyroclastic deposits (Fig. 2A).

12 March—9 May 1955. A new submarine source, cone 6, became active about 550 m south of cone 5. Varying rates of lava flow were indicated by the changes in volume of material reaching sea-surface; the flows continued to 9 May. The cone did not reach sea level.

16 May—26 June 1955. Minor explosions, rumbling, and the usual surface manifestations of previous submarine activity were observed in the vicinity of cone 1. A lava dome reached sea level by the morning of 6 June, and a brief period of explosions followed. Steady vapour emission continued from 20th to 26th, when the dome subsided below sea level. Eruptions of mud were observed at cone 2 (later mud eruptions also took place spasmodically at cone 4 between July and October).

26-28 September and 3-7 October 1955. Short periods of submarine lava effusion were indicated by the appearance of lava at sea level east of cone 2. This is attributed to early activity from cone 7.

7 November 1956—28 January 1957. Following increased vapour emission from north of cone 2 a small cone-shaped island, cone 7A, was built to 3 m above sea level on 28 November (Fig. 4). Spasmodic explosions and jet-like blasts of vapour and dust characterised this phase of the eruption. On 30 November a short period of lava flow began between the new island and cone 2. This was from another vent — probably belonging to the same cone as 7A, and designated 7B. Mud eruptions were also seen at cone 4 at this time. Activity at 7A continued until about 11 December. As this phase declined, extrusion of lava flows resumed at 7B and gradually filled the area between the islands of 2 and 7A above sea level (Fig. 2B).

Fig. 5: The Tuluman islands, 12 January 1957, with lava from cone 7B joining islands of cones 2 and 4; island of cone 3 in foreground. Aerial view from northwest.

Spasmodic explosions continued into January 1957, but the eruption had ceased by the 28th. The lava pile above sea level was approximately 430 m by 300 m, and 30 m high at the crater rim; the average height of the lava surface above sea level was 20 m (Fig. 5).

Subsequent tilting to the northeast, and rapid erosion, resulted in the early removal of the remnants of cone 5, and finally left a large island made of the products of cones 2, 4, and 7, and a small lava island above cone 3 (cf. Fig. 2C).

The main activity outlined above, and some other minor events, are listed on an individual cone basis in Table 1.

TYPES OF ACTIVITY

The eruption consisted of three main types of activity: (1) the extrusion of flows onto the sea-floor, building cones, most of which reached the sea surface; (2) explosions from subaerial vents; and (3) the extrusion of subaerial lava flows.

Submarine extrusion

This submarine activity produced gas-charged masses of lava which floated to the surface. These lava masses were highly vesiculated, and had red-hot interiors and chilled peripheries; they presumably flaked off from submarine lava flows during cone-building on the seabed. The arrival of the masses at the surface could usually be expected at intervals of about ten minutes, although there was no regular periodicity.

The reduction in hydrostatic pressure when the masses reached sea level permitted vesiculation, sometimes explosively. The water vapour generated, and the escaping gases, formed clouds above the remnants of the masses. Most of the

Table 1. Phases of activity at Tuluman volcano, 1953-57

Cone No.	Period	Weeks	Type of activity
1	27 June—6 July 1953	1½	Submarine*
	20 Oct.—6 Nov. 1954	2½	Submarine
	16 May—6 June 1955	3	Submarine to surface
	6-20 June 1955	2	Minor explosive & vapour
	20 September 1955		Vapour emission
2	14 Nov. 1953—18 Feb. 1954	14	Submarine (intermittent)
	9-11 & 21-27 July 1954	1½	Submarine to surface
	10-12 Feb. 1955	½	Minor explosive
3	11-14 July 1954	½	Surface explosive
4	11-15 Feb. 1955	½	Surface explosive
	21 July—2 Aug. 1955	2	Minor explosions & vapour
	Oct.—Nov. 1955	1 to 4	Minor explosions & vapour
	24 Nov.—4 Dec. 1956	1	Minor explosions & vapour
5	11-13 & 16-23 Feb. 1955	1½	Submarine to surface
	24 Feb.—11 Mar. 1955	2	Surface explosive & effusive
6	12 Mar.—9 May 1955	9	Submarine (intermittent)
7?	26-28 Sept. & 3-7 Oct. 1955	1	Submarine
7A	7-26 Nov. 1956	3	Mainly vapour emission
	27 Nov.—11 Dec. 1956	2	Surface explosive
7B	30 Nov.—2 Dec. 1956	½	Submarine
	11 Dec. 1956—28 Jan. 1957	7	Surface effusive & minor explosions

* All submarine activity was effusive; stronger phases are underlined.

lava masses sank within a few minutes of reaching the surface, but some drifted as vesicular lava or pumice fragments for many kilometres. The greater part of the floating masses was submerged.

Submarine activity was probably restrained by the rapid cooling of the lava and the confining hydrostatic pressure of the sea, and possibly by the load effects imposed by the sea-floor sediments and lava accumulation on the flanks of the cone. Submarine activity occurred as one or two individual phases at each of the main vents until the cones reached the surface of the sea. The lava flows appear to have built up to 3 to 4 metres above sea level by the close of each submarine phase.

Surface explosive activity

A phase of explosive activity took place at each centre after the cone had reached the surface (except for the extrusive phase at cone 7B towards the end of the eruption). Clouds of vapour and ash ascended to heights of 2000 m above sea level or up to 5000 m in a calm atmosphere; larger debris (mainly red-hot bombs, coarse glassy and pumiceous fragments) fell around the cone, and lapilli rained down over a radius of more than 750 m. Bright yellow, orange, and brick-red plumes commonly accompanied the vapour clouds. Gases such as sulphur dioxide and hydrogen sulphide were recognised in small quantities, but never in sufficient amounts to cause discomfort near the active centres. A faint smell like burning tar was noticed occasionally.

The most intense period of explosive activity at Tuluman was from February to March 1955, when explosions ranged from continuous to intermittent with a period of about five minutes. The explosive phase at cone 7 in November-December 1956, when the main outbursts had a frequency of about one per hour, was not as strong. However, it allowed clearer observation of the following sequence of events, which probably characterised the earlier explosive phases:

(a) in a 'static' condition, the summit of the cone was above sea level and composed of late lava flows, and the vent was sealed by a plug of brown to red scoriaceous rock; vapour and gas emission was constant, but some jet-like emissions with finely comminuted material took place at times;

(b) commencement of explosions of varying intensity took place, and the plug and sometimes the upper part of the cone disintegrated; ejectamenta ranged from fine ash to bombs several cubic metres in volume; sharp concussive effects were noticed;

(c) re-emplacement of a plug of red-hot, mainly solid lava in the vent reformed the cone; jet-like emissions of steam, gas and dust were common;

(d) return to static condition.

Only the more intense explosive phases of 1954 and 1955, with frequent outbursts, produced accumulations of pyroclastic materials and formed islands. These were at cone 3, and cones 2, 4 and 5 (contiguous). Erosion of the unconsolidated finer material was rapid, and the only islands that survived were those with summits composed of lava (island of cone 3, and the island of cones 2 and 4 joined by the lava of cone 7).

Although the summit of cone 7 emerged above sea level at centre 7A a number of factors combined to preclude the formation of a prominent island: the summit was small by comparison with those of other cones, the volume of ejected material from the 7A vent was not as great, and much longer periods elapsed between explosions. The activity from November to December 1956 at cone 7 represented a declining phase of the explosive type of activity.

Subaerial extrusion

The final and main surface effusive phase of the Tuluman eruption began in late December 1956 between the island of cones 2 and 4 and the peak of cone 7. The actual time at which activity changed to effusive is not clear. The explosive phase at cone 7 declined after 11 December, but according to observations made from Baluan Island, vapour emission appeared to increase. By 21 December it was apparent that a separate crater (crater B) of cone 7 had formed to the south of the explosive crater (crater A), and that lava was flowing from this new source.

Observations were impaired by the generation of dense clouds of water vapour, and the vent was obviously close to sea level at this time. By 28 January 1957 the lava island embracing cone 7 and the island of cones 2 and 4 had formed. Local observations during and after the effusive phase, and aerial inspections indicated that the outpouring the lava was intermittent and that some minor spasmodic explosions occurred.

Later inspections showed that the flows consisted of hard, slightly vesicular obsidian and dark grey vesicular rhyolite. Blocks of lava up to 2 m across were strewn over the surface of the island; these may have been ejectamenta, but some may have spalled off from the flows. Masses of grey pumice, with black obsidian bands which show flow lamination and in parts marked convolution, were common. The products of the surface effusive activity were apparently very similar to those of earlier submarine phases and to the bombs ejected during the major explosive periods.

The brief period of lava flow at sea level at cone 2 from 21 to 27 July 1954 is regarded as the last phase of effusive activity which, up to that time, had been entirely submarine.

Although not observed at the time, a brief period of surface effusive activity at cone 5 took place in early March 1955 during what is regarded as the most violent explosive phase of the eruption. An inspection of the Tuluman Islands shortly afterwards (24 March) showed remnants of an original crater wall of cone 5 composed of pyroclastic debris, and a smaller inner crater whose sides contained agglomerate interbedded with lava flows.

PATTERN OF ERUPTION

The most violent phase of activity appears to have been the surface explosive period from cones 4 and 5 in February to early March 1955. This seems to have been the zenith of the eruption and was preceded by uplift of cone 2 — the only premonitory warning of any of the major phases of activity. This climax took place 21 months after the beginning of the eruption in June 1953. The first 18 months of the eruption produced four major detectable submarine effusive phases, and a further period of submarine activity accompanied the eruption at cone 5; the first subaerial explosive phase took place in mid-1954.

Apart from the submarine lava flows from cone 6, which immediately followed the violent explosive phase at cone 5, other activity in 1955 was of much lower intensity. A quiescent period of 12 months preceded the final stage of the eruption at cone 7, which began on 7 November 1956. The major subaerial effusive phase occurred in this final period.

The overall pattern of the eruption, as witnessed, therefore appears to have been a build-up period as the volcano approached the surface of the sea (shown mainly by submarine effusive phases for some 20 months), an apparently climatic explosive phase over 2 months, and a period of decreasing intensity for the next 22 months. Although prolonged, the eruption was mild compared with some of the other eruptions with known histories elsewhere in Papua New Guinea.

ACKNOWLEDGEMENTS

The history of the 1953-57 eruption of Tuluman volcano could not have been written without the invaluable data supplied by village people, airline pilots, trawler masters, and particularly the Administration officers (Messrs J. Landman, E.G. Hicks, and W. Murdoch) who were stationed at Baluan Patrol Post during that period. Permission to publish this summary has been given by the Director of the Bureau of Mineral Resources (Canberra).

We are both grateful for the privilege of working with Tony Taylor in the period following the 1951 Lamington eruption, and in the formative years of the Vulcanological Observatory at Rabaul. His enthusiasm was an inspiration, and many of our ideas were guided by his concepts of the mechanics of eruption in island-arcs.

REFERENCES

Johnson, R.W., & Smith, I.E., 1974: Volcanoes and rocks of St Andrew Strait, Papua New Guinea. *J. geol. Soc. Aust., 21(3),* pp. 333-351.

Miklouho-Maclay, N. de., 1884-5: On volcanic activity on the islands near the north-coast of New Guinea and evidence of rising of the Maclay coast in New Guinea. *Proc. Linn. Soc. N.S.W., 9,* pp. 963-967.

Page, R.W., & Johnson, R.W., 1974: Strontium isotope ratios of Quaternary volcanic rocks from Papua New Guinea. *Lithos, 7,* pp. 91-100.

Reynolds, M.A., & Best, J.G., 1957: The Tuluman Volcano, St Andrew Strait, Admiralty Islands. *Bur. Miner. Resour. Aust. Rep.* 33.

Reynolds, M.A., & Best, J.G., in prep.: The 1953-1957 eruption of Tuluman Volcano, St Andrew Strait (northern Bismarck Sea) Papua New Guinea. *Bur. Miner. Resour. Aust.*

FELDSPATHOID-BEARING POTASSIC ROCKS AND ASSOCIATED TYPES FROM VOLCANIC ISLANDS OFF THE COAST OF NEW IRELAND, PAPUA NEW GUINEA: A PRELIMINARY ACCOUNT OF GEOLOGY AND·PETROLOGY

R.W. JOHNSON[1], D.A. WALLACE[2], and D.J. ELLIS [1,3]

Bureau of Mineral Resources, P.O. Box 378, Canberra City, A.C.T. 2601.

[2]Geological Survey of Papua New Guinea, Volcanological Observatory, P.O. Box 386, Rabaul, Papua New Guinea.

ABSTRACT

The volcanic rocks of the Tabar, Lihir, Tanga, and Feni Islands range from pre-Middle Miocene to Pleistocene, and show an exceptionally wide range of compositions. Strongly silica-undersaturated types are basanites, tephrites, olivine nephelinites, phonolitic tephrites, and tephritic phonolites; these are especially common on the Tanga and Feni Islands. Less undersaturated and silica-saturated types are alkali and 'transitional' basalts, trachybasalts, and trachyandesites; these are well-represented on the Tabar and Lihir Islands. Quartz trachytes have been found on all the island groups except Lihir, and undersaturated trachytes and phonolitic trachytes are present in the Tanga group. Tholeiitic basalts, andesites, and dacites all appear to be absent. K_2O/Na_2O values are mainly between 0.5 and 1, and TiO_2 contents are notably low (mainly less than 1 percent). The tectonic conditions which led to the development of these magmas are uncertain. The magmas may be related to a slab of downgoing lithosphere, but the islands are not underlain by a zone of intermediate- or deep-focus earthquakes at the present day. An alternative model involving an isostatically controlled fault zone beneath the islands is being explored.

INTRODUCTION

Strongly alkaline rocks are rare among the Late Cainozoic volcanoes of Papua New Guinea, and appear to be restricted to the Tabar-to-Feni Islands, which form a narrow, 260 km-long chain off the coast of New Ireland (Fig. 1). The rocks of these islands show an extremely wide range of compositions (bearing in mind the limited size and number of the islands, Fig. 1), ranging from the strongly undersaturated types (containing up to almost 25 percent normative nepheline) to those near the critical plane of silica undersaturation in the 'basalt tetrahedron' (Yoder & Tilley, 1962), to quartz-normative salic compositions.

The islands comprise four groups — Tabar, Lihir, Tanga, and Feni — which are equally spaced about 75 km apart (Fig. 1). The Tabar group in the northwest is closest to New Ireland (about 30 km away), and the Tanga and Feni Islands are farthest away (about 55 km). The Green Islands (Nissan and Pinipel) lie 75 km to the southeast on the extension of the Tabar-to-Feni line (Fig. 1), but they are raised coral atolls and show no volcanic rocks (except for drift pumice on beaches). The line continues southeastwards to Bougainville Island, whose Late Cainozoic volcanoes have rocks of calcalkaline compositions but apparently none of alkaline type (Blake & Miezitis, 1967; Taylor et al., 1969; Bultitude, this volume). To the northwest, the Tabar-to-Feni line is roughly co-linear with the St Matthias Group, which consists mainly of limestones although volcanic rocks are known from Mussau Island (White & Warin, 1964).

[3] Present address: Research School of Earth Sciences, Australian National University, P.O. Box 4, Canberra, A.C.T. 2600.

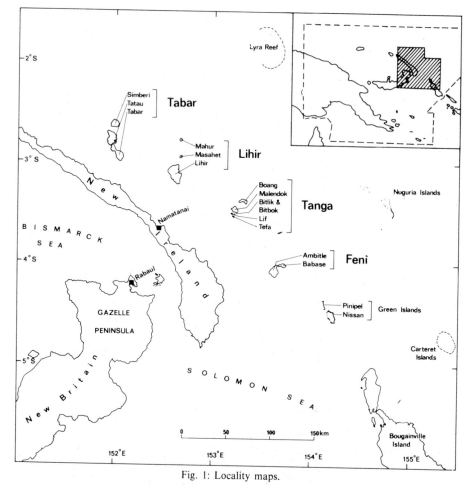

Fig. 1: Locality maps.

GEOLOGICAL INVESTIGATIONS

Glaessner (1915) was the first to draw attention to the alkaline character of rocks from the Tabar-to-Feni Islands. He identified a leucite-like mineral in a 'trachydolerite' from Lihir Island, nosean and aegirine-augite in 'glass-rich trachydolerites' from Ambitle Island, and aegirine-augite in a 'limburgite trachydolerite' of high total-alkali content, also from Ambitle Island. It appears that for fifty years Glaessner's paper was the only record of alkaline rocks in the Melanesian region — until Allen & Deans (1965) described rocks of alnöitic and kimberlitic affinities from Malaita Island, British Solomon Islands.

G.A.M. Taylor was the first to undertake systematic geological work on the islands. He surveyed them in 1969 and 1971, made comprehensive collections of rocks from all the island groups, and obtained many major-element whole-rock analyses (see, for example, Johnson et al., 1973). Wallace and I.A. Crick resurveyed the islands in 1973, and Wallace and Johnson revisited the islands in 1974.

This paper discusses 44 major-element analyses of rocks collected by Taylor, and 40 analyses of rocks from the Wallace-and-Crick surveys. It briefly describes the geology of the islands and their environment, and discusses the mineralogy of 11 Wallace-and-Crick rocks. Johnson is responsible for the processing and plotting of the whole-rock analyses, Wallace is primarily responsible for the geology, and Ellis obtained the mineral analyses. Other reports are in preparation which will provide detailed descriptions of geology and petrology and give more detailed attention to problems of magma genesis.

PLATE BOUNDARIES AND CRUSTAL STRUCTURE

The tectonic setting of the Tabar-to-Feni Islands is uncertain, despite recent geological mapping and geophysical surveying of the surrounding region (BMR, 1972; Finlayson & Cull, 1973; Murauchi et al., 1973; Furumoto et al., in prep.; Gulf, 1973).

Each of three recent papers on plate tectonics in Papua New Guinea has provided a different interpretation of present-day tectonic activity in the New Ireland region. Johnson & Molnar (1972) proposed that a left-lateral transcurrent plate boundary runs west of, and parallel to, the west coast of New Ireland. Curtis (1973), on the other hand, believed a present-day plate boundary closely follows a narrow zone of deep water that runs west-east north of Manus Island and then swings down the east side of New Ireland — between the Tabar-to-Feni Islands and the line joining Lyra Reef, the Nuguria Islands, and the Carteret Islands (Fig. 1; Mammerickx et al., 1971). However, present-day seismicity east, and immediately west, of New Ireland is very infrequent (see Denham, 1969, 1973), and it is therefore considered that Krause (1973) was probably justified in choosing not to recognise a present-day plate boundary in this region.

Seismic refraction studies in the region report the following crustal thicknesses: 28-30 km beneath the Feni group (Finlayson & Cull, 1973); 18-19 km half-way between the Green Islands and the southern tip of New Ireland (Murauchi et al., 1973) where the crust is up to 35 km thick (Finalyson & Cull, 1973); 16.2-17.4 km in the channel between New Ireland and the Tanga and Feni Islands (Furumoto et al., in prep.). Farther east, refraction profiles across parts of the Ontong Java Plateau (which begins about 190 km east of the Feni group) give depths to the Moho from 35 km to well over 40 km (Furumoto et al., 1973, in prep.). These data are insufficient to characterise the thickness and structure of the crust beneath the entire Tabar-to-Feni chain, although they suggest the crust there is thinner than that beneath New Ireland, and thicker than that of true oceanic crust found in the Pacific basin.

A southwest-northeast seismic reflection profile run between the southern end of New Ireland and the Ontong Java Plateau by Furumoto et al. (in prep.) shows a series of horst- and graben-like structures for much of its length. It suggests that the Green Islands and the narrow zone of deep water east of them occupy a zone controlled by normal faulting (see also Gulf, 1973). The seismic reflection profiles by Gulf (1973) indicate substantial accumulations of sediment in the channel between New Ireland and the Tabar-to-Feni Islands, and faulting and a dome-like 'basement' structure in the vicinity of the Tabar group.

SUMMARY OF GEOLOGY

The volcanic rocks of the Tabar-to-Feni Islands are Tertiary and Pleistocene. Holocene rocks have yet to be identified, and no eruptions have been reported from any of the islands, although thermal areas are present in all the island groups (especially on Lihir and Ambitle Islands; see Fisher, 1957). The islands of both the Tabar and Lihir groups form north-south lines, whereas those of the Tanga and Feni groups form roughly arcuate patterns. Many of the Tabar-to-Feni Islands have coral terraces which indicate periods of uplift and tilting throughout the Cainozoic.

The volcanic rocks of Simberi Island, the northernmost of the Tabar group, are overlain by Middle Miocene reef limestone (micropalaeontological determinations by D.J. Belford, pers. comm., 1975). Rocks of Tertiary age are probably also present on Tatau Island, and at the northwestern end of Tabar Island, but the southern part of Tabar Island is a dissected volcano which is probably Pleistocene. The K-Ar age of one rock from the northwestern part of Tabar Island is 0.986 ± 0.08 m.y. (determination on a plagioclase concentrate)*. Faulting is common in the Tabar group; one north-south fault which is especially prominent on Tabar Island may continue northwards onto Tatau Island, and even to Simberi Island. Alluvial gold was mined on a small scale in the 1920s from a stream on the northeast side of Tatau Island.

The two northernmost islands of the Lihir group — Mahur and Masahet — are raised coral reefs (with volcanic cores), but Lihir Island itself is a subaerial volcanic complex consisting of at least three major centres, all of which are probably Pleistocene. Thermal activitiy is widespread on the floor of a breached caldera on the east side of the island.

Except for Boang Island, which consists of raised coral reef, the islands of the Tanga group appear to be parts of the same volcano. The walls of a submerged caldera are preserved on Malendok, Tefa, and Lif Islands, and Bitlik and Bitbok Islands appear to be post-caldera extrusions. K-Ar ages of two rocks from the Tanga group are 1.14 ± 0.08 m.y. (determined on biotite — a second determination gave 1.08 ± 0.08 m.y.) and 0.187 ± 0.02 m.y. (whole-rock age).

Ambitle (or Feni, or Anir) and Babase Islands comprise the Feni (or Anir) group. The two islands define a rough semicircle, suggesting their volcanism may have been controlled by a ring fracture. Oligocene limestone is present on Ambitle Island, but the volcanic rocks appear to be mainly Pleistocene. Two K-Ar determinations on biotite from one sample gave ages of 0.68 ± 0.10 and 0.49 ± 0.10 m.y. Ambitle Island shows an eroded central crater, open to the southwest, which contains a quartz-normative salic extrusion. Thermal areas are found throughout the central area and in the southwest of Ambitle Island. Babase Island consists of a low cone and crater (in the east), and limestone (in the west) which overlies an extrusion of similar type to the one inside the central crater of Ambitle Island. The K-Ar age of one Babase rock is 1.53 ± 0.15 (hornblende concentrate).

Xenoliths, consisting of clinopyroxene and smaller amounts of opaque minerals, amphibole, and zoned plagioclase, are found in the lava flows of several

*All K-Ar dates supplied by the Australian Mineral Development Laboratories, Adelaide.

islands. These pyroxenite inclusions range from a few millimetres to over 20 cm in diameter, and appear to be crystal cumulates.

One striking feature of the geology of some islands is the abrupt change in composition between older and younger rocks. On Tabar Island, in the Tanga group, and on Ambitle Island, earlier rocks are mainly silica-undersaturated (strongly undersaturated in Tanga and Ambitle), but the later rocks are salic and strongly quartz-normative. In these three areas, at least, the process of magma generation must have changed drastically to produce such divergent rock types.

CHEMICAL ANALYSES

Variations in chemical compositions of the 84 analysed rocks are illustrated in Figures 2 to 7.

Some of the analysed rocks are chemically altered — especially those from the older Tertiary centres of the Tabar group — and a measure of this alteration is indicated on the variation diagrams. Sample points without a vertical dash on top of the appropriate symbol represent 'fresh' rocks which contain less than 1 percent H_2O-, and less than 1 percent CO_2. The vertical dash indicates that the rock contains more than 1 percent H_2O-, or more than 1 percent CO_2, and therefore is possibly altered. These 'altered' rocks contain less than 2 percent H_2O-, or less than 2 percent CO_2, except for three samples from the Tabar group which have CO_2 values of 2.40, 2.45 and 2.65 percent. This division into 'fresh' and 'altered' rocks is largely arbitrary, and may obscure the fact that some of the strongly alkaline lavas could have been volatile-rich, containing relatively high proportions of 'primary' CO_2 and H_2O.

Seventy-three of the 84 chemical analyses have Fe_2O_3 values equal to, or greater than, those of FeO. This appears to correspond with the exceptionally low ulvöspinel contents and consequently, the high Fe_2O_3/FeO values of the homogenous titanomagnetite grains in these rocks (see Appendix). In order that the chemistry of these rocks can be compared on the same basis with the compositions of others in which Fe_2O_3/FeO values are much lower, the CIPW norms of the 84 samples have been calculated using the oxidation adjustment recommended by Irvine & Baragar (1971) — i.e., if Fe_2O_3 is greater than TiO_2 + 1.5 wt percent, sufficient Fe_2O_3 is converted to FeO so that $Fe_2O_3 = TiO_2 + 1.5$. All the variation diagrams in Figures 2 to 7 show the values of the whole-rock analyses recalculated to 100 percent on a volatile-free basis (H_2O+, H_2O-, CO_2, S, SO_3, and Cl, all excluded) after use of the Irvine & Baragar transform.

PETROCHEMISTRY

Alkalinity

Most of the analysed rocks from the Tabar-to-Feni Islands are nepheline-normative, as shown in Figure 2. In Figure 3, the broad spread of points in the Hawaiian alkaline field shows a positive slope, but at the silica-poor end the points define a nearly vertical array that cuts upward through the field boundary from the Hawaiian tholeiitic field.

There is an absence of rocks in the silica range 59-64 percent, but all those

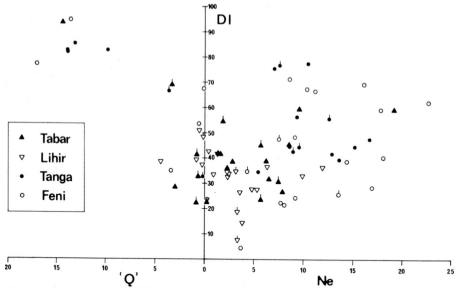

Fig. 2: Differentiation Indices (Thornton & Tuttle, 1960; DI = $\Sigma\,Q + ab + or + ne + lc$) plotted against wt percent normative nepheline (Ne), or normative 'quartz' ('Q'), where 'Q' is normative quartz plus the silica of normative hypersthene. The values of Ne and 'Q' are those obtained after using the Irvine & Baragar (1971) transform (see text). When the Fe_2O_3/FeO values of the original chemical analyses are used in the norm calculation, the points are displaced towards the left, and some nepheline normative rocks become 'quartz' normative. Vertical dash on symbol indicates 'altered' rocks (see text).

rocks showing silica contents greater than 64 percent fall in the Hawaiian tholeiitic field (Fig. 3). As shown in Figure 2, these rocks have high Differentiation Indices (DI), and more than 9 percent normative 'quartz' (see Fig. 2 for definition of 'quartz'). The extremely wide range in compositions of the Tabar-to-Feni rocks is also illustrated in Figure 2: the rocks contain between 23 percent normative nepheline and 17 percent normative 'quartz', and show a range of DI values between 4 and 95.

Potash : soda ratios

Most analysed rocks from the Tabar-to-Feni Islands have K_2O/Na_2O values greater than 0.5, and a few have values greater than 1 (Fig. 4). These values are typical of rocks from a wide range of tectonic settings, including rocks of the 'shoshonite association' (Joplin, 1968) which are believed to be characteristic of island-arcs and recently stabilised orogenic areas (cf. Jakeš & White, 1972: Mackenzie & Chappell, 1972; Johnson et al., 1973).

Potash : silica relationships

The ratios of K_2O to SiO_2 in the Tabar-to-Feni rocks show a wide range (Fig. 5), and the pattern of variation is similar to that for Na_2O+K_2O versus SiO_2 (cf. Fig. 3). Silica-poor rocks, especially, show a wide range of K_2O values, and form a steep overall trend whose slope decreases as silica increases to about 59 percent. Most of the quartz-normative rocks (greater than 64 percent SiO_2) contain only moderate amounts of K_2O.

$K_2O:SiO_2$ relationships are used widely for purposes of rock classification,

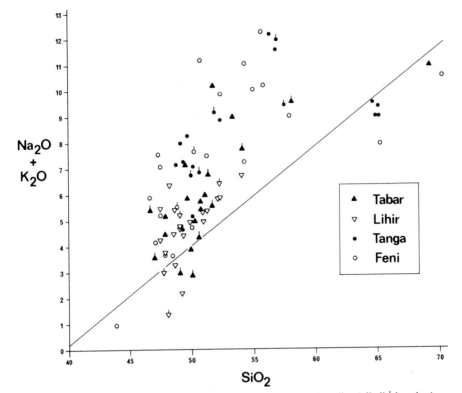

Fig. 3: Na2O + K2O versus SiO2 diagram. Straight line separates Hawaiian 'alkaline' rocks (upper field) from 'tholeiitic' ones (Macdonald, 1968), and has been extrapolated to high silica values. The line is not an effective field boundary for the Tabar-to-Feni rocks, as some nepheline-normative (alkaline) rocks fall below the line, and some hypersthene-normative (tholeiitic) ones fall above it. Vertical dash on symbol indicates 'altered' rocks (see text).

and in Figure 5 the Tabar-to-Feni samples are compared with two taxonomic schemes proposed recently for island-arc rocks. A comparison of these schemes illustrates the confusion that exists in current nomenclature for potassic rocks, particularly in island-arc regions. Some volcanic rocks from oceanic islands and continental areas also plot in the same parts of the K2O:SiO2 diagram as do many Tabar-to-Feni rocks, and this adds further confusion, as many of them have been given names different from those normally used in island-arc areas (e.g. the basalt — trachybasalt — tristanite — trachyte series of Gough, Tristan da Cunha, and Nightingale Islands; Tilley & Muir, 1964). While recognising the importance of K2O contents as possible genetic indicators, the taxonomic scheme adopted in this paper uses K2O values only as qualifiers to a system or rock names based primarily on normative mineralogy (see below).

Magnesium : iron ratios

In Figures 2 to 5 the Tabar-to-Feni rocks show broad fields of chemical variation, but in the MgO versus FeO + Fe2O3 diagram in Figure 6 they show a relatively narrow trend. None of the rocks of intermediate MgO content have the high total-iron values of tholeiitic rocks, and the general trend of slight iron-

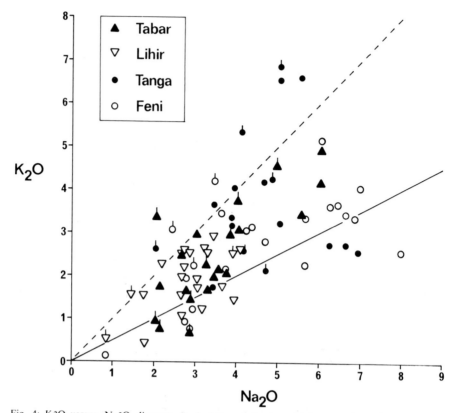

K_2O

Na_2O

Fig. 4: K_2O versus Na_2O diagram. Dashed lines indicates K_2O/Na_2O = 1. Solid line indicates K_2O/Na_2O = 0.5. Vertical dash on symbol indicates 'altered' rocks (see text). Note that more than half of the rocks in which K_2O/Na_2O is greater than 1 are 'altered' (see text).

enrichment is that shown by alkaline associations and by many hypersthene-normative island-arc associations (Fig. 6).

Titania and alumina contents

The histogram in Figure 7A shows that most rocks from the Tabar-to-Feni Islands contain less than 1 percent TiO_2, and that only one sample contains more than 1.5 percent TiO_2. It also shows that most of the Tabar and Lihir rocks contain more than 0.8 percent TiO_2, whereas most of those from Tanga and Feni have less than 0.8 percent TiO_2.

Rocks containing less than 1.5 percent TiO_2 are characteristic of basaltic rocks (DI less 50, SiO_2 less than 54 wt percent) from circumoceanic regions (island-arcs, continental margins, etc.), whereas those from oceanic islands generally contain more than 2 percent TiO_2 (Chayes, 1965). The consistently low TiO_2 contents of the Tabar-to-Feni rocks are therefore a persuasive reason for regarding these rocks as 'circumoceanic' types.

The Al_2O_3 contents of the Tabar-to-Feni rocks have a wide range — from less than 7 percent in pyroxenite inclusions to more than 21 percent in three rocks from the Tanga group (Fig. 7B). In general, however, Al_2O_3 values are high — mainly

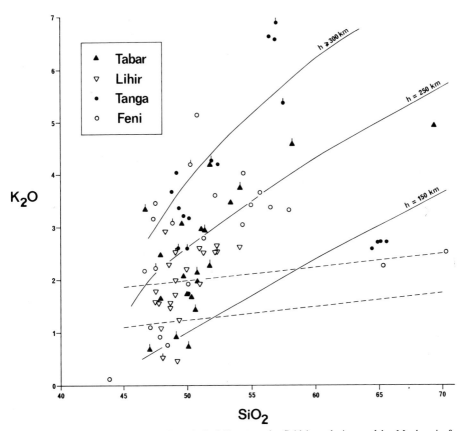

Fig. 5: K2O versus SiO2 diagram. Two dashed lines are the field boundaries used by Mackenzie & Chappell (1972) to define a lower field of 'calcalkaline' rocks, a middle one of 'high-K calcalkaline' rocks, and an upper one of 'shoshonitic' rocks; rocks containing more than about 3 percent K2O fall outside the scope of this classification. The three solid lines are schematic boundaries which correspond to depths (h) to Benioff zones beneath volcanoes in Alaska, the Izu-Marianas arc, Chile, and Indonesia, and which were proposed by Ninkovich & Hays (1972) to define a 'calcalkaline suite' (below h = 150 km), an 'Atlantic alkaline suite' (between h = 150 km and h ≥ 300 km), and a 'Mediterranean alkaline suite' (above h ≥ 300 km). Vertical dash on symbol indicates 'altered' rocks (see text).

between 16 and 20 percent — and again typical of circumoceanic volcanic rocks (cf. Chayes, 1965).

Geochemistry

Page & Johnson (1974) presented Sr^{87}/Sr^{86}, strontium, and rubidium values for 10 Tabar-to-Feni rocks. The Sr^{87}/Sr^{86} values are uniform (0.7040-0.7044) and only slightly higher than those of other Late Cainozoic rocks in the Bismarck Archipelago. They are typical of values for many other island-arc rocks, and are too low to suggest that the compositions of the magmas have been changed significantly by contamination from old radiogenic crust (Page & Johnson, 1974). Strontium contents are notably high (1223-1667 ppm). Rubidium values range between 17 and 124 ppm.

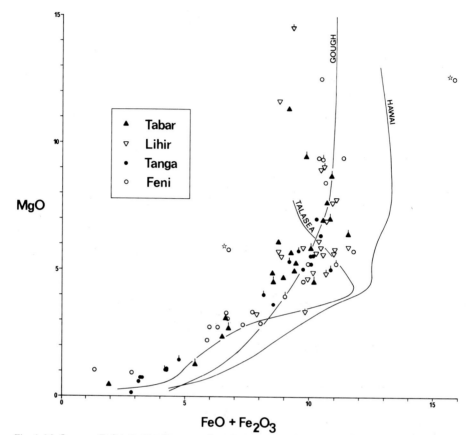

Fig. 6: MgO versus FeO+ Fe2O3 diagram, where FeO and Fe2O3 values are those obtained after use of the Irvine & Baragar (1971) transform (see text). Solid lines for Gough Island and Hawaii are drawn through points representing the 'average' rock compositions given by LeMaitre (1962) and Macdonald (1968), respectively. The 'Talasea' line is defined by 11 rocks analysed by Lowder & Carmichael (1970) from the northern part of Willaumez Peninsula, New Britain, and illustrates a trend of 'mild' iron-enrichment for a hypersthene-normative island-arc association (basalts, andesites, dacites, and rhyolites). Note that except for the two rocks from the Feni group marked with a star, the Tabar-to-Feni rocks constitute a relatively narrow band of chemical variation. Vertical dash on symbol indicates 'altered' rocks (see text).

TAXONOMY AND DISTRIBUTION OF ROCK TYPES

The rock names used in this study are based almost entirely upon whole-rock chemical compositions. The rock names are derived initially from the grid shown in Figure 8, and are assigned the prefixes 'potassic' and 'sodic' when K_2O/Na_2O values are greater than, or less than, 0.5, respectively. The amount of modal olivine distinguishes basanites (olivine greater than 1 percent by volume) from tephrites (olivine less than 1 percent). The term 'olivine nephelinite' is used here in a purely chemical sense for basaltic rocks rich in normative nepheline and without normative albite (cf. Green & Ringwood, 1968), and its use does not necessarily imply the presence of modal nepheline (cf. samples 1 and 4, Tables 1 and 2). The 'transitional' basalts (Coombs, 1963) contain low amounts of normative

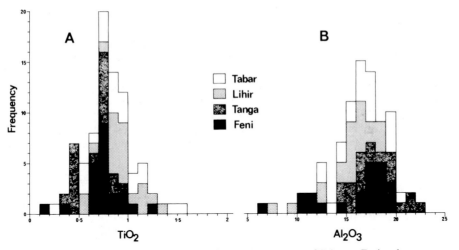

Fig. 7: Histograms for (A) TiO 2 and (B) Al2O3 contents of Tabar-to-Feni rocks.

hypersthene but show the mineralogy of alkali basalts (e.g. sample 3, Tables 1 and 2); none of the analysed basaltic rocks are true tholeiites.

Although this taxonomic system provides an internally consistent series of well-known rock names, it is essential to point out that some of these names have synonyms; for example, the 'potassic *ne*-trachyandesites' are roughly equivalent to

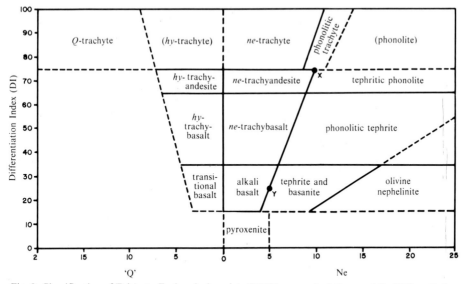

Fig. 8: Classification of Tabar-to-Feni rocks based on CIPW norm calculations and the Differentiation Index of Thornton & Tuttle (1960). Solid lines define field boundaries between rocks represented on the Tabar-to-Feni Islands. Dashed lines are approximate outer boundaries of compositional fields. The horizontal lines correspond to the Differentiation Indices used by Baker et al.'(1964) for the rocks of Atlantic oceanic islands. X-Y is the arbitrary line used by Coombs & Wilkinson (1969) to separate mildly undersaturated rock associations from strongly undersaturated ones. Q-, hy-, and ne-, indicate that quartz (*sensu stricto*), hypersthene, and nepheline, respectively, appear in the CIPW norm.

the 'tristanites' of Tilley & Muir (1964). In particular, it should be noted that the 'potassic transitional basalts', 'potassic alkali basalts', and 'potassic trachybasalts', are the same as many rocks grouped by some petrologists in the 'shoshonite association' (i.e. the absarokites, shoshonites, banakites, etc.; Joplin, 1968). However, unlike the rocks of shoshonite association, many Tabar-to-Feni rocks are strongly silica-undersaturated, and they cannot be assigned familiar names which relate them to the shoshonite association. The decision not to use rock names of the shoshonite association has also been influenced by recent accounts which ignore, side-step, or recommend abandoning these names (Streckeison, 1967; Coombs & Wilkinson, 1969; Nicholls & Carmichael, 1969; Irvine & Baragar, 1971).

The following generalisations on rock type distribution can be made from a comparison of Figures 2, 4 and 8.
1. 'Potassic' rocks appear to be more common than 'sodic' ones in all the island groups (Fig. 4).
2. The most common analysed rocks from the Tabar and Lihir Islands are low in normative 'quartz', or low in normative nepheline. Potassic trachybasalts appear to be common on both island groups, and potassic alkali basalts seem to be common on Lihir Island.
3. Rocks showing high normative nepheline values appear to be the most common types on the Tanga and Feni Islands. Potassic phonolitic tephrites seem to be especially abundant.
4. Undersaturated salic rocks (DI more than 75) have been found only on Malendok Island, in the Tanga group (ne-trachytes and one phonolitic trachyte).
5. Q-trachytes are present on Tabar Island, on Bitlik and Bitbok Islands (Tanga group), and on Ambitle and Babase Islands (Feni group).

PETROGRAPHY AND MINERALOGY

The Tabar-to-Feni rocks show an exceptionally wide range of petrographic types. In this short report it is not possible to give a comprehensive account of this mineralogical diversity, but by way of illustration the petrography and mineralogy of 11 rocks are summarised in Table 1. As shown in Table 2, these 11 rocks cover a wide range of the chemical spectrum. There is, however, a bias towards the rocks of Ambitle Island where strongly undersaturated types rich in feldspathoid and analcite phenocrysts are especially well developed. Additional mineralogical notes are provided in the Appendix.

MODELS FOR PETROGENESIS

There appears to be no simple explanation for the wide range of rock compositions represented on the Tabar-to-Feni Islands. The broad compositional spectrum of the basic rocks suggests that a wide range of primary magmas may have been produced by partial melting in the upper mantle, and the even wider range of intermediate and salic rocks suggests that, if these are differentiates of the basic rocks, many different liquid lines of descent were followed. The Q-trachytes appear to form a distinct chemical group; they could have had an origin not directly connected with the development of the more basic rocks, and of all the rock types seem to be the most likely candidates for possible derivation by crustal anatexis.

It is difficult to establish a feasible model to explain the petrogenesis of primary magmas beneath the Tabar-to-Feni Islands. This difficulty results largely from the uncertain nature of the tectonic setting that gave rise to the volcanism, and to the absence of relevant experimental data on rocks of these compositions. There are, however, no compelling geological or petrological reasons to suggest that the magmas are associated with a seafloor spreading regime, or related to relative movements of lithospheric slabs over asthenospheric 'plumes' (Morgan, 1972). At the present time, therefore, it is considered that the most feasible models are those which recognise the Tabar-to-Feni magmas as broadly 'circumoceanic' in type — i.e., they have developed in regions where plate convergence and subduction have taken place, and where partial melting in the upper mantle may have taken place under hydrous conditions.

Curtis (1973) suggested that in late Eocene to Oligocene times a submarine trench (a subduction zone) existed on the west side of a New Ireland island-arc (see Hohnen, in press), and that left-lateral transcurrent movement along the trench in the late- and post-Miocene translated New Ireland to its present position northeast of New Britain. This mechanism would destroy the trench and disconnect the downgoing part of the lithosphere beneath New Ireland. It is possible, therefore, that this disconnected slab could act as a potential source of water that led to the generation of primary magmas in the overlying mantle during the descent of the slab into the asthenosphere.

The narrowness of the Tabar-to-Feni chain, and the $K_2O:SiO_2$ relationships shown in Figure 5, could be explained by proposing a near-vertical subducted slab of lithosphere beneath the volcanic chain. The K_2O contents of island-arc rocks are believed to be a function of depth to underlying Benioff zone (e.g. Dickinson, 1970; Nielson & Stoiber, 1973; Whitford & Nicholls, this volume), and the wide range of K_2O values shown in Figure 5 might be due to primary magmas of·different compositions (1) originating from a wide range of depths immediately above a near-vertical slab, (2) rising more or less vertically through common channels, (3), fractionating, and (4) erupting from the same volcanoes in a narrow zone above the slab (note how in Fig. 5 the trend of points transects the 'h' lines of Ninkovich & Hays, 1972).

If the zone of deep water north and east of the Tabar-to-Feni Islands is the remnant of a submarine trench, a possible alternative subduction model is that a plate was consumed from the northeast, and that the downgoing slab was not disconnected by faulting. However, this model becomes unnecessarily complicated when it is developed to explain the Tertiary evolution of the New Ireland island-arc at a site farther from the zone of deep water than the volcanic chain. In addition, the seismic reflection profile presented by Furumoto et al. (in prep.) does not provide evidence that the zone of deep water is necessarily a submarine trench.

There is, in any case, no evidence for the existence of a downgoing slab — disconnected or otherwise — beneath the Tabar-to-Feni Islands. Intermediate- and deep-focus earthquakes have not been reported from beneath the volcanic chain; the raised coral reefs of the islands indicate uplift, rather than sinking of seafloor over a downgoing slab of lithosphere; and no tholeiitic basalts, andesites, and dacites (the rocks most characteristic of circum-Pacific island-arc volcanoes) appear to be present on the islands.

Table 1. Summary of petrography and mineralogy of eleven rocks from Tabar-to-Feni Islands.

Sample[a]	Rock-type[b]	Phenocrysts[c]	Groundmass[c]
1.	Potassic olivine nephelinite	31% Cpx, <1% Ol (Fo86), <1% Tm (extremely rare)	Hypocrystalline. Rounded patches of alkali-bearing chabazite (6% of total rock, up to ca. 0.4 mm across, apparently of primary origin — identification confirmed by X-ray diffraction), in matrix of Cpx, Tm, and brown glass.
2.	Potassic basanite	24% Cpx, 8% Ol (Fo90), 2% Tm, <1% Pl, <1% pseudomorphs, probably after Amph.	Hypocrystalline. Pl (An58-69), Cpx, Tm, analcite (patches), Bi, Ap, and turbid brown glass.
3.	Potassic transitional basalt	26% Pl (zoned, An66-81 in one grain), 3% Ol (Fo86, partly pseudomorphed by "iddingsite"), 2% Tm.	Holocrystalline. Pl (An62, An65), Cpx, Tm, Ap, some alteration to secondary minerals.
4.	Potassic olivine nephelinite	31% Cpx, 1% Ol (Fo84), 1% Tm.	Holocrystalline. <1% leucite microphenocrysts in matrix of Cpx, Tm, Pl (An41, An64, interstitial), and leucite.
5.	Sodic tephrite	33% Cpx, 4% Tm, 3% Pl (An76, An82), 2% analcite (n = 1.492). <1% Ol.	Holocrystalline. Pl, Cpx, Tm, and analcite.
6.	Potassic phonolitic tephrite	17% Cpx, 3% Pl (An82, An87), 3% Tm, (these 3 phenocryst minerals also in aggregates), 2% analcite (n = 1.489), <1% Amph (basaltic hornblende), <1% Ap, <1% Bi.	Holocrystalline. Cpx, Pl (An85), Bi, Tm, Amph, analcite, minor calcite, and zeolites (possible chabazite and mesolite).
7.	Potassic phonolitic tephrite	30% Pl (strong oscillatory zoning, cores of An75 to margins of An42) rimmed by Kf (Or51 Ab46 An3), 14% Cpx, 4% Tm, 1% haüyne (brownish or bluish due to alteration of pyrite inclusions, some with colourless margins), <1% Ap, <1% pseudomorphs, probably after Amph.	Holocrystalline. Pl (An31). A'clase (Or20 Ab75 An5), Tm, Cpx, rare sodalite, chalcopyrite, and secondary calcite.
8.	Potassic tephritic phonolite	11% Pl (An36-40), 6% Cpx, 4% haüyne (similar in appearance to those in sample 7), 2% Tm, 2% pseudomorphs, probably after Amph, <1% Ap, 1% sodalite.	Holocrystalline. A'clase microphenocrysts in matrix of A'clase (Or30 Ab66 An4, Or23 Ab71 An6), Tm, Cpx, and possible analcite.

9.	Sodic *ne*-trachyandesite	19% Pl (Or4 Ab54 An42 to Or7 Ab69 An24), 3% Cpx, 2% haüyne (n = 1.5045, colourless where fresh — especially at margins, greyish or brownish where altered pyrite inclusions present), 1% Tm, 1% pseudomorphs, probably after Amph, <1% Ap (dusty brown-grey, containing fine-grained inclusions).	Holocrystalline. Ab, A'clase (Or23 Ab73 An4 to Or12 Ab79 An9), Cpx, Tm, Ap, and Bi.
10.	Potassic *ne*-trachyte	5% Pl (An58), 3% haüyne (mainly colourless cores, bluish margins, and turbid brownish rims; some mainly brown, where altered pyrite inclusions present; rare crystals have rims of Kf — Or85 Ab14 An1), 2% Cpx, <1% Tm, <1% Bi.	Holocrystalline. Pl microphenocrysts in matrix of Pl (An36), Tm, Cpx, Kf and A'clase (Or55 Ab44 An1 to Or29 Ab70 An1), analcite.
11.	Potassic *Q*-trachyte	57% alkali feldspar (one analysed grain = Or7 Ab89 An4; another = Or39 Ab61), <1% Amph (tremolite), <1% Bi (largely pseudomorphed by Tm), <1% Tm.	Holocrystalline. Quartz, alkali feldspar, Amph (pale blue riebeckite), Tm, and sericite.

a Locality descriptions (numbers in parentheses are BMR registered sample numbers)
1. In situ outcrop, 1.5 km west of Nanum River outlet, southwest coast of Ambitle Island, Feni group (73680007).
2. Boulder in stream, 2 km southeast of Napekur village, Simberi Island, Tabar group (73680045).
3. In situ lava flow, stream 1.5 km south of Datawa village, Tabar Island, Tabar group (73680034).
4. In situ outcrop, Waramung Bay, west coast of Ambitle Island, Feni group (74400047).
5. Boulder near head of stream draining southwest part of caldera at Luise Harbour, Lihir Island, Lihir group (74400042).
6. In situ outcrop, 1.5 km up Niffin River, southwest side of Ambitle Island, Feni group (73680010).
7. Boulder on track south of Balum Plantation, Babase Island, Feni group (73680005).
8. In situ outcrop, 0.8 km south of Nanum River outlet, south coast of Ambitle Island, Feni group (74400048).
9. Boulder by roadside, 1.2 km south of Nanum River outlet, south coast of Ambitle Island, Feni group (73680011).
10. In situ outcrop, summit of peak above Put Plantation, southwest coast of Malendok Island, Tanga group (73680013).
11. In situ outcrop, near head of stream 1 km southeast of Tiripats Plantation, Tabar Island, Tabar group (73680037).

b Samples listed in order of increasing Differentiation Indices.
c Abbreviations for common rock-forming minerals: Ol = olivine, Fo = fosterite, Cpx = clinopyroxene, Amph = amphibole, Bi = biotite, Tm = titanomagnetite, Pl = plagioclase, An = anorthite, Ab = albite, Or = orthoclase, A'clase = anorthoclase, Kf = potash feldspar, Ap = apatite. (*ne* and *Q* refer to *normative* nepheline and quartz).

JOHNSON AND OTHERS

Table 2. Chemical analyses [a] and CIPW norms of eleven rocks listed in Table 1.

	1	2	3	4	5	6	7	8	9	10	11
SiO_2	45.8	46.6	49.0	45.9	46.6	46.0	50.3	54.5	55.8	54.9	68.5
TiO_2	0.67	0.75	1.18	0.80	0.95	0.70	0.75	0.80	0.68	0.39	0.30
Al_2O_3	10.2	12.9	13.8	12.1	15.4	17.4	18.7	17.90	17.7	21.1	16.3
Fe_2O_3	5.60	3.95	4.50	5.80	6.30	5.65	4.95	4.35	3.90	2.40	1.46
FeO	4.90	5.10	6.15	5.00	5.40	4.15	3.20	2.50	2.20	1.71	0.46
MnO	0.20	0.18	0.20	0.20	0.22	0.24	0.26	0.13	0.12	0.18	0.04
MgO	9.05	11.0	7.50	8.30	5.80	4.40	2.85	3.05	2.70	1.01	0.42
CaO	14.7	12.3	11.80	14.4	11.80	11.1	9.55	5.75	5.30	3.15	0.55
Na_2O	2.85	2.70	2.10	3.65	3.60	3.50	4.60	6.55	6.85	4.85	5.95
K_2O	2.15	1.60	1.70	2.15	1.75	3.35	2.75	3.40	3.35	6.65	4.85
P_2O_5	0.58	0.38	0.29	0.53	0.60	0.64	0.48	0.43	0.40	0.19	0.01
H_2O+	1.54	1.79	1.08	1.29	1.30	2.25	0.67	0.25	0.15	1.50	0.29
H_2O-	1.70	0.35	0.80	0.03	0.36	0.59	0.67	0.09	0.13	1.44	0.33
CO_2	0.05	0.25	0.10	0.05	0.05	0.15	0.20	0.05	0.05	0.05	0.10
Cl	0.06	0.01	0.01	n.d.	n.d.	0.03	0.18	n.d.	0.36	0.21	0.01
SO_3	0.01[b]	0.005	0.04[b]	0.03[b]	0.04[b]	0.005	0.04[b]	0.07[b]	0.11	0.04[b]	0.04[b]
S	n.d.	0.023	n.d.	n.d.	n.d.	0.028	n.d.	n.d.	—	n.d.	n.d.
sub-total	100.06	99.88	100.24	100.23	100.17	100.18	100.15	99.77	99.80	99.77	99.61
lessO=Cl	0.01	—	—	—	—	0.01	0.04	—	0.08	0.05	—
Total	100.05	99.88	100.24	100.23	100.17	100.17	100.11	99.77	99.72	99.72	99.61
D.I.	25.92	26.52	28.36	28.39	33.19	38.90	48.49	66.71	70.25	76.83	93.64

CIPW norms [d]

	1	2	3	4	5	6	7	8	9	10	11
Q	—	—	—	—	—	—	—	—	—	—	13.74
C	—	—	—	—	—	—	—	—	—	0.68	0.29
or	9.43	9.72	10.24	6.01	10.55	20.45	16.56	20.26	20.03	40.72	28.99
ab	—	8.91	18.12	—	12.70	4.11	22.80	35.31	40.30	28.53	50.91
an	9.02	18.87	23.67	10.45	21.11	22.60	22.69	9.49	7.76	14.91	2.69
lc	2.94	—	—	5.40	—	—	—	—	—	—	—
ne	13.55	7.89	—	16.98	9.95	14.34	9.13	11.14	9.93	7.58	—
di ⌠wo	26.19	17.24	14.23	24.45	14.44	12.51	9.35	6.87	6.76	—	—
di ⌡en	16.05	11.81	8.61	14.61	7.38	6.18	4.42	3.95	3.97	—	—
di ⌡fs	8.65	4.06	4.85	8.57	6.70	6.09	4.81	2.60	2.47	—	—
hy ⌠en	—	—	3.48	—	—	—	—	—	—	—	1.06
hy ⌡fs	—	—	1.96	—	—	—	—	—	—	—	—
ol ⌠fo	5.14	11.45	4.87	4.46	5.15	3.60	1.97	2.60	1.99	1.83	—
ol ⌡fa	3.05	4.34	3.02	2.89	5.15	3.91	2.37	1.89	1.36	1.69	—
mt	3.27	3.35	3.96	3.39	3.62	3.30	3.32	3.36	3.20	2.84	0.75
hm	—	—	—	—	—	—	—	—	—	—	0.96
il	1.32	1.46	2.29	1.54	1.84	1.37	1.45	1.53	1.31	0.77	0.58
ap	1.43	0.93	0.70	1.28	1.45	1.57	1.16	1.03	0.96	0.47	0.02

[a] Supplied by Australian Mineral Development Laboratories, Adelaide.
[b] SO_3 value is a total sulphur ($SO_3 + S$ as SO_3) determination.
[c] n.d. = not determined.
[d] calculated using Irvine & Baragar transform (see text) and on volatile- & S-free basis.

The possibility of another model is being explored — that the volcanic chain overlies a zone of faulting which marks the boundary of a region that has undergone isostatic readjustments. Talwani & Eldholm (1973) identified major faults at the boundaries of continental and oceanic crust in various parts of the world, attributing them to isostasy. Analogous faults may exist beneath the Tabar-to-Feni Islands between the Pacific oceanic crust in the northeast and the thicker crust of the Bismarck Sea/New Ireland region in the southwest. Evidence of these faults is provided by the seismic reflection profile given by Furumoto et al. (in prep.), and evidence for isostatic instability of the islands is provided by the raised reefs, and by Bouguer gravity anomalies (+60 and +180 mGal) which, according to Finlayson & Cull (1973), are probably due 'to the crust's supporting accumulated volcanic piles without isostatic compensation'.

In this isostatic model it is proposed that, during the early Tertiary, northeastward subduction beneath the New Ireland island-arc introduced water and partial melts of the downgoing slab to the mantle peridotite above the slab, producing hydrous minerals such as amphibole (e.g. Green, 1972), and probably an uneven spatial distribution of incompatible elements such as potassium. During the late Tertiary and Quaternary, an isostatically controlled Tabar-to-Feni faulted zone extended to depths where partial melting could take place under hydrous conditions by releases of total-pressure caused by the faulting. This hydrous melting produced a wide variety of basic primary magmas which had the low titania contents of other circumoceanic basalts, and which fractionated to give rocks containing hydrous minerals (zeolites, amphibole, and biotite).

This model is unorthodox, but is perhaps justified in that the compositions of the Tabar-to-Feni rocks are unusual in the context of Late Cainozoic volcanism in the southwest Pacific region.

ACKNOWLEDGEMENTS

Sole credit for the initiation and early development of this study is due to the late G.A.M. Taylor. We gratefully acknowledge the field work of I.A. Crick, and critical reviews of the draft manuscript by I.A. Nicholls and J.F.G. Wilkinson. The surveys by Wallace and Crick were undertaken as part of a program of regional geological mapping by the Geological Survey of Papua New Guinea. This paper is published with the permission of the Director of the Bureau of Mineral Resources, and of the Chief Geologist of the Geological Survey of Papua New Guinea.

REFERENCES

Allen, J.B. & Deans, T., 1965: Ultrabasic eruptives with alnöitic-kimberlitic affinities from Malaita, Solomon Islands. *Miner. Mag., 34,* pp. 16-34.

Baker, P.E., Gass, I.G., Harris, P.G. & LeMaitre, R.W., 1964: The volcanological report of the Royal Society expedition to Tristan da Cunha, 1962. *Phil. Trans. R. Soc. Lond., A, 256,* pp. 439-578.

Blake, D.H., & Miezitis, Y., 1967: Geology of Bougainville and Buka Islands, New Guinea. *Bur. Miner. Resour. Aust. Bull.* 93.

BMR, 1972: 1:1 million scale geological map of Papua New Guinea. *Bur. Miner. Resour. Aust.*

Chayes, F., 1965: Titania and alumina content of oceanic and circumoceanic basalt. *Miner. Mag., 34,* pp. 126-131.

Coombs, D.S., 1963: Trends and affinities of basaltic magmas and pyroxenes as illustrated on the diopside-olivine-silica diagram. *Miner. Soc. Am. spec. Pap. 1,* pp. 227-250.

Coombs, D.S. & Wilkinson, J.F.G., 1969: Lineages and fractionation trends in undersaturated volcanic rocks from the east Otago volcanic province (New Zealand) and related rocks. *J. Petrol., 10,* pp. 440-501.

Curtis, J.W., 1973: Plate tectonics and the Papua New Guinea — Solomon Islands region. *J. geol. Soc. Aust., 20,* pp. 21-36.

Denham, D., 1969: Distribution of earthquakes in the New Guinea — Solomon Islands region. *J. geophys. Res., 74,* pp. 4290-4299.

Denham, D., 1973: Seismicity, focal mechanisms and the boundaries of the Indian-Australian plate. *in* Coleman, P.J., (Ed), *The Western Pacific: island arcs. marginal seas. geochemistry.* Perth, Univ. W. Aust. Press, pp. 35-53.

Dickinson, W.R., 1970: Relations of andesites, granites, and derivative sandstones to arc-trench tectonics. *Rev. Geophys. Space. Phys., 8,* pp. 813-860.

Finlayson, D.M. & Cull, J.P., 1973: Structural profiles in the New Britain — New Ireland region. *J. geol. Soc. Aust., 20,* pp. 37-48.

Fisher, N.H., 1957: Catalogue of the active volcanoes of the world. Part 5, Melanesia. *Int. Assoc. Volc., Naples.*

Furumoto, A.S., Wiebenga, W.A., Webb, J.P., & Sutton, G.M., 1973: Crustal structure of the Hawaiian Archipelago, northern Melanesia, and the central Pacific basin by seismic refraction methods. *Tectonophysics, 20,* pp. 153-164.

Furumoto, A.S., Webb, J.P., Odegard, M.E., & Husson, D.M., in prep.: Seismic studies on the Ontong Java plateau, 1970.

Glaessner, R., 1915: Beitrag sur kenntnis der eruptivegesteine der Bismarck-Archipelo und der Salomon Inseln. *Beitr. geol, Erforsch. dtsch. Schutzgeb, 10,* pp. 1-85.

Green, D.H., 1972: Magmatic activity as the major process in the chemical evolution of the Earth's crust and mantle. *Tectonophysics, 13,* pp. 47-71.

Green, D.H. & Ringwood, A.E., 1967: The genesis of basalt magmas. *Contr. Miner. Petrol., 15,* pp. 103-190.

Gulf, 1973: Regional marine geophysical reconnaissance of Papua/New Guinea. *Rep. Gulf Research and Development Co., Sydney.*

Hohnen, P.D., in press: Geology of New Ireland. *Bur. Miner. Resour. Aust. Rep.*

Irvine, T.N., & Baragar, W.R.A., 1971: A guide to the chemical classification of the common volcanic rocks. *Can. J. Earth Sci., 8,* pp. 523-548.

Jakeš, P., & White, A.J.R., 1972: Major and trace element abundances in volcanic rocks of orogenic areas. *Bull. geol. Soc. Am., 83,* pp. 29-40.

Johnson, R.W., Mackenzie, D.E., Smith, I.E., & Taylor, G.A.M., 1973: Distribution and petrology of late Cenozoic volcanoes in Papua New Guinea. *in* Coleman, P.J., (Ed.), *The Western Pacific: island arcs, marginal seas, geochemistry.* Perth, Univ. W. Aust. Press, pp. 523-533.

Johnson, T., & Molnar, P., 1972: Focal mechanisms and plate tectonics of the southwest Pacific. *J. geophys. Res., 77,* pp. 5000-5032.

Joplin, G.A., 1968: The shoshonite association: a review. *J. geol. Soc. Aust., 15,* pp. 275-294.

Krause, D.C., 1973: Crustal plates of the Bismarck and Solomon Seas. *in* Frazer, R., (Ed.), *Oceanography of the South Pacific 1972.* N.Z. Nat. Comm. UNESCO, Wellington, pp. 271-280.

Le Maitre, R.W., 1962: Petrology of volcanic rocks, Gough Island, South Atlantic. *Bull. geol. Soc. Amer., 73,* pp. 1309-1340.

Lowder, G.G., & Carmichael, I.S.E., 1970: The volcanoes and caldera of Talasea, New Britain: geology and petrology. *Bull. geol. Soc. Amer. 81,* pp. 17-38.

MacDonald, G.A., 1968: Composition and origin of Hawaiian lavas. *Mem. geol. Soc. Amer., 116,* pp. 447-522.

Mackenzie, D.E., & Chappell, B., 1972: Shoshonitic and calc-alkaline lavas from the Highlands of Papua New Guinea. *Contr. Miner. Petrol., 35,* pp. 50-62.

Mammerickx, J., Chase, T.E., Smith, S.M. & Taylor, I.L., 1971: Bathymetry of the south Pacific, chart No. 11. *Scripps Instn Oceanogr. and Inst. Mar. Res.*

Morgan, W.J., 1972: Deep mantle convection plumes and plate motions. *Bull. Am. Ass. Petrol. Geol, 56,* pp. 203-213.

Murauchi, S., Ludwig, W.J., Den, N., Hotta, H., Asanuma, T., Yoshii, T., Kubotera, A., & Hagiwara, K., 1973: Seismic refraction measurements on the Ontong Java plateau northeast of New Ireland. *J. geophys. Res., 78,* pp. 8653-8663.

Nicholls, J., & Carmichael, I.S.E., 1969: A commentary on the absarokite-shoshonite-banakite series of Wyoming, U.S.A. *Schweize. Min. Petrog. Mitt., 49,* 47-64.

Nielson, D.R., & Stoiber, R.E., 1973: Relationships of potassium content in andesitic lavas and depth to the seismic zone. *J. geophys, Res., 78,* pp. 6887-6892.

Ninkovitch, D., & Hays, J.D., 1972: Mediterranean island arcs and origin of high potash volcanoes. *Earth. Plan. Sci. Lett., 16,* pp. 331-345.

Page, R.W. & Johnson, R. W., 1974: Strontium isotope ratios of Quaternary volcanic rocks from Papua New Guinea. *Lithos, 7,* pp. 91-100.

Streckeison, A.L., 1967; Classification and nomenclature of igneous rocks (final report of an inquiry). *N. Jb. Mineral. Abh. 107,* pp. 144-240.

Talwani, M., & Eldholm, O., 1973: Boundary between continental and oceanic crust at the margin of rifted continents. *Nature, 241,* pp. 325-330.

Taylor, S.R., Capp, A.C., Graham, A.L., & Blake, D.H., 1969: Trace element abundances in andesites. II Saipan, Bougainville and Fiji. *Contr. Miner. Petrol, 23,* pp. 1-26.

Thornton, C.P., & Tuttle, O.F., 1960: Chemistry of igneous rocks. I Differentiation Index. *Am. J. Sci, 258,* pp. 664-684.

Tilley, C.E., & Muir, I.D., 1964: Intermediate members of the oceanic basalt-trachyte association. *Geol. For. Stockh. Forh., 85,* pp. 436-444.

White, W.C., & Warin, O.N., 1964: A survey of phosphate deposits in the south-west Pacific and Australian waters. *Bur. Miner. Resour. Aust. Bull, 69.*

Yoder, H.S. Jr., & Tilley, C.E., 1962: Origin of basalt magmas: an experimental study of natural and synthetic rock systems. *J. Petrol., 3,* pp. 342-532.

Appendix. *Mineralogical Notes*

Haüyne

Phenocrysts of haüyne in samples 7, 8, 9, and 10 (Table 1) have been partly analysed with the electron microprobe (S and Cl are undetermined). As shown in Figure A, there are pronounced chemical variations between, and within, the phenocrysts, and the most common type of substitution appears to be a replacement of K by Na and Ca. In the trend of Ca substitution, electrostatic neutrality is maintained by a decrease in the ideal 1:1 Al-Si substitution, or else the Al-Si substitution is unaffected, and neutrality is apparently restored by SO4-Cl(or S) paired substitutions. (Many of the haüyne crystals contain zones of minute inclusions of pyrite which are, perhaps, evidence for S-Cl zoning).

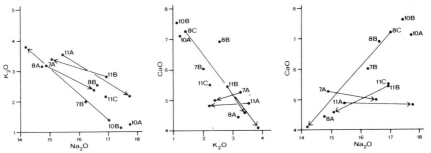

Fig. A: Weight percent K2O-Na2O-CaO variations in haüyne phenocrysts. Tie lines connect core-rim pairs (arrow heads point to rims). 7, 8, 9, and 10 are rock sample numbers (cf. Table 1), and A, B, and C are different phenocrysts in the same rock.

Analyses of the cores and rims of four haüyne phenocrysts in three different samples indicate two different patterns of zoning. One crystal in sample 8 and another in sample 10 show an increase in K and a decrease in Na and Ca from core to rim. A second crystal in sample 10, however, and one in sample 7, show a different pattern: from core to rim, K decreases, Na increases, and Ca shows only a light decrease (i.e., more or less a simple K-Na substitution). These preliminary results indicate complexities in the types of possible substitutions in haüyne, and the need for more detailed studies with the microprobe.

Analcite

Twelve phenocrysts and groundmass grains of analcite in five rocks have been analysed with the microprobe. They show a restricted and erratic compositional variation, even though they are present in a wide variety of rock types (e.g. samples, 2, 5, 6, and 10). The analcite phenocrysts in any one rock do not necessarily show identical compositions (e.g. sample 6).

The most obvious chemical variation is a slight Al-Si substitution (SiO2 = 50-57%, Al2O3 = 27-23%). Wilkinson (1965) noted that the principal replacement of Si by Na and Al in analcite from the Square Top intrusion (New South Wales) decreases as temperature of crystallisation decreases. In contrast, the compositions of the analcite in the Tabar-to-Feni rocks do not appear to vary significantly from one rock to another, and therefore appear to be independent of temperature.

Iron-titanium oxide

Optically homogeneous titanomagnetite is the only iron-titanium oxide found in the eleven rocks examined with the microprobe. The ulvöspinel contents (calculated by the method of Carmichael, 1967) of these titanomagnetite crystals are much lower than in those of other basic igneous rocks: in some speciments (e.g. no. 4) there is insufficient Ti to even satisfy the Mn, Mg, and Ca components of the ulvöspinel molecule ($2R''O.TiO_2$), so that a small amount of Mg-Al spinel is also present in solid solution.

The analysed Tabar-to-Feni rocks have high Fe_2O_3 and TiO_2 contents. As these are also the characteristics of the analysed, optically homogeneous titanomagnetite crystals, it seems likely that the apparently high oxidation state of the rocks is not due to secondary oxidation processes, but rather is a reflection of a high oxidation state in the magmas before their eruption.

Clinopyroxene

The analysed clinopyroxene phenocrysts of all the rocks are pale yellow-green salite. Many of them show weak pleochroism, and oscillatory zoning is common. The margins are more Fe-rich than the cores, but crystals in sample 4 have margins which are *lower* in Fe, Ti and Al, and higher in Si and Mg, compared to the cores, because of the presence of titanomagnetite grains in the margins.

References

Carmichael, I.S.E., 1967: The iron-titanium oxides of salic volcanic rocks and their associated ferromagnesian silicates. *Contr. Miner. Petrol, 14,* pp. 36-34.

Wilkinson, J.F.G., 1965: Some feldspars, nephelines, and analcimes from the Square Top intrusion, Nundle, N.S.W. *J. Petrol., 6,* 420-444.

ERUPTIVE HISTORY OF BAGANA VOLCANO, PAPUA NEW GUINEA, BETWEEN 1882 AND 1975

R.J. BULTITUDE

Bureau of Mineral Resources, P.O. Box 378, Canberra City, A.C.T. 2601

ABSTRACT

Bagana is a highly active volcano on Bougainville Island and, unlike the other active volcanoes in Papua New Guinea, appears to be built mainly of thick steep-sided blocky andesitic lava flows with rounded steep-sided fronts up to about 150 m high; pyroclastic deposits are relatively minor. The eruptive history of Bagana before about 1950 is poorly documented and largely unknown. Since 1947 the volcano has been almost continuously active, the longest interval with no reports of any abnormal activity being about 6 years (from late 1953 to late 1959). Eruptions are marked mainly by the effusion of thick, slow-moving, blocky lava flows from the summit area and violently explosive activity which produces nuées ardentes and voluminous ash clouds. Major eruptions occurred in 1950, 1952, and 1966. Lava extrusion may continue for months, even years, and is not always accompanied by explosive activity. Lava was extruded more or less continuously from 1972 to early 1975, but no reports of any explosive activity were received at the Central Volcanological Observatory, Rabaul. The characteristic activity between eruptions is the almost continuous emission of voluminous white vapours from the summit area.

INTRODUCTION

Bagana volcano (06°09'S; 155°11'E) rises about 1,750 m above sea level (Branch, 1967), 19 km northeast of Torokina, in central Bougainville (Figs 1, 2). With the possible exception of Manam, Bagana has been the most consistently active volcano in Papua New Guinea (Fisher, 1954, 1957). However, very little information has been published on the observed eruptive histories of either volcano. This paper is the first attempt to comprehensively collate observations of eruptions from Bagana, and to a large extent it parallels a similar paper on Manam volcano (Palfreyman & Cooke, this volume). Sixteen new analyses of rocks from Bagana are also presented; many of the specimens come from lava flows erupted since 1950.

GEOLOGICAL SETTING

Bougainville Island is about 203 km long, up to 62 km wide, and up to about 2,590 m high and is the largest island of the Solomon Group, a northwesterly-aligned island chain on the southwestern margin of the Pacific Ocean. Bougainville is made up of mainly Cainozoic volcanic rocks, sedimentary rocks derived from the volcanics, and subordinate Lower Miocene and Pleistocene limestones; it is dominated by an axial chain of high mountains formed mainly on post-Miocene volcanoes and derived pyroclastic and sedimentary deposits (Blake & Miezitis, 1967; Speight, 1967; Blake, 1968). There are also several dioritic to granodioritic intrusions, some of which may be the cores of deeply eroded volcanoes. A large porphyry-copper type deposit, associated with the intrusion of quartz diorite and granodiorite into andesite, is being mined at Panguna, in southern Bougainville (Blake & Miezitis, 1967; Macnamara, 1968). Pliocene to Recent volcanics cover more than half the island and are the products of several named and unnamed volcanoes (Fig. 2), of which the best known are the active Bagana volcano and the

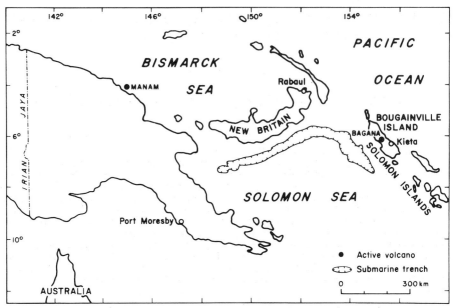

Fig. 1: Papua New Guinea and the northwestern part of the British Solomon Islands.

dormant Balbi and Loloru volcanoes (Blake & Miezitis, 1967; Blake, 1968). The other volcanoes on the island are considered to be extinct.

    Bougainville occupies an island-arc environment (see Coleman, 1970). A deep submarine trench, thought to mark the boundary between converging plates, lies south of New Britain and west of Bougainville (Fig. 1; Denham, 1969, 1973; Johnson & Molnar, 1972; Curtis, 1973a), where it reaches a depth of about 9,100 m (Denham, 1969). Earthquake activity is high in the vicinity of Bougainville Island, and many intermediate-focus earthquakes have been reported (Denham, 1969, 1973; Curtis, 1973a). A well-defined Benioff zone dips northeastwards at between 80^0 and 90^0 beneath Bougainville (Denham, 1969). The orientation of the Benioff zone and the focal-mechanism solutions for this boundary suggest that the floor of the Solomon Sea underthrusts Bougainville in a northeasterly direction (Denham, 1969, 1973; Ripper, 1970; Johnson & Molnar, 1972; Curtis, 1973a, b; Krause, 1973). Estimates of the rate of subduction of the underthrust slab range from about 7.8 to 11 cm per year (Curtis, 1973b; Krause, 1973). Cross-sections presented by Denham (1969) and Curtis (1973a) show that very few earthquakes with epicentres at depths between about 200 and 400 km have been recorded in the Bougainville region. The apparent gap in activity may indicate that the lower part of the subducted slab beneath Bougainville has become detached from the major part of the plate and has sunk under its own weight (see Oliver et al., 1973; also see Halunen & von Herzen, 1973, for an alternative explanation) and that the deep-focus earthquakes are caused by the disjoined section of lithosphere.

    The post-Miocene volcanoes of Bougainville are built up of calcalkaline lavas and pyroclastic deposits, mostly of andesitic composition, although some dacitic rocks are also present (Blake, 1968). The chemistry and mineralogy of the rocks

Fig. 2: Sketch map of Bougainville Island.

have not been studied in great detail, but Taylor et al. (1969) described low-Si and high-K andesites from the island and postulated that they originated by processes involving seafloor spreading and transportation of oceanic crust down the steeply dipping seismic plane beneath Bougainville, where it was remelted to form andesites.

BAGANA VOLCANO

The volcano forms a roughly symmetrical cone with slopes of about 30°, rising above deeply dissected mountains to the south, north, and east (Fig. 3). The volcano is situated at the head of the drainage basins of the Torokina and Saua Rivers and is flanked to the west and southwest by a series of low-lying, coalescing, volcano-alluvial fans made up mainly of volcaniclastic debris from Bagana. The superficial parts of the cone are built predominantly of thick, blocky andesitic lava

Fig. 3: Bagana volcano from the north on 5 May 1960, showing some of the lava flows erupted during the previous seventeen years (see Fig. 4), and voluminous white vapours issuing from the summit area. Photograph taken by G.A.M. Taylor.

flows, but volcaniclastic debris is interspersed between the flows. The surfaces of the flows consist of a jumble of loose, angular, smooth-sided blocks of massive to moderately vesicular andesite up to several metres across and are commonly characterised by hummocks and hollows and prominent transverse, arcuate pressure ridges. Older, deeply eroded lava flows have crudely jointed, massive interiors up to about 30 m thick. The more extensive flows have reached the base of the cone, and some abut on and partly overlie the foothills of the adjacent mountains.

The flows are generally bounded for most of their lengths by prominent marginal levées which project up to about 60 m above the central axes of the flows. The depths of the troughs below the enclosing levées decrease down slope, and near the base of the cone the levées lose their identities and the lavas spread out to form broad, steep-sided flows with rounded flow-fronts up to about 150 m high and several hundred metres wide; most flows are less than 50 m wide near the summit. The inner walls of the levées are vertical or dip steeply inwards towards the central axes of the flows and are generally worn smooth as a result of abrasion by lava debris carried down by the hotter, more mobile, central parts of the flows.

Violent explosive activity has also been a feature of many eruptions. The northern flanks of the cone are littered with breadcrust bombs (up to about 1 m long) and large blocks of lava (up to about 1 m across), and commonly show impact craters up to about 1 m deep and 2 m in diameter.

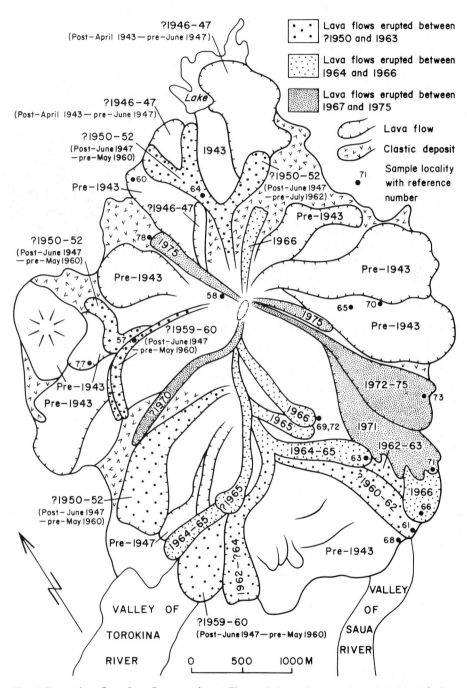

Fig. 4: Recent lava flows from Bagana volcano. The small dome shown partly encircled by lava flows at the foot of the cone on the northwestern side of the volcano is densely vegetated and poorly exposed; it may have formed relatively early in the development of Bagana volcano or it may be due to unrelated volcanic activity. The sample numbers shown are all preceded by the prefix 744000-.

SUMMARY OF ERUPTIVE HISTORY BETWEEN 1882 and 1975

The following account is a summary of published data, of unpublished reports of the Bureau of Mineral Resources, and of data in the files of the Central Volcanological Observatory, Rabaul.*

The data on eruptions from Bagana are not comprehensive; many of the records are incomplete, and some are conflicting. Bagana is isolated geographically, and the area around the volcano is only sparsely populated by Europeans. Furthermore, the upper part of the cone is shrouded in clouds for a large part of most days, and much minor activity has probably not been reported. Most of the reports of activity came from Europeans living on Bougainville and from the crews of boats and aircraft in the area. Several photographs of the volcano taken between 1943 and 1974 have also been examined by the writer, and form the basis of the sequence of lava flows shown in Figure 4. Because of inadequate records, difficulty was found in correlating some lava flows with reported effusive activity, particularly in the 1950s and 1960s, when the volcano was not frequently inspected. Much of the apparent conflict in the reports probably stems from the fact that slow extrusion of lava sometimes continued intermittently for years over much the same path.

Activity prior to 1948

Guppy (1887a, p.VIII) reported that in 1882 Bagana was the only volcano in 'active eruption' in the Solomon Islands. He also reported being told by a native chieftain of a 'great explosion' (p. 22) from Bagana in December 1883 or January 1884 which killed 'several natives'. According to local informants, the volcano had been in more or less continuous eruption during at least the previous 15 to 20 years (Guppy, 1887b).

In September 1937, loud explosions from Bagana were heard at Kieta, about 50 km away (Fig. 1). Ash ejected during a severe eruption on 15 May 1938 fell on Kieta (Fisher, 1939), and the upper part of the cone became mantled by loose, hot lava blocks; a large area southwest of the mountain was covered by ash. Water vapour issued from the flanks of the volcano in numerous places, and boiling springs and hot quicksands were reported from around the base of the cone.

Aerial photographs taken in 1943 show abundant white vapour issuing from the summit area of Bagana, and those taken on 7 April 1943 show a newly erupted lava flow extending down the north-northeastern flank of the volcano (Fig. 5). A later wartime aerial photograph shows the lava flow extending to the base of the cone and bounded by prominent marginal levées for most of its length.

In 1945-46 'light' explosions accompanied by ejection of 'scoria' were reported (Fisher, 1957, p. 73). Aerial photographs taken on 20 June 1947 show voluminous white vapour being emitted from the summit area, and another new lava flow with particularly prominent marginal levées extending down the north-northeastern side of the cone. This flow was extruded over the top of the 1943 flow on the upper part of the cone, but bifurcated about half-way down the north-northeastern flank (Fig. 4).

* The writer acknowledges the use of unpublished reports by G.A.M. Taylor, J.G. Best, M.A. Reynolds, G.W. D'Addario, R.F. Heming, I.E. Smith, M.G. Mancini, W.D. Palfreyman, R.A. Davies, and R.J.S. Cooke of the Volcanological Observatory.

Fig. 5: Bagana from the north on 7 April 1943, showing emission of white vapours from the summit area and a newly erupted lava flow extending more than half way down the northern side of the cone. Note that lava in the axial parts of the flow is darker than lava along the margins and forming the surface of the adjacent older flows. In the foreground is the crater of the extinct Billy Mitchell volcano. The deeply dissected extinct Reini volcano lies between, and to the left of, Billy Mitchell and Bagana volcanoes. Behind Bagana are the channels of the Saua (left) and Torokina (right) Rivers. Photograph taken by U.S. Air Force.

Activity between 1948 and 1953

Towards the end of 1948, spasmodic explosive activity commenced and continued for about 12 months (Fisher, 1957), after which the frequency and intensity of eruptions increased. Eruptions during 1950 produced ash clouds, nuées ardentes (the first reported from Bagana) and lava flows, and a lava dome was extruded in the summit crater. The most violent eruptions were from June to early October, and the height of the activity took place at the end of this period when there were up to five eruptions a day, each accompanied by a loud roaring noise. Ash was ejected up to about 10,000 m, and new lava flows were reported. The nuées ardentes completely devastated the upper Saua River valley (Fig. 4). Subsequent mudflows filled the channel of the river, and much of the low land towards the coast was flooded.

In December 1950, activity was confined to summit vapour emission, extremely slow movement of a lava flow down the south-southwestern flank, and spasmodic rumblings (Taylor, 1956). Forest had been flattened by the passage of recent nuées ardentes up to about 3 km from the base of the mountain on the southern and western sides. The fallen tree trunks were orientated radially away from the crater. Bark on standing stumps was bruised and pitted on the sides facing the volcano but was not charred.

Bagana was 'active throughout most of 1951' (Fisher, 1957, p. 73), but activity intensified on 29 February 1952 and continued until October, producing audible rumblings, earth tremors, explosions, nuées ardentes, and lava flows. A new crater formed on the northern side of the summit area, and by 10 March 1952 a new lava flow extended down the southwestern flank of the volcano to the base of the cone (Fig. 4). J.G. Best observed 15 nuées ardentes and 45 explosive eruptions between 18 March and 31 March 1952; most of the nuées ardentes failed to reach the base of the volcano, but new vegetation in part of the area devastated by nuées ardentes in 1950 was stripped and flattened up to about 1 km from the base of the volcano on the western side. Ash from one of the nuées was still hot (90°C) when collected about an hour after deposition. Three of the explosive eruptions observed by Best were extremely violent; ash-laden clouds were ejected to about 10,000 m above sea level. Incandescent blocks of lava were commonly observed at night tumbling down the western flank of the cone from the summit area.

Ash clouds, explosive eruptions, possible nuées ardentes, and rumblings were produced between June and September 1953. A new crater, reported to have formed on the northern side of the summit during June, produced a series of powerful explosions. Particularly violent eruptions took place on 10 and 25 July and during August 1953, but there were no further reports of activity until 1959.

Activity between 1959 and 1961

In January 1959 G.A.M. Taylor reported that the viscous lava flows produced during the 1948-53 activity had descended the northwestern and southwestern flanks to the foot of the cone. The summit crater was reported to be completely filled by an irregular, hummocky lava dome. In September frequent earth tremors were felt in the vicinity of the volcano, and continuous emission of vapour clouds, frequent audible explosions, and a glow in the summit area at night were reported up to early May 1960, when Bagana again erupted with the ejection of ash clouds to about 3,000 m.

During an aerial inspection on 5 May 1960 G.A.M. Taylor observed a lava flow which appeared to be hot and moving down the southwestern flank of the volcano. Photographs taken during the inspection show what appears to be the new flow extending down the side of the cone towards the head of the Saua River and another recent flow extending down the southwestern flank of the volcano to the head of the Torokina River (Fig. 4).

The only reported activity in 1961 was on 26 July when brown, ash-laden clouds were seen rising above the summit area.

Activity between 1962 and 1965

Bagana next erupted on 15 February 1962. Major activity — marked by earth tremors, emission of ash clouds, and effusion of lava — continued until April. Part of the eastern rim of the crater was reported to have been destroyed. Many of the streams issuing from the base of the cone were hot, and dead fish were common in some.

In July 1962 lava was seen descending the south-southwestern flank of the cone. A divided vapour column suggested that the lava was issuing from two

Fig. 6: Vertical aerial photograph of Bagana volcano taken on 24 June 1963, showing recent lava flows on the southern and southwestern flanks of the volcano (see Fig. 4). Photograph taken by Royal Air Force.

separate vents within the crater. The southeastern part of the crater was reported to be growing and blocks of lava were seen tumbling from this part of the summit. Lava was extruded from the southwestern part of the crater for some time and branched into two separate flows on the lower slopes of the volcano (Figs 4, 6). By 24 June 1963 the two flows had extended to the foot of the mountain — one on the southern flank, the other on the south-southwestern flank (Figs 4, 6).

Comparison of aerial photographs taken in 1960, 1962, and 1963 with those taken in 1947 confirm that in the intervening years several lava flows had been extruded. Their distribution is shown in Figure 4.

Extensive minor damage was caused in southern Bougainville by earth tremors between 17 and 23 April 1964, and on 24 April loud explosions and vast 'mushroom-shaped' eruption clouds emanated from Bagana. Between 6 and 9 October the most prominent activity was the continuous emission of voluminous white vapour clouds from the summit area, with 'mild steam' explosions from a small crater in the western side of the summit area about every 10 minutes (Branch, 1967, p. 11). A deep red glow was seen in the crater at night. A new lava flow, most probably initiated in late April 1964, had descended the south-southwestern flank of the volcano (Fig. 4) and was slowly advancing at the rate of a few centimetres per hour (Branch, 1967). There was no evidence of any recent nuées ardentes.

Activity in 1966 and 1967

Build-up of activity early in 1966 culminated in a large eruption at the end of May. Increased vapour emissions from fumaroles on the western flank of Bagana, and fluctuating glows at night in the summit area and from several places on the flanks of the volcano, were reported on 20 March. Incandescent blocks of lava were observed tumbling down the southwestern slopes of the volcano between 20 and 25 March, resulting in several reports that a new vent had opened up on the western flank of the volcano.

On 1 April an aerial inspection by mining company geologists showed that the summit crater was full (presumably due to growth of a lava dome), but no lavas were reported flowing down the outer flanks. No abnormal activity was detected by a volcanologist from the Observatory during an aerial inspection of the volcano on 4 April.

Observed activity throughout May included: lavas flowing down the southern, southwestern, and eastern flanks of Bagana; severe explosions on 5 and 7 May; red glows in the summit area on nights between 16 and 26 May; and lava flowing on 26 May on the southern and northwestern flanks (with minor extrusion of lava on the eastern flank). Voluminous dark grey ash-laden clouds were erupted on 27 and 29 May, and a new vent was reported at the southwestern end of the summit.

On 30 May a large eruption sent a nuée ardente down the southern slopes of the volcano. A northeast-trending ridge at the foot of the cone diverted most of the dense basal part into the Saua River valley, where the nuée reached a point about 9.5 km from the crater. At least part of the ash cloud rising above the basal part of the pyroclastic flow jumped the ridge and continued southwards, destroying a fan-shaped area of forest extending up to about 6 km from the base of the volcano. Up to 4 m of pyroclastic and lava debris were deposited by the nuée on the floor of the upper Saua River valley; the sides of the valley were almost completely stripped of vegetation and most trees were flattened. Movement over irregular terrain produced vortices in the nuée so that, in places, trees were felled in the direction from which the nuée descended. The larger of the two newly erupted lava flows extending down the southern side of the cone had reached the lower slopes of the volcano by the time of the ground inspection between 10 and 13 October 1966. The lava flow was reported to have been extruded through a prominent breach in the southern wall of the crater and was still slowly moving. By 12 April 1967 the front of the larger of the 1966 lava flows on the southern flanks of the volcano had reached the head of the Saua River valley (Fig. 4).

Fig. 7: Close-up of hummocky protuberance in the more extensive of the 1966 lava flows on the southern flank of Bagana volcano. Photograph taken by R.F. Heming on 23 September 1967.

Bagana was inspected from the air by R.F. Heming on 23 September 1967, following a report of emission of a dark brown ash cloud. The summit area was described as containing a clearly defined crater, about 200 m in diameter, with a small lava dome which spilled through a break in the south-southeastern rim of the crater. Dense white vapour clouds were being continuously emitted from the surface of the dome and from numerous small vents around the crater rim. A new lava flow about 160 m long was observed to the east of the larger 1966 flow on the southern flank. Heming postulated that lava had been extruded more or less continuously since May 1966 and that more mobile lava had recently broken through the solidified crust of the 1966 flow in the summit area and started a new lava flow. The front of the 1966 flow at the head of the Saua River valley had not moved since the previous April, but the steep upper parts of the flow had been mobile in the interim. Withdrawal of more mobile lava from the axial parts had caused the centre of the flow to subside, producing prominent marginal levées. Lava had piled up on the lower slopes of the cone and formed a large protuberance near where the lava flow made a fairly sharp bend and flattened out (Fig. 7).

Emission of black ash-laden clouds was reported on 23 October and 22, 23, and 30 November 1967.

Activity between 1969 and 1974

Activity during an aerial inspection on 4 May 1970 consisted of emissions of dense white vapour from many parts of the summit crater, including the surface of a lava dome. This dome, which began growing after the May 1966 eruption, appeared to fill the old crater and almost to have reached the level of the crater rim. It was reported that there was very little change in the state of activity since the last inspection on 3 December 1969.

A fresh overflow of blocky lava, less than 300 m long, was observed blocking the southeastern breach in the crater rim during an aerial inspection on 6 June 1970, and some fresh lava was also noted on the northwestern side of the cone. The lava dome was reported to be at a lower level than when inspected in December 1969. Continuous emission of white vapour from the crater rim and lava dome and intermittent ash-laden vapour ejections, some accompanied by rumbling explosions, took place during the latter half of 1970 and the first half of 1971. On 11 October 1970 a blocky lava flow of recent origin was reported to extend down the western side of the cone to about 1,000 m above sea level. Incandescent blocks of lava were observed cascading down the front of the flow, and white vapour was being discharged from its upper surfaces. This flow had not been seen during the last inspection of Bagana in June 1970, and was not mentioned in later Observatory reports.

Explosive volcanic activity was renewed in March 1971, reaching a climax late in the month. A new blocky lava flow was observed on the southern flank of the cone on 23 March and 2 April, and was reported to have been extruded through a breach in the southeastern section of the crater wall; it followed much the same path as, and extensively covered, the larger of the flows extruded in 1966 down the southern flanks (Fig. 4). High marginal levées and transverse pressure ridges were well developed. On 23 March the crater appeared to be completely filled with blocky lava and a vent had formed near the centre of the crater, but by 2 April some of the lava had spilled through the breach in the crater rim.

Activity during late March and early April 1971 consisted of single explosive events which produced voluminous 'cauliflower-shaped' ash-laden clouds up to about 5,500 m above sea level. The frequency and magnitude of these explosive events slowly diminished during April, and activity had ceased altogether by 24 April.

A new blocky lava flow extending to the base of the cone was observed during the aerial inspection on 24 February 1973. A single report of a red glow in the summit one night, a vague story of a possible explosion heard from a remote area, and second-hand reports (undated) of red glows seen by villagers are the only events reported in 1972 and early 1973. The volcano had been previously inspected on 9 December 1971, but no new lava flows were reported. However, photographs taken during the inspection show a recent lava flow at the foot of the cone on the eastern side of the main 1971 lava flow (Fig. 4). It is highly likely that this flow was a forerunner to, and forms part of, the voluminous lava flow observed in February 1973 (and labelled 1972-75 in Fig. 4). A huge cloud of brownish 'smoke' had been reported billowing out of the side of the new lava flow, about 150 m below the summit, on 20 February 1973. In the light of the writer's observations between 15

Fig. 8: Upper part of Bagana on 18 November 1973 from the eastern side of the 1972-75 lava flow, showing: (1) recently erupted dark lava in the central parts of the flow and dull grey, older, more weathered lava on the margins of the flow; (2) voluminous white vapours emanating from the southeastern part of the summit area and from an apron of recently erupted lava on the upper southeastern part of the cone; (3) the distinct notch in the rim of the summit area to the left of the 1972-75 flow which appears to have been extruded over a high part of the crater rim; and (4) the grey streak (arrowed) down the side of the 1972-75 lava flow marking the site of rock falls from the top of the flow.

and 27 November 1973 it seems certain that the cloud was produced by lava debris avalanching down the side of the lava flow. Most of the vapour emissions observed on 24 February 1973 originated from a lava dome which did not appear to overlap the crater rim. It was reported that the newest flow and the other recent flows did not appear to have flowed through the prominent breach in the southern rim of the crater, but to have overflowed relatively high parts of the southeastern rim.

Lava was still being slowly extruded from the summit when the writer visited the area in November 1973. Blocks in the upper central part of the new flow reported on 24 February 1973 had fresh black surfaces whereas earlier extruded lava blocks along the margins of the flow and on the surface near the front of the flow had dull grey weathered surfaces (Fig. 8). Lava in the central channel was at about the same height as the stationary margins. Many of the blocks at and near the front of the flow were warm; a few were too hot to handle. White vapour issued under low pressure from numerous points along the length of the flow. Smells of hydrogen sulphide were particularly strong down-wind from the summit crater. On several nights a dull red glow was observed in the summit area. Glimpses of a lava dome were obtained from a vantage point about 100 m below the summit on the southeastern flank of the volcano, and extensive deposits of sulphur (cf. Branch, 1967) were observed on the upper part of the cone from high on the northern flank of the volcano.

Avalanches of loose lava blocks were common from the summit area, down the steep sides of the new flow, and off the face of the flow, particularly where it took a sharp downward bend about half way up the cone. The cascading blocks generated large billowing clouds of pale brown dust (finely comminuted lava debris) formed by attrition between blocks in the slow-moving lava. Blocks of incandescent lava tumbling down the flow face gave rise to red streaks in the night and generated voluminous dust clouds that obscured the sky.

Quiet effusive activity continued during 1974 and early 1975 resulting in the building of a very thick broad lava flow on the lower slopes of the cone. However, the lava flow appeared inactive when inspected from the air by R.J.S. Cooke on 26 June 1975. Lava in the central channel of the flow had drained downhill leaving high marginal levées. A small flow of recent origin was reported on the southeastern side of the cone, adjacent to the 1972-75 flow, and lava was observed flowing down the north-northwestern flank (Fig. 4). The prominent lava dome observed by Cooke on 24 February 1973 had been destroyed and a new dome was reported to be building up in the summit area.

PETROLOGY

Sixty thin sections of volcanic rocks from Bagana have been examined. Chemical analyses of 16 specimens are presented in Table 1. In common with most volcanic rocks from island-arcs, those of Bagana are highly porphyritic, especially in plagioclase (Table 1). Clinopyroxene (probably mainly augite) is the next most abundant ubiquitous phenocryst, and all the rocks examined contain minor amounts of iron-titanium oxide phenocrysts. Some rocks contain rare small olivine phenocrysts (less than 1 percent by volume) which, in a few samples, are mantled by small grains of calcium-rich clinopyroxene; coronas of calcium-poor pyroxene are absent. The majority of the lavas contain no olivine but minor orthopyroxene as small, faintly pleochroic phenocrysts and granules, and sparse, but large and conspicuous, phenocrysts of brown hornblende, generally surrounded by reaction rims of mainly opaque oxide commonly with clinopyroxene and plagioclase. Groundmasses range from glassy to microcrystalline. The most readily identified minerals in the micro-crystalline groundmasses are plagioclase, clinopyroxene, iron-titanium oxides, orthopyroxene or olivine (or both), and apatite. At least one specimen contains rare prominent anhedral grains of groundmass quartz.

Some lavas contain scattered inclusions, the most common type bearing abundant large white plagioclase phenocrysts. These inclusions superficially resemble a porphyritic microdiorite intrusive exposed beneath the lavas near the head of the Saua River. Rare inclusions of andesite and layered gabbro were also found.

The rocks display only a very limited range in major-element abundances (Table 1; Figs 9a, b) and are similar in composition to those previously described

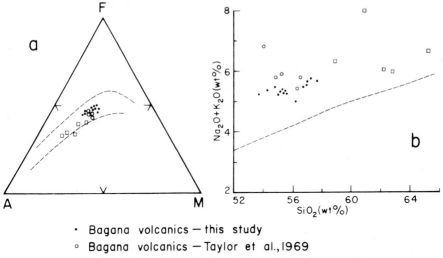

- Bagana volcanics — this study
- Bagana volcanics — Taylor et al.,1969
- Other Bougainville volcanics—Taylor et al.,1969

Fig. 9: Chemical variation diagrams for Cainozoic volcanic rocks from Bougainville Island. a: AFM (total iron as FeO) diagram. The limits of Kuno's (1968; p. 649) 'hypersthenic rock series' (calcalkaline series) are shown by the broken lines. b: Total-alkalis versus silica diagram. The boundary between Kuno's (1968; p. 627) 'pigeonitic rock series' (tholeiitic series — below) and his 'high-alumina basalt series' (above) is shown by the broken line.

from Bagana by Blake & Miezitis (1967) and Taylor et al. (1969). They are quartz-normative, predominantly low-silica (53-57 percent, Taylor et al., 1969) andesites with major-element abundances which appear to be typical of calcalkaline andesites from island-arcs. In recent years, the cause of variation in potassium content of volcanic rocks in island-arc regions has received much attention. Rocks with similar K_2O contents (for comparable SiO_2 contents) have been reported from several parts of Papua New Guinea (Jakeš & White, 1969; Jakeš & Smith, 1970; Mackenzie & Chappell, 1972; Johnson et al., 1973), but the significance of the chemistry of the andesites from Bagana cannot be fully evaluated until the chemistry of volcanic rocks from other post-Miocene volcanoes on Bougainville has been studied in detail. Andesites with K_2O contents greater than 2.5 percent have been reported from Balbi volcano and Malabita Hill on Bougainville (Blake & Miezitis, 1967; Taylor et al., 1969).

Cainozoic volcanic rocks so far analysed from Bougainville fall in a straight band extending from near the centre of the AFM diagram towards the alkalis apex (Fig. 9a), although they are more enriched in iron than calcalkaline volcanics from the eastern Papua New Guinea mainland (Jakeš & White, 1969; Johnson et al., 1973). They plot mainly in the field of the 'high-alumina basalt series' of Kuno (1968, p. 627) on a total-alkalis versus silica diagram (Fig. 9b), and show only a gradual overall increase in total alkalis with increasing SiO_2 content. Analyses of rocks from Bagana listed by Blake & Miezitis (1967) and Taylor et al. (1969) tend to have slightly higher total-alkalis than those listed in Table 1 (also see Fig. 9b). The reason for this is not known, but it could possibly be due to systematic analytical errors. Strontium isotope analyses of four of their specimens (from

Table 1. Chemical* and modal analyses of 16 rocks from Bagana volcano.

	1	2	3	4	5	6	7
SiO_2	53.7	53.7	54.3	54.8	55.0	55.2	55.3
TiO_2	0.96	0.88	0.85	0.83	0.85	0.81	0.82
Al_2O_3	17.8	17.1	17.6	17.8	17.8	17.9	18.2
Fe_2O_3	4.10	4.15	3.90	4.35	4.25	3.45	4.45
FeO	5.10	4.95	4.80	4.10	4.25	4.85	3.70
MnO	0.18	0.17	0.17	0.17	0.17	0.18	0.18
MgO	3.85	4.25	3.85	3.65	3.45	3.35	3.40
CaO	8.30	8.55	8.50	8.05	8.20	8.10	7.95
Na_2O	3.60	3.60	3.75	3.80	3.70	3.75	3.75
K_2O	1.60	1.61	1.61	1.65	1.50	1.57	1.61
P_2O_5	0.35	0.36	0.36	0.35	0.34	0.36	0.35
H_2O+	0.02	0.17	0.02	<0.01	0.06	0.04	<0.01
H_2O-	0.08	0.07	0.06	0.06	0.06	0.08	0.05
CO_2	<0.05	0.05	0.05	0.05	0.05	0.05	<0.05
Total S	0.01	0.02	0.015	0.01	0.01	0.01	0.01
Total	99.65	99.63	99.84	99.67	99.69	99.70	99.77

CIPW norms

	1	2	3	4	5	6	7
Q	4.41	4.11	4.33	5.86	7.00	6.12	7.26
or	9.50	9.57	9.54	9.79	8.90	9.32	9.54
ab	30.59	30.64	31.80	32.27	31.43	31.85	31.81
an	27.82	25.91	26.51	26.75	27.66	27.50	28.16
di ⎰ wo	4.70	5.88	5.46	4.48	4.45	4.25	3.80
di ⎨ en	2.95	3.83	3.48	3.12	2.99	2.51	2.76
di ⎱ fs	1.46	1.64	1.63	0.99	1.12	1.52	0.68
hy ⎰ en	6.68	6.81	6.13	6.01	5.63	5.86	5.73
hy ⎱ fs	3.29	2.91	2.88	1.90	2.10	3.55	1.42
il	1.83	1.68	1.62	1.58	1.62	1.54	1.56
mt	5.97	6.06	5.67	6.33	6.19	5.02′	6.47
ap	0.83	0.86	0.86	0.83	0.81	0.86	0.83
he	0.00	0.00	0.00	0.00	0.00	0.00	0.00
cc	0.00	0.11	0.11	0.11	0.11	0.11	0.00

Volume percent phenocrysts

	1	2	3	4	5	6	7
Plagioclase	28	22	25	27	24	32	38
Olivine	<1	<1	—	—	—	—	<1
Orthopyroxene	—	—	—	<1	1	<1	2
Clinopyroxene	10	13	10	8	4	7	7
Opaque oxides	3	4	3	3	2	4	5
Hornblende	1	1	4	6	1	<1	<1
Total	42	39	42	44	32	43	52

*Samples analysed at Australian Mineral Development Laboratories, Adelaide.

1. Post ?1950-52 lava flow on northwestern flank of volcano (74400057)
2. Western limb of ?1950-52 lava flow, northern flank (74400064)
3. Relatively old lava flow on southern flank (74400069)
4. Relatively old lava flow from head of clastic deposit on southern flank (74400072 — overlies 74400069)
5. Bomb on northern flank (74400078)
6. Near front of the more extensive of the 1966 lava flows on southern flank (74400066)
7. Front of ?1960-62 lava flow, southern flank of volcano (74400061)

Table 1 cont.

8	9	10	11	12	13	14	15	16
55.4	55.5	55.5	56.2	56.7	56.9	56.9	57.2	57.6
0.85	0.80	0.81	0.73	0.74	0.73	0.67	0.67	0.68
17.9	18.0	17.9	17.3	18.0	18.1	17.9	18.1	18.1
4.15	3.95	3.95	5.05	4.35	3.65	5.65	3.30	3.95
4.30	4.30	4.25	2.80	3.15	3.50	1.90	3.80	2.90
0.17	0.17	0.17	0.16	0.16	0.16	0.16	0.16	0.16
3.35	3.25	3.30	3.80	2.70	2.70	2.75	2.65	2.55
8.10	7.90	8.10	8.10	7.40	7.45	7.55	7.30	7.30
3.75	3.70	3.75	3.60	3.85	4.00	3.95	4.05	3.95
1.51	1.61	1.51	1.40	1.66	1.66	1.61	1.70	1.70
0.35	0.37	0.35	0.31	0.35	0.32	0.35	0.34	0.34
0.01	0.07	0.04	0.04	0.50	0.19	0.09	0.15	0.21
0.05	0.03	0.06	0.06	0.10	0.13	0.17	0.15	0.11
0.05	0.05	<0.05	0.05	<0.05	0.05	0.05	0.05	0.05
0.01	0.02	0.01	0.01	0.01	0.01	0.01	0.01	0.015
99.95	99.72	99.70	99.61	99.67	99.55	99.71	99.63	99.62

8	9	10	11	12	13	14	15	16
7.24	7.50	7.30	10.30	10.22	9.02	10.52	8.77	10.83
8.93	9.55	8.96	8.31	9.90	9.88	9.57	10.11	10.12
31.75	31.42	31.85	30.60	32.87	34.10	33.60	34.49	33.65
27.59	27.87	27.67	27.05	27.19	26.74	26.51	26.37	26.84
4.19	3.65	4.33	4.58	3.15	3.37	3.56	3.14	2.96
2.77	2.35	2.82	3.94	2.38	2.27	3.08	1.94	2.23
1.12	1.05	1.21	0.03	0.45	0.85	0.00	1.01	0.43
5.58	5.77	5.42	5.57	4.40	4.50	3.81	4.70	4.17
2.26	2.59	2.32	0.04	0.83	1.68	0.00	2.45	0.81
1.62	1.53	1.54	1.39	1.42	1.40	1.28	1.28	1.30
6.02	5.75	5.75	7.36	6.37	5.33	4.73	4.82	5.77
0.83	0.88	0.83	0.74	0.84	0.76	0.83	0.81	0.81
0.00	0.00	0.00	0.00	0.00	0.00	2.42	0.00	0.00
0.11	0.11	0.00	0.11	0.00	0.11	0.11	0.11	0.11

8	9	10	11	12	13	14	15	16
29	33	33	28	33	35	29	27	25
—	—	—	—	—	—	—	—	—
1	1	2	1	1	<1	1	<1	<1
11	8	7	9	7	10	6	7	8
3	3	3	3	2	4	3	3	3
1	1	3	6	2	3	3	2	<1
45	46	48	46	45	52	42	39	36

8. Near front of 1971 lava flow (74400071)
9. 1964-65 lava flow, southern flank of volcano (74400063)
10. Front of 1972-75 lava flow (in November 1973 — 74400073)
11. Relatively old (pre-1943) lava flow near head of Saua River (74400068)
12. Thin, relatively old (pre-1943) lava flow at head of clastic deposit on northwestern flank of volcano (74400058) — may be now covered by new lava flow observed on 26 June 1975 (see Fig. 4)
13. Pre-1943 flow, southeastern flank (74400070)
14. Pre-1943 lava flow, northwestern flank (74400077)
15. Part of large block on southeastern flank (74400065)
16. Pre-1943 lava flow, northern flank (74400060)

Bagana, Balbi, and Takuan volcanoes and Malabita Hill) yielded Sr^{87}/Sr^{86} ratios ranging from 0.7038 to 0.7040 (Page & Johnson, 1974). Page & Johnson interpreted these uniform and relatively low ratios as indicating that the lavas were derived from relatively homogeneous upper mantle sources with little (if any) contamination by old radiogenic crustal material.

DISCUSSION

Bagana vies with Manam for being the most active volcano in Papua New Guinea, but because of its isolated position the eruptive history is poorly documented, especially before about 1950. Since the Second World War the volcano has been almost continuously active, the longest interval with no reports of any abnormal activity being about 6 years, from late 1953 to late 1959. Major eruptions, characterised by violent explosive activity, occurred in 1950, 1952, and 1966, during which voluminous ash clouds, nuées ardentes, and lava flows were produced; fifteen nuées ardentes were observed in a 14-day period in March 1952. However, effusion of lava is not always accompanied by explosive activity. Lava was extruded more or less continuously from 1972 to early 1975 but there were virtually no reports of any associated explosive activity.

Bagana differs from the other active volcanoes in Papua New Guinea in that the principal eruptive activity has been the extrusion of viscous, sluggish, blocky lava flows of predominantly low-silica andesite composition; pyroclastic deposits are relatively minor. Slow extrusion of lava commonly continues for months, in some cases (for example, the 1972-75 lava flow) for years, and steep-sided flows up to about 150 m thick are produced. Since 1943 at least 18 lava flows with fronts at the base of the cone or on the lower and intermediate slopes of the volcano have been extruded. It has not been possible to determine any cycle or periodicity in the pattern of eruptive events, possibly because of the general lack of comprehensive observations; nor is there any apparent link between eruptions from Bagana and eruptions from volcanoes along the southern margin of the Bismarck Sea, six of which erupted between 1972 and 1975 (Cooke et al., this volume). Of the lava specimens analysed, none (with the possible exception of 74400065, whose age is not precisely known) of those erupted since 1950 have SiO_2 contents greater than 55.5 percent. It is the relatively old lava flows that are characterised by relatively low total-iron, CaO, MgO, and TiO_2, and high SiO_2, contents.

Since the first recorded observations in 1882 the main activity from Bagana between eruptive episodes has been the almost continuous emission of dense white vapours from the summit area. Marked fluctuations (generally of short duration) in the volume of vapours emitted, and wisps of pale blue 'smoke', were observed by the writer and have been noted by other observers.

Bagana is far removed from any major centres of population, but there are more than 30 villages within a radius of about 20 km of the volcano. Except for falls of ash, 'normal' eruptions from Bagana are unlikely to seriously affect many areas beyond the base of the volcano. The lava flows are slow-moving and none have reached far beyond the foot of the cone. However, nuées ardentes and mudflows pose a potential threat to areas far beyond the base of the volcano. The area most likely to be affected is the low open country drained by the Torokina and Saua Rivers on the western and southwestern sides of the volcano. In May 1966 a nuée ardente swept down the valley of the Saua River for about 9.5 km from the crater. In 1950 nuées completely devastated the upper part of the Saua River valley. The data therefore indicate that centres near or in the valleys of the Saua and Torokina Rivers and within 15-20 km of the volcano are likely to suffer in a very major or catastrophic eruption.

ACKNOWLEDGEMENTS

The writer acknowledges the assistance provided by the Geological Survey of Papua New Guinea in making the study possible, and thanks R.J.S. Cooke, C.O. McKee, and R. Bill of the Central Volcanological Observatory, Rabaul, for help received during the field work and preparation of the paper. He also thanks his wife, Joyce, for drawing the diagrams. This paper is published with the permission of the Director, Bureau of Mineral Resources, Canberra.

REFERENCES

Blake, D.H., 1968: Post Miocene volcanoes on Bougainville Island, Territory of Papua New Guinea. *Bull. volcanol., 32*, pp. 121-138.

Blake, D.H., & Miezitis, Y., 1967: Geology of Bougainville and Buka Islands, New Guinea. *Bur. Miner. Resour. Aust. Bull. 93.*

Branch, C.D., 1967: Short papers from the Vulcanological Observatory, Rabaul, New Britain. *Bur. Miner, Resour. Aust. Rep. 107.*

Coleman, P.J., 1970: Geology of the Solomon and New Hebrides Islands, as part of the Melanesian re-entrant, southwest Pacific. *Pacific Science, 24*, pp. 289-314.

Curtis, J.W., 1973a: Plate tectonics and the Papua New Guinea-Solomon Islands region. *J. geol. Soc. Aust., 20*, pp. 21-36.

Curtis, J.W., 1973b: The spatial seismicity of Papua New Guinea and the Solomon Islands. *J. geol. Soc. Aust., 20.* pp. 1-20.

Denham, D., 1969: Distribution of earthquakes in the New Guinea-Solomon Islands region. *J. geophys. Res., 74*, pp. 4290-4299.

Denham, D., 1973: Seismicity, focal mechanisms and the boundaries of the Indian-Australian Plate. *in* Coleman, P.J. (Ed.), *The Western Pacific: island arcs, marginal seas, geochemistry*, pp. 35-53. Perth, Univ. West. Aust. Press.

Fisher, N.H., 1939: Report on the volcanoes of the Territory of New Guinea. *Terr. N. Guinea geol. Bull. 2.*

Fisher, N.H., 1954: Report of the sub-committee on vulcanology, 1951. *Bull. volcanol., 15*, pp. 71-79.

Fisher, N.H., 1957: *Catalogue of the active volcanoes of the world. Part V. Melanesia.* Naples, Int. volcanol. Assoc.

Guppy, H.B., 1887a: *The Solomon Islands and their Natives.* London, Swan, Sonnenschein, Lowrey & Co.

Guppy, H.B., 1887b: *The Solomon Islands: their Geology, General Features, and Suitability for Colonization.* London, Swan, Sonnenschein, Lowrey & Co.

Halunen, A.J., & von Herzen, R.P., 1973: Heat flow in the western equatorial Pacific Ocean. *J. geophys. Res., 78*, pp. 5195-5208.

Jakeš, P., & White, A.J.R., 1969: Structure of the Melanesian arcs and correlation with distribution of magma types. *Tectonophysics, 8*, pp. 223-236.

Jakeš, P., & Smith, I.E., 1970: High potassium calc-alkaline rocks from Cape Nelson, eastern Papua. *Contr. Mineral. & Petrol., 28*, pp. 259-271.

Johnson, R.W., Mackenzie, D.E., Smith, I.E., & Taylor, G.A.M., 1973: Distribution and petrology of late Cenozoic volcanoes in Papua New Guinea. *in* Coleman, P.J. (Ed.), *The Western Pacific: island Arcs, marginal seas, geochemistry*, pp. 523-533. Perth, Univ. West. Aust. Press.

Johnson, T., & Molnar, P., 1972: Focal mechanisms and plate tectonics of the southwest Pacific. *J. geophys. Res., 77*, pp. 5000-5032.

Krause, D.C., 1973: Crustal plates of the Bismarck and Solomon Seas. *in* Brodie, J.W., Burns, D.A., Dawson, E.W., Shirtcliffe, T.G.L., & Stanton, B.R. (Eds), *Oceanography of the South Pacific 1972*, pp. 271-280. Wellington, New Zealand National Commission for UNESCO.

Kuno, H., 1968: Differentiation of basalt magmas. *in* Hess, H.H., & Poldervaart, A. (Eds), *Basalts. The Poldervaart Treatise on Rocks of Basaltic Composition, 2.* pp. 623-688. New York, John Wiley & Sons.

Mackenzie, D.E., & Chappell, B.W., 1972: Shoshonitic and calc-alkaline lavas from the highlands of Papua New Guinea. *Contr. Mineral. & Petrol., 35*, pp. 50-62.

Macnamara, P.M., 1968: Rock types and mineralization at Panguna porphyry copper prospect, upper Kaverong valley, Bougainville Island. *Proc. Aust. Inst. Min. Metall., 228*, pp. 71-79.

Oliver, J., Isacks, B., Barazangi, M., & Mitronovas, W., 1973: Dynamics of the down-going lithosphere. *Tectonophysics, 19*, pp. 133-147.

Page, R.W., & Johnson, R.W., 1974: Strontium isotope ratios of Quaternary volcanic rocks from Papua New Guinea. *Lithos, 7*, pp. 91-100.

Ripper, I.D., 1970: Global tectonics and the New Guinea-Solomon Islands region. *Search, 1*, pp. 226-232.

Speight, J.G., 1967: Geology of Bougainville and Buka Islands. *CSIRO Land Res. Ser., 20*, pp. 71-77.

Taylor, G.A., 1956: Review of volcanic activity in the Territory of Papua New Guinea, the Solomon and New Hebrides Islands, 1951-53. *Bull. volcanol., 18*, pp. 25-37.

Taylor, S.R., Capp, A.C., Graham, A.L., & Blake, D.H., 1969: Trace element abundances in andesites. II. Saipan, Bougainville and Fiji. *Contr. Mineral. & Petrol., 23*, pp. 1-26.

DEVELOPMENT OF PORPHYRY COPPER AND STRATIFORM VOLCANOGENIC ORE BODIES DURING THE LIFE CYCLE OF ANDESITIC STRATOVOLCANOES

C.D. BRANCH

School of Applied Geology, South Australian Institute of Technology, North Terrace, Adelaide, S.A. 5000

ABSTRACT

Andesitic stratovolcanoes are large complex structures ranging up to 4 km high and 30 km in diameter. As a result of successive eruptive cycles a complex of sills, laccoliths, dykes, and pipes is formed near the base of the volcano. Magma injected in this zone which is not erupted may crystallise to form a stock with a halo of hydrothermal alteration; if the magma is enriched in copper a porphyry copper orebody may be produced. Fumarolic activity, particularly in crater areas, may give rise to gold, mercury, and sulphur deposits. Marine and lacustrine environments both on the flanks of the volcano and in central craters and calderas are favourable sites for the generation of stratiform lead, zinc, and copper ore bodies.

INTRODUCTION

Rocks formed in modern and ancient volcanic environments are considered by most geologists to have a high potential for the discovery of copper, lead, zinc, gold, mercury, and sulphur orebodies (e.g., Tatsumi et al., 1970). However, many exploration geologists have had little or no experience in active volcanic areas and hence are unaware of the many igneous and stratigraphical complications that may exist.

I consider myself fortunate to have been a volcanologist in Papua New Guinea, and in particular at an early stage in my career there to have had the company of Tony Taylor for several months in the field. From discussions with him I was able to absorb some of his deep understanding of volcanic behaviour and share in his insight into volcanic processes. This has considerably influenced my thinking in later years. This volcanological experience (Branch, 1967c), combined with my research on volcano-plutonic provinces (Branch, 1966, 1967a, b, 1969), has enabled me to develop my concept of volcanic acitivity which has been of value in geological exploration.

FEATURES IN THE EVOLUTION OF A STRATOVOLCANO

Many examples exist in Papua New Guinea of basaltic to andesitic stratovolcanoes in various stages of evolution, and from these it is possible to make the following observations and inferences.

1. Most active and dormant volcanoes are aligned along lineaments (Taylor, 1969) that are probably associated with plate boundaries and are generally related to underlying Benioff zones (Johnson et al., 1971).

2. The volcanic centres along a lineament may be widely spaced (e.g., Manam and Karkar volcanoes are about 120 km apart), or clustered so that their superstructures overlap (e.g., Bougainville).

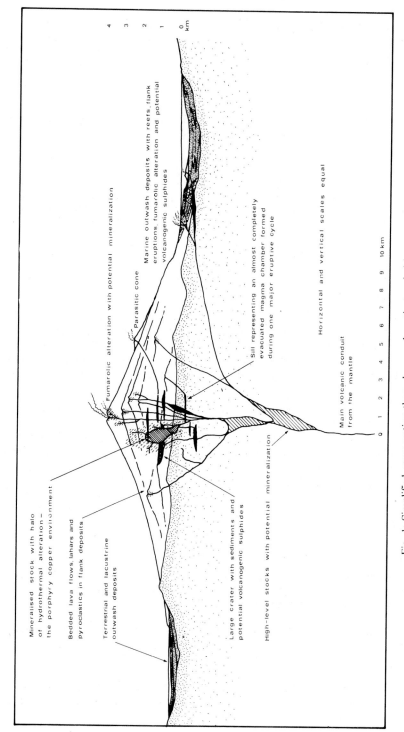

Fig. 1: Simplified cross-section through a nearly extinct andesitic stratovolcano.

3. Adjacent volcanic centres need not be at similar stages of growth or of the same magma type (e.g., Manam and Boisa volcanoes).

4. During the life of a major volcanic centre — probably less than a million years and possibly only 100,000 years or so (Ruxton, 1966; Blake & McDougall, 1973) — the magma available to a volcano commonly evolves from basic to intermediate or acid.

5. Most andesitic volcanoes are stratovolcanoes constructed of lava flows, lahar deposits, and pyroclastic beds. The volcano superstructure may attain a height of 2.5 to 4 km above its basement, with an apron 15 to 30 km across, before the volcano becomes extinct (Fig. 1). The profile of the volcano depends on the dominant magma type; as a general rule, the more acid the magma the steeper the profile unless nuée ardente eruptions are common, in which case the profile tends to be low with an apical dome. An appreciation of the size of a stratovolcano is important, because many geologists make their reconstruction of ancient centres more like small cinder cones (e.g., Rickard, 1966, Fig. 1). An excellent model for considering the size of volcanoes is the map of Bougainville prepared by Blake & Miezitis (1967).

6. During a single eruptive cycle, which may last a few months to a few years, magma rises along conduits from the mantle and accumulates in a subvolcanic magma chamber. As magma accumulates the volcano superstructure inflates (e.g., Branch, 1967c) indicating that the magma chamber is located either just below or within the volcano superstructure (Eaton & Murata, 1960). In the case of mature stratovolcanoes the magma chambers are probably at a depth of 2 to 4 km below the summit crater. From here the magma generally erupts through the central vent vertically above the magma chamber, and the volcano superstructure deflates: remnants of the evacuated magma chamber probably form sills and laccoliths (Fig. 1).

7. In successive eruptions the formation of new magma chambers and the associated inflation and deflation of the volcano cause intense shattering of the rocks around the zone where the magma chambers are formed.

8. In some cases much of the shattered material is rapidly evacuated during an exceedingly large and explosive eruption, and the volcano superstructure collapses as huge chaotic blocks to form a caldera 5 to 20 km across and generally more than 300 m deep. Under these circumstances any further volcanic activity is scattered, not localised, and generally on a minor scale (e.g., Rabaul caldera — Heming, 1974).

9. In some cases magma is moved sideways along dykes in a subsurface rift zone, and flank eruptions are generated: these may occur in terrestrial, lacustrine, or marine environments.

10. Eruptive cycles may be one year to 10,000 years apart. In the later, intermediate to acid, stage of the development of a volcano, remnants of the subvolcanic magma chambers for each eruption form an irregular complex of coalescing sills, laccoliths, pipes, and dykes of slightly different composition and texture in a zone 3 to 5 km across, intruding both the basement to the volcano and

the volcanics within the volcano itself. (In more stable continental areas similar volcanic complexes have a more regular, ring shape — Branch, 1966.)

11.   It is possible that one or two eruptive cycles will be abortive; the magma will rise and form a subvolcanic magma chamber, but will not erupt. Should this occur, then:

(a)   a small stock 0.5 to 2 km across will be formed;

(b)   the magma in the stock will differentiate and crystallise (Phillips, 1973);

(c)   late-stage hydrothermal fluids will soak into the surrounding country rock, which may be either comagmatic intrusions, shattered lavas and tuffs of the volcanic pile, or basement rocks below the volcano; the area soaked will be irregular and may extend for 1 or 2 km away from the intrusion (Macnamara, 1968; Lowell & Guilbert, 1970; Fountain, 1972; Sillitoe, 1973);

(d)   if there is more than one period of intrusion and hydrothermal activity the paragenetic sequence of minerals formed during the early phase may be strongly distorted by the later phases.

12.   The volcanic pattern may be complicated by several factors.

(a)   Alteration due to fumarolic activity. Two types must be distinguished: first, extensive alteration around and below a crater; and second, superficial alteration related to one flow or ash bed.

(b)   Overlapping and intertonguing volcanic aprons, originating both along the same lineament and from adjacent lineaments.

(c)   Freshwater and marine deposits in intervolcanic basins, in craters, and more importantly in large calderas breached to the sea.

(d)   Flank eruptions, particularly below sea level when pillow lavas are generated.

(e)   Post-caldera eruptions, especially when the caldera is filled by a lake or is open to the sea. Under these conditions sediments, fossils, pillow lavas, and subaerial deposits are intermingled within a steep-sided basin up to 20 km across and 300 m or more deep (e.g., Rabaul caldera).

(f)   Vertical tectonics, whereby in tectonically active areas it is possible for large crustal blocks to be depressed, or elevated, or tilted by several hundreds of metres in a relatively short period. For example, Pliocene limestone has been elevated 3000 m above sea level in the highlands of New Guinea (J.E. Thompson, pers. comm.). As a consequence, long-lived volcanoes may grow from below to above sea level, subside below sea level (possibly as a result of isostatic loading) and then either grow or be elevated above sea level again. Such volcanoes would provide a wide range of potentially mineralised environments, but could be difficult to recognise in the field.

(g)   Rapid erosion of unconsolidated sediment or volcanic ash and breccia, particularly when deposits are uplifted. Hence (i) volcano flank deposits may be rapidly stripped in the form of lahars within hours or days of being erupted, deposited in local basins 5 to 20 km away, uplifted, re-eroded, and redeposited elsewhere in a few thousand or million years. (ii) The surfaces on which detritus is laid down will generally have high relief, thus accentuating the lack of continuity between contemporaneous deposits. For example, on Manam volcano the southern vent, about 1800 m above sea level, is about 0.5 km away from the top of a cliff 600 m high at the head of the southwest avalanche valley.* (iii) A volcano, especially in its waning stages, may be moderately eroded during the thousand or more years between eruptions (often with the development in the crater area of an erosion caldera or an amphitheatre-headed valley) but sporadic eruptions may construct small cones on the degraded or scree-covered surfaces.

ENVIRONMENTS WITHIN A STRATOVOLCANO WITH A POTENTIAL FOR MINERALISATION

Porphyry copper environment

Porphyry copper mineralisation is generally associated with intermediate to acid subvolcanic stocks probably intruded late in the construction of a stratovolcano (Thompson & Fisher, 1967; Fisher, 1970; Sillitoe, 1973). The intrusions with the greatest potential for mineralisation are those stocks formed during abortive eruptions which intrude either the volcano superstructure or the basement to the volcano. In these cases, if the magma is rich in copper, as the magma differentiates and crystallises, late-stage hydrothermal fluids could generate a porphyry copper deposit. Alternatively, if the rocks around the intrusion are rich in copper (e.g., Cu in the ferromagnesian minerals in the volcanics) then the hydrothermal fluids could concentrate this copper into an orebody. The richest orebodies should be formed when these alternatives are combined and the copper-rich fluids soak into shattered, porous, copper-rich volcanics within the volcano superstructure; enriched zones can occur where the fluids encounter beds of limestone either in the basement rocks or in sediments preserved within the volcanic pile (Fig. 1).

Hence, in exploration for porphyry copper orebodies, you are looking for an extinct major stratovolcano which has evolved through the intermediate to acid stage, and whose central area has been eroded about 1 to 4 km below the original summit to expose the upper part of a copper-rich hydrothermal alteration zone.

Gold, mercury, sulphur environment

In zones of fumarolic activity gold (e.g., Wau), mercury and sulphur may be deposited. These minerals are more likely to occur in crater regions where the fumarolic activity is deepseated, and both juvenile water and percolating groundwater are involved.

Lead, zinc, copper environments

Two volcanic environments occur in which lead, zinc, and copper orebodies may be formed. First, after caldera-forming eruptions, subsequent volcanic activity is scattered, not localised, and any mineralising fluids will be dispersed through the fractured rocks beneath and around the caldera. As a result, scattered vein deposits similar to those associated with the Silverton caldera, U.S.A. (Varnes, 1963) will be formed.

* See Palfreyman & Cooke (this volume). Ed.

Second, where volcanic activity occurs in a marine or lacustrine environment it is probable that stratiform lead, zinc, and copper orebodies will be generated (Anderson, 1969; Stanton, 1972). Generally, favourable environments would exist around the outer apron of the stratovolcano, where fumarolic activity and flank eruptions could occur either beside or within a lake or the sea (Fig. 1): these environments have the best chance of being preserved when the volcano superstructure is eroded. However, aqueous environments with the greatest potential for mineralisation would be either crater lakes or calderas located near the centre of the volcano where magmatic activity is concentrated: large calderas breached to the sea provide the optimum conditions.

REFERENCES

Anderson, C.A., 1969: Massive sulfide deposits and volcanism. *Econ. Geol., 64,* pp. 129-143.
Blake, D.H., & McDougall, I., 1973: Ages of the Cape Hoskins volcanoes, New Britain, Papua New Guinea. *J. geol. Soc. Aust., 20,* pp. 199-204.
Blake, D.H. & Miezitis, Y., 1967: Geology of Bougainville and Buka Islands, New Guinea. *Bur. Miner. Resour. Aust. Bull. 93.*
Branch, C.D., 1966: The volcanic cauldrons, ring complexes, and associated granites of the Georgetown Inlier, Queensland. *Bur. Miner. Resour. Aust. Bull. 76.*
Branch, C.D., 1967a: The source of eruption for pyroclastic flows: cauldrons or calderas. *Bull. volc., 30,* pp. 41-50.
Branch, C.D., 1967b: Genesis of magma for acid calc-alkaline volcano-plutonic formations. *Tectonophysics, 4,* pp. 83-100.
Branch, C.D., 1967c: Short papers from the Vulcanological Observatory, Rabaul, T.P.N.G. *Bur. Miner. Resour. Aust. Rep. 107,* pp. 1-39.
Branch, C.D., 1969: Phanerozoic volcanic history of northern Queensland. *Geol. Soc. Aust. spec. Publ. 2,* pp. 177-182.
Eaton, J.P., & Murata, K.T., 1960: How volcanoes grow. *Science, 132,* pp. 925-938.
Fisher, N.H., 1970: Rock weathering, anatexis and ore deposits. *Search, 1,* pp. 111-119.
Fountain, R.J., 1972: Geological relationships in the Panguna porphyry copper deposit, Bougainville Island, New Guinea. *Econ. Geol., 67,* pp. 1049-1064.
Heming, R.F., 1974: Geology and petrology of the Rabaul caldera, Papua New Guinea. *Geol. Soc. Amer. Bull., 85,* pp. 1253-1264.
Johnson, R.W., Mackenzie, D.E., & Smith, I.E., 1971: Seismicity and late Cenozoic volcanism in parts of Papua New Guinea. *Tectonophysics, 12,* pp. 15-22.
Lowell, J.D., & Guilbert, J.M., 1970: Lateral and vertical alteration-mineralization zoning in porphyry ore deposits. *Econ. Geol., 65,* pp. 373-408.
Macnamara, P.M., 1968: Rock types and mineralization at Panguna porphyry copper deposit, upper Kaverong valley, Bougainville Island. *Proc. Aust. Inst. Min. Metall., 228,* pp. 71-79.
Phillips, W.J., 1973: Mechanical effects of retrograde boiling and its probable importance in the formation of some porphyry ore deposits. *Trans. Inst. Min. Metall., (Sec. B: App. Earth Sci.), 82,* pp. B90-B98.
Rickard, M.J., 1966: Reconnaisance geology of Vanua Levu, Fiji. *Geol. Surv. Fiji Mem. 2.*
Ruxton, B.P., 1966: Correlation and stratigraphy of dacite ash-fall layers in northeastern Papua. *J. geol. Soc. Aust., 13,* pp. 41-67.
Sillitoe, R.H., 1973: The tops and bottoms of porphyry copper deposits. *Econ. Geol., 68,* pp. 799-815.
Stanton, R.L., 1972: *Ore petrology.* New York McGraw-Hill.
Tatsumi, T., Sekine, Y., & Kanehira, K., 1970: Mineral deposits of volcanic affinity in Japan: Metallogeny. *in* Tatsumi, T.(Ed.), *Volcanism and ore genesis.* Tokyo, Univ. Tokyo Press, pp. 3-47.
Taylor, G.A.M., 1969: Post-Miocene volcanoes in Papua-New Guinea. *Geol. Soc. Aust. spec. Publ. 2,* pp. 205-208.
Thompson, J.E., & Fisher, N.H., 1967: Mineral deposits of New Guinea and Papua, and their tectonic setting. *in* Woodcock, J.T., Madigan, R.T., & Thomas, R.G. (Eds), *Proceedings-general: Eighth Commonwealth Mining and Metallurgical Congress, Vol. 6,* Aust. Inst. Metall., Melb., pp. 115-148.
Varnes, D.J., 1963: Geology and ore deposits of the South Silverton mining area, San Juan County, Colorado. *U.S. geol. Surv. prof. Pap. 378-A.*

RESIDUAL VOLCANIC EMANATIONS FROM THE BRITISH SOLOMON ISLANDS

G.R. TAYLOR

Mineral and Water Resources Division, Ministry of Trade Industry and Labour, Honiara, British Solomon Islands

ABSTRACT

The Paraso thermal area is the latest feature along a linear progression of volcanic activity within the Vella Graben which strikes northeast-southwest across the Vella Lavella Island, British Solomon Islands. The thermal area has a natural heat flow of 36×10^3 k.cals/sec which is of the same order of magnitude as some geothermal power producing areas. Both acid-sulphate mudpools and neutral hot springs occur. Analyses from Paraso and Simbo Island demonstrate the precipitation of Fe and Mn hydroxides from volcanic exhalations on entering an oxidising environment. Derived subsurface temperatures for the Peléan volcano, Savo Island, are shown to be a possible means of volcano surveillance. Minor copper mineralisation, pyritisation, and wall-rock alteration seen on Savo are the result of a vapour dominated hydrothermal system. The composition of a mineralised brine emanating from an unroofed diorite intrusion on Guadalcanal indicates contemporary argillic alteration of the underlying rock. It is concluded that the data presented support the hypothesis for the penecontemporaneous occurrence of volcanic exhalative and sub-volcanic porphyry copper mineralisation in an evolving active island-arc.

INTRODUCTION

Volcanism has occurred throughout the main Solomon Islands chain from the early Tertiary to the present time. Within the historical record there have been eruptions, largely andesitic in character, on Savo Island and the submarine volcanoes of Kavachi and Cook (Fig. 1). Volcano morphology and regular seismic disturbances on Vella Lavella and Simbo suggest that these islands are potentially active. Fumaroles and hot springs in thermal areas associated with the andesite volcanoes are predominantly of the acid-sulphate type, but near-neutral, high-chloride springs are found on Vella Lavella. The chemistry of these exhalations contrasts with that of warm brines emanating from dioritic plutons, thought to be the eroded cores of older andesitic volcanic complexes on Guadalcanal. This paper presents preliminary data on these various emanations, and their significance to metallogenesis is briefly discussed.

PARASO THERMAL AREA, VELLA LAVELLA

Geology of Vella Lavella

The petrology of the New Georgia islands (Fig. 1) has been described by Stanton & Bell (1969). Vella Lavella is the northernmost island of the group, and an extensive thermal area lies in the hinterland of Paraso Bay on the island's northeast coast (Figs 2, 3).

A photo-interpretation of the geology of part of Vella Lavella is shown in Figure 2. The island was formed by the coalescence of several cones of volcaniclastic material which shows a gradational change from older pyroxene basalt to younger hornblende andesite. Minor outcrops of limestone sediment and

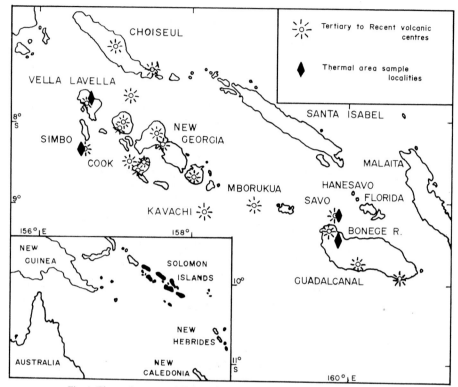

Fig. 1: Thermal areas and Cainozoic volcanic centres in the Solomon Islands.

coral reef platforms have been elevated above sea level during Pleistocene and recent times.

The oldest pyroxene basalts form a heavily dissected complex characterised by a mature relief and drainage pattern in which no relict volcanic centres can be distinguished (V3, Fig. 2). This unit is succeeded by hornblende andesite lavas and agglomerates forming eroded cones within which heavily dissected craters may still be recognised (V2, Fig. 2). The most recent volcanism produced hornblende andesite lava flows and volcanic agglomerates of the lahar type (V1a, Fig. 2). Subsequent uplift has exposed peripheral deposits of submarine tuff and tuffaceous mudstone which occasionally possess a calcareous matrix. These sediments show a juvenile topography of monotonous relief. The drainage is not yet incised and several recent fault scarps trending northeast have disrupted river courses with the resultant deposition of alluvium in small temporary lakes.

The map and cross-section (Fig. 2) show the recent volcanic centres within a northeast-trending fault-controlled graben — here termed the Vella Graben. A linear progression of volcanic activity along the graben is evident. The two oldest craters in the southwest are heavily dissected whereas the Nonda Volcano, lying within a subcircular erosional caldera, has clearly visible crater walls and a prominent central dome. There is no historical record of eruption from this volcano, but during a major earthquake in 1959 the local inhabitants reported

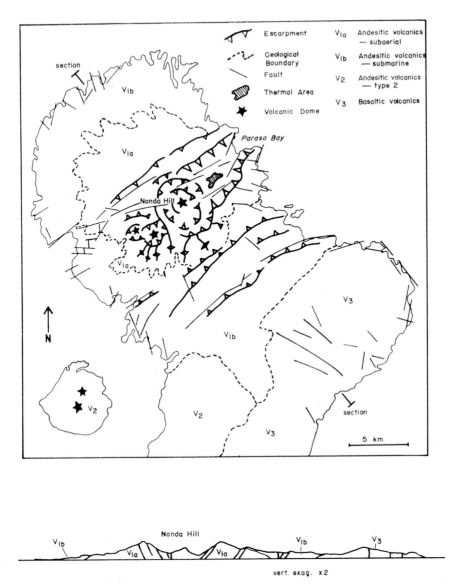

Fig. 2: Photogeological interpretation of part of Vella Lavella Island.

smoke and explosive activity in the vicinity of Nonda Hill (Grover, 1959). The Paraso thermal area forms the most recent feature along the line of volcanic progression (Fig. 2).

The crust in this region of the Solomon Islands is abnormally thin, as indicated by a mantle depth of 8 km (Furumoto et al., 1970). The structural features described above suggest that the area is one of marked crustal attenuation in a northwest-southeast direction, i.e. normal to the alignment of the Vella Graben. This northeast-southwest structural trend is parallel to the alignment of volcanic centres in the New Georgia group noted by Stanton & Bell (1969). These structures are in accordance with Hackman's (1973) analysis of the regional plate tectonics, and are readily explained as the result of crustal stretching between major east-west transcurrent faults or 'megashears' as described by Hackman.

Paraso thermal area

The floor of the Ngokosole River Valley (Fig. 3) exposes well bedded coarse volcanic tuff and agglomerate of hornblende andesite composition. Toward the seaward end of the valley these are covered by a thick mantle of recent alluvium. The steep sides of the flat-bottomed valley are fault controlled and the valley is essentially a tectonic depression within the Vella Graben proper (Fig. 2). A zone of anomalous heat flow (Fig. 3) is marked by a distinctive flora of pandanus trees and tall grass, and is estimated from aerial photography to cover about 1.7 km². Surface indications of anomalous heat flow can be seen in the Ngokosole and Ulo Rivers for distances of over 3 km.

A preliminary heat flow survey was carried out in the valley using over 350 measuring stations at 50 m intervals along 18 km of widely separated traverse lines. Temperatures were measured at the surface and at 15 cm and 1 m depths using a portable electronic thermometer with stainless steel temperature probes. Figure 3 shows the isotherm for up to 25°C increase of temperature between the surface and 1 m depth. The distribution of the +25°C isotherm, too small to be shown in Figure 3, indicates that the areas of highest heat flow are localised and occur around structurally controlled lineations.

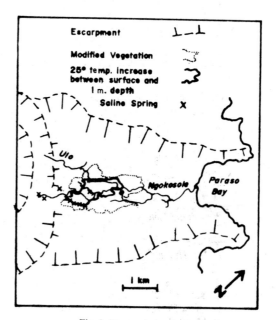

Fig. 3: Paraso thermal area.

Table 1. Water analyses from Paraso.

Sample Number	MUDPOOLS		FAULT SPRINGS				NGOKO-SOLE R.
	10	12	26	27	28	30	5
pH ⎫ at	2.2	2.5	5.0	6.7	7.9	7.0	7.5
Temp°C ⎰ collection	38	48	49	85	48	79	32
SiO_2	100.0	168.0	192.8	148.3	97.5	84.6	57.2
Ca	3.2	2.0	46.7	17.0	27.5	22.2	42.4
Mg	2.9	1.8	16.4	9.8	16.5	6.1	13.9
Cl	11	11	206	69	45	149	263
Na	2.8	1.3	131.0	70.3	48.4	97.0	150.0
K	1.50	0.60	11.37	15.0	6.8	13.5	17.4
CO_3(total)	83.5	59.2	9.5	35.6	0.2	0.0	26.8
H_2S	5.96	1.70	nd	1.87	6.48	4.77	2.22
SO_4	1054	1288	nd	195.3	nd	224.1	62.9
Fe	18.09	6.7	7.70	1.50	0.50	2.90	0.00
Mn	0.00	0.00	0.48	0.85	0.03	0.17	0.00
Derived equilibrium temperature °C	nd	nd	224	289	240	259	247

Analyses in mg/l   nd = not determined

The main thermal zone is a flat, open area generally at an elevation of about 5 m above the meandering courses of the Ngoksole and Ulo Rivers. Volcanic tuff and agglomerate exposed in the river banks are highly altered and show pervasive disseminated pyrite. There are several hundred individual hot pools, and it has not yet been possible to accurately survey them. The area of barren sinter and mudpools is 0.19 km^2, and it is estimated that about 20 percent of this is occupied by active mudpools. These are of the acid-sulphate type, but a series of near-neutral, low sulphate hot springs occurs at the southern margin of the thermal area (Fig. 3).

Analyses of filtered mudpool samples (Table 1) are characteristic of the acid-sulphate pools frequently found above vapour dominated hydrothermal systems occurring within recently active volcanic areas (White et al., 1971). The pool temperatures, the rate of ebullition, and the consistency of the mud are highly variable. Pools with the thickest mud often have spatter cones and miniature mud volcanoes. The nature of the pools varies with the rainfall, and the level of the local water-table is highly variable as there may be as much as 3 m difference between the levels of standing water in adjacent pools. Gas condensates were collected using techniques similar to those of Mahon & Glover (1965) and CO_2:H_2S molar ratios have been determined. Condensate from a hot (94°C) vigorous mudpool had a CO_2:H_2S ratio of 63.0 whilst that from a cool (55°C) quiescent mudpool had a CO_2:H_2S ratio of 28.1. Following the reasoning of Mahon & Glover (1965), this suggests that the vapour from the hotter pool has only recently separated from the liquid phase and therefore probably is closer to a feeder conduit.

Near-neutral, very hot and vigorous saline springs are located on the southern boundary of the thermal area (Fig. 3) along a linear fault-controlled feature. Samples from these springs were collected and analysed using procedures described by Ritchie (1961); the results are shown in Table 1. The low sulphate and moderately high chloride and silica contents of these waters suggest the presence of a hot-water-dominated hydrothermal system. However, the total dissolved solids and silica contents are significantly lower than those of water samples from economically important geothermal areas elsewhere (Ellis, 1968). Derived equilibrium temperatures using the Na-K-Ca thermometer of Fournier & Truesdell (1973) suggest a maximum subsurface temperature of between 220-290°C. But the silica contents are below the solubility of amorphous silica at these temperatures, which would appear to invalidate these derived equilibrium temperatures. As the hot springs are over 3 km distant from the sea, and are between 20 and 50 m above sea-level, it is not thought likely that any sea water is drawn into the system, and the anomalous chloride values are therefore considered to be significant. Most previous work on hot-spring chemistry has been done in areas where the aquifer rocks are richer in silica than those of Vella Lavella. At Wairakei, New Zealand, quartz-bearing ignimbrites and rhyolites predominate (Healy, 1965). It is tentatively suggested that the low dissolved-solids and silica contents of the Paraso springs are a function of the aquifer lithology. Alternatively, the passage of water through the system may be so rapid that the time available for rock/hot-water reaction is relatively short. It is therefore apparent that the permeability and fracturing of the underlying rock are critical in assessing the potential of the areas as a source of geothermal power.

The hot springs all contain significant amounts of dissolved iron and manganese. Near its mouth the waters of the Ngokosole River retain the major characteristics of a part diluted and cooled hot-spring water, but have deposited all the iron and manganese (Table 1). From flow measurements of the river it is estimated that 570 t/year of chemical sediment is deposited downstream of the thermal area. The volume of alluvium fill in the lower reaches of the Ngokosole valley is judged to be about 10^8 m^3. If the component due to normal terrigenous sediments is ignored, then an approximate figure of 600,000 years is arrived at for the age of the Paraso thermal area. The precipitation of the ferro-manganese hydroxides will be due largely to atmospheric oxidation and dilution by meteoric water, and is accompanied by a rise in pH of the river water (Table 1).

Heat flow estimation

Using techniques developed in New Zealand (Robertson & Dawson, 1964; Dawson & Fisher, 1964; Dawson, 1964) the following estimate of the amount of natural heat flow from the Paraso thermal area has been made:

	k.cal/sec
By conduction through the soil	581
— convection through the soil	1,271
— evaporation and radiation from pools	21,700
— carried by the Ngokosole River	12,600
Total heat flow	36,152

In making the calculations it was assumed that rock conductivities would be similar to those of Wairakei. The above estimate should be considered a minimum figure as it was not possible to measure some significant categories of heat loss: e.g. no estimation could be made of the energy involved in the heat and mass transfer from fumaroles and geysers emitting gases under pressure. The natural heat flows from the Wairakei and Taupo thermal areas of New Zealand were estimated to be 160,000 k.cal/sec and 50,000 k.cal/sec respectively before major interference with the systems by drilling and subsequent power generation (Banwell, 1964). The Paraso thermal area is seen to be of a similar order of magnitude to these major geothermal regions.

SIMBO ISLAND

An age of 2.3 ± 1 m.y. has been determined for the pyroxene basalt lavas of the Simbo volcanic complex (Snelling, 1969). Iron hydroxide gels precipitated from submarine volcanic exhalations around the coast of Simbo Island were described by Stanton & Baas Becking (1962). Analyses of samples collected from Kosinge (Fig. 4) in 1970 by the present author showed up to 20 percent iron and 10 percent manganese, and an analysis of condensate from an active fumarole (temp. 100°C), just above the high-tide level, showed a $CO_2:H_2S$ molar ratio of 18.6. This molar ratio is typical of fumaroles on recently active volcanoes (White & Waring, 1963).

Lake Ove is a salt-water lagoon with hot springs at its southern end (Fig. 4). Ten water and twenty sediment samples collected in 1974 by I.J. Tedder were analysed by the present author. The lake water is partly modified sea water. Figure 4 summarises the significant data. Water from close to the spring outlets possesses a low pH (minimum value 2.5) and contains both iron and manganese in solution. Away from the springs the buffering effect of the sea water causes a rise in the pH which corresponds to a decrease in the iron and manganese concentrations. This is directly related to an increase in these constitutents in precipitated sediments collected from the sea-floor. The manganese contents of these sediments are very

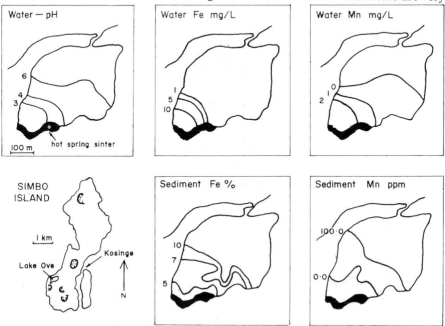

Fig. 4: Distribution of pH, iron, and manganese in the waters and sediments of Lake Ove, Simbo Island.

low. Under the pH conditions in Lake Ove, manganese oxides should not precipitate at all, and their direct association with the variations in iron contents suggest that the manganese is included within iron hydroxide gels which have acted as collecting agents. Minor amounts of zinc also follow the same trend as the iron distributions in both the water and sediments, but no copper was recorded in either. Similar metal-rich sediments in the environment surrounding volcanic exhalations at Matupi, New Britain, have been described by Ferguson & Lambert (1972).

SAVO VOLCANO

Savo Island is a volcanic cone composed of lava flows, volcanic agglomerates, lahar deposits, and nuée ardente deposits of pyroxene and hornblende andesite. It has been fully described by Grover (1958) and G.A.M. Taylor (1959). Violent eruptions in 1568 and between 1830 and 1840 produced nuees ardentes. Attempts to maintain regular surveillance of the volcano have been of limited success owing to the paucity of stable sites for tiltmeters and the tendancy for hot springs to drastically alter their character as old conduits are closed by sinter deposition. The measurement of surface temperatures at specific points is therefore thought to be of limited significance. Partial analyses of several hot-spring water samples were carried out to determine the suitability of the dissolved silica and the Na-K-Ca geothermometers for assessing the subsurface temperatures.

Hot water samples from three minor thermal areas near the coast have silica concentrations below the solubility of amorphous silica at their surface temperatures (SiO_2 65-190 mg/l) and sodium/potassium ratios approaching that of sea water. They are considered to be locally heated groundwaters that have some slight sea-water contamination. The two most active thermal areas, Voghala and Toakomata, lie on a conspicuous northeast-trending fault that affects the central area occupied by the youngest crater (Fig. 5). Partial analyses of hot-spring waters from these areas are given in Table 2. These springs are of the acid-sulphate category and are composed largely of locally heated meteoric water. Derived subsurface temperatures using the silica method of Fournier & Rowe (1966) show a fair degree of consistency over a one-year period, but they contrast with the derived temperatures calculated from the Na-K-Ca method of Fournier & Truesdell (1973). Several authors, including Fournier & Truesdell, have emphasised the errors involved in applying either method to the estimation of subsurface temperatures in vapour-dominated acid-sulphate systems. In the absence of any alternative it is suggested that the silica method may still provide a readily applicable method for assessing *relative* variations in subsurface temperatures of dormant volcanoes such as Savo, where the water-table is fairly constant owing to the regular high rainfall.

There is abundant evidence that vapour-dominated hydrothermal systems occuring on quiescent volcanoes are capable of effecting considerable hydrothermal alteration on the reservoir rocks, and the leached constituents may result in minor accumulations of base metals. In the vicinity of most thermal areas on Savo the country rock is frequently altered to a grey clay pug containing pervasive pyrite disseminations. Over 60 m of outcrop in the upper Mbazo River the pyritisation is especially intense. Individual hand specimens contain up to 13,300 ppm copper in the form of chalcopyrite, but a channel sample over the

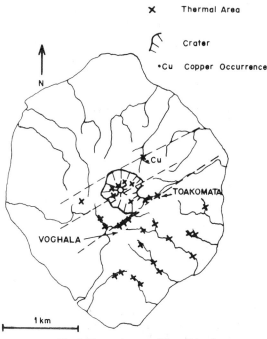

Fig. 5. Thermal areas of Savo Island.

whole outcrop assayed only 850 ppm copper. Microscopic examination of the gangue shows it to be composed of mainly fine-grained anhydrite with disseminated sulphide and fracture fillings of crystalline gypsum. The mineralised zone is related to a fault parallel to the major northeast shear controlling the Voghala-Toakomata thermal areas.

BONEGE BRINES, GUADALCANAL

Hackman (1971) listed 24 saline, warm and hot spring occurrences on the island of Guadalcanal. There is an apparent increase in the temperature of the springs from east to west, the hottest springs occuring within Pliocene to Recent andesitic lavas of the Gallego Volcanics. Potassium-argon age determinations (Hackman, 1971) show the latter to be 6.4 $\pm$ 0.2 million years old. Dioritic plutons such as the Koloula intrusive complex are considered to be the eroded basements of andesitic volcanoes belonging to the same phase of volcanism that produced the Gallego Volcanics. Dates of 7.8 $\pm$ 0.8 m.y. and less than 1.6 m.y. have been obtained from different intrusive phases of the Koloula pluton. The diorites contain base- and precious-metal mineralisation of the porphyry-copper type. Older intrusions that occur in the Poha and Bonege Rivers show distinct petrochemical affinity with the Koloula diorite, and although of probable Oligocene age may date from the commencement of Tertiary to Recent andesitic volcanism in Guadalcanal. The analysis of water collected by B.D. Hackman from a spring within the Bonege diorite is presented in Table 2.

Determinations of the base-metal contents of this sample are as follows: Fe

Table 2. Water analyses from Savo and Guadalcanal.

	TOAKOMATA		VOGHALA		BONEGE
	1	2	1	2	
pH ⎫ at	5.4	6.6	6.4	6.3	8.0
Temp°C ⎭ collection	50	80	65	67	47
SiO2	370	349	264	276	38
Ca	35.8	20.8	12.8	18.6	2,760
Mg	9.5	48.7	15.9	7.8	2.3
Cl	14.8	47.5	7.9	4.9	8,100
Na	85.5	145.0	36.0	60.0	2,935
K	15.3	17.0	10.4	11.4	39.5
SO4	57.0	972	453	726	1,026
Alkalinity as CaCO3					37.0
Derived temp (Silica)°C	208(213)	204(209)	183(187)	187(189)	—
Derived temp (Na-K-Ca)°C	267	256	295	270	140

Analyses in mg/l.

Values in brackets are silica-derived temperatures for samples collected approximately one year later.

0.50, Mn 0.07, Zn 0.50, Cu 0.00 mg/l. The analyses show that the Bonege spring is a highly mineralised brine. It may be argued that the occurrence together of saline spring and diorite pluton is entirely fortuitous, but the frequent association on Guadalcanal of saline springs with diorite plutons implies that the brines eminate from an aquifer that draws its heat from the original intrusion and not from some later feature. The moderate value of the derived subsurface temperature and the high concentration of calcium in the solution would suggest that argillic alteration is occurring in the underlying rocks. The evidence therefore suggests that low-temperature hydrothermal alteration can take place during a period measured in tens of millions of years after the emplacement of the intrusion.

CONCLUSIONS

From the results of preliminary investigations it is apparent that both liquid- and vapour-dominated hydrothermal systems associated with residual volcanism have an important bearing on the genesis of mineral deposits. Stanton (1955) stressed the association of stratiform ore deposits and volcaniclastic sediments developed during the eugeosynclinal stage of island-arc formation. Stanton (1958) and Oftedahl (1958) have postulated a direct volcanic exhalative origin for this class of ore. Taylor (1974) described a complex massive sulphide/ferro-manganese oxide ore deposit from a submarine volcanic exhalative environment within an ophiolite suite in the Solomon Islands. The analytical data presented above are strong evidence for the penecontemporaneous formation of the volcanic exhalative and the sub-volcanic porphyry-copper categories of mineralisation during the evolution of an active island-arc.

Whilst it has been shown that both vapour-dominated and liquid-dominated hydrothermal systems may be closely associated with anomalous base-metal concentrations, it is suggested that the chloride-rich solutions of the latter, such as the Bonege brine, more closely approach the composition of a mineralising fluid. The base- and precious-metal content of such a solution will be a function of both its temperatures, Eh and pH conditions, and the availability of these metals in the rocks with which the solution has had sufficient time to reach equilibrium. In normal circumstances the first three factors could be expected to change significantly as the solution approached the atmospheric conditions of the surface and result in the precipitation of its metal content before debouching as a saline spring. Only the existence of atypical conditions at the point of examination, such as exists in the case of the Red Sea and Salton Sea brines, will provide for our examination the true mineralising fluid.

ACKNOWLEDGEMENTS

The author is indebted to B.D. Hackman and R.L. Stanton for much help and advice. Thanks are due also to I.B. Lambert, P.R.L. Brown and G.W. Grindley for constructive criticism made during review of the first draft. W. Stanley and P. Samane provided willing assistance in the field. This paper was partly written during the tenure of a Conzinc Riotinto of Australia Post Graduate Scholarship at the University of New England, Armidale, N.S.W.

REFERENCES

Banwell, C.J., 1963: Thermal energy from the Earth's crust. *N.Z. J. Geol. Geophys, 6(1)*, pp. 52-69.

Dawson, G.B., 1964: The nature and assessment of heat flows from hydrothermal areas. *N.Z. J. Geol. Geophys, 7(1)*, pp. 155-171.

Dawson, G.B. & Fisher, R.G., 1964: Diurnal and seasonal ground temperature variations at Wairakei. *N.Z. J. Geol. Geophys, 7(1)*, pp. 144-154.

Ferguson, J. & Lambert, I.B., 1972: Volcanic exhalations and metal enrichments at Matupi Harbour, New Britian, T.P.N.G. *Econ. Geol., 67*, pp. 25-37.

Fournier, R.O & Rowe, J.J., 1966: Estimation of underground temperatures from the silica content of water from hot springs and wet steam wells. *Amer. J. Sci., 264*, pp. 685-697.

Fournier, R.O. & Truesdell, A.H., 1973: An empirical Na-K-Ca geothermometer for natural waters. *Geochimica et Cosmochimica Acta, 37*, pp. 1255-1275.

Furumoto, A.S., Hussong, D.M., Campbell, J.F., Sutton, G.H., Malahoff, A., Rose, J.C., & Woollard, G.P., 1970: Crustal and upper mantle structure of the Solomon Islands as revealed by seismic refraction survey of November-December, 1966. *Pacific Science, 24(3)*, pp. 315-332.

Grover, J.C., 1958: Savo volcano — A potential danger to its inhabitants. *The Solomon Islands — Geological Exploration and Research 1953-1956*. Memoir of the B.S.I.P. Geological Survey Dept, pp. 102-108.

Grover, J.C., 1959: The great earthquake in the western Solomons on 18 August, 1959. *The British Solomon Islands Geological Record, Vol. II — 1959-62*. pp. 174-182.

Hackman, B.D., 1971: The regional geology of Guadalcanal: a contribution to the geology of fractured island arcs. *Ph.D. thesis, Univ. W. Aust.*

Hackman, B.D., 1973: The Solomon Island fractured arc. *in* Coleman, P.J. (Ed.), *The Western Pacific, island arcs, marginal seas, geochemistry*. Perth, Univ. West. Aust. Press, pp. 179-191.

Healy, J., 1965: The geology of the Wairakei geothermal field. *8th Comm. Min. and Metall. Cong., Aust. and N.Z. 1965. N.Z. Sect. paper 218.*

Mahon, W.A.J. & Glover, R.B., 1965: The chemistry of geothermal fluids discharged from drill holes at Wairakei New Zealand. *8th Comm. Min. and Metall. Congr. Aust. and N.Z. 1965. N.Z. Section.*

Oftedahl, C., 1958: A theory of exhalative-sedimentary ores. *Geol. Foren Stockholm Forh., 80*, pp. 1-19.

Ritchie, J.A., 1961. Chemical analysis and laboratory requirements: Experience in New Zealand's hydrothermal areas. *Working paper for U.N. Conf. on new sources of energy.* E/conf. 35/9/29.

Robertson, E.I. & Dawson, G.B., 1964: Geothermal heat flow through the soil at Wairakei. *N.Z. J. Geol. Geophys., 7, 1*, pp. 134-143.

Snelling, N.J., 1969: Unpublished age determinations for the B.S.I.P. Geological Survey Dpt. Published by permission of the Director, Institute of Geological Science, London, U.K.

Stanton, R.L., 1955: Lower Palaeozoic mineralisation near Bathurst, New South Wales. *Econ. Geol., 50,* pp. 681-714.

Stanton, R.L., 1958: Abundances of copper, zinc, and lead in some sulphide deposits. *J. Geol. 66,* pp. 484-502.

Stanton, R.L. & Baas Becking, L.G.M., 1962: The formation and accumulation of sedimentary sulphides in seaboard volcanic environments. *Proc. Koninkl. Nederl. Akademie van Wetenschappen, Series B, No. 3,* pp. 236-243.

Stanton, R.L & Bell, J.D., 1969: Volcanic and associated rocks of the New Georgia Group, British Solomon Islands Protectorate. *Overseas Geol. and Miner. Resour. 10(2),* pp. 113-145.

Taylor, G.A., 1965: Notes on Savo volcano, 1959: *The British Solomon Islands Geological Record,* Vol. II — 1959-62. Report 56, pp 168-173.

Taylor, G.R., 1974: Volcanogenic mineralisation in the islands of the Florida Group, B.S.I.P. *Trans. Inst. Min. Metall. Sect. B, 83, B120-B130.*

White, D.E., Muffler, L.J.P., & Truesdell, A.H., 1971: Vapour-dominated hydrothermal systems compared with hot-water systems. *Econ. Geol., 66(1),* pp. 75-98.

White, D.E., & Waring, G.A., 1963: Volcanic emanations, chap. K in data of geochemistry, 6th ed. *U.S. Geol. Survey prof. Pap.* 440K.

VARIATIONS IN STRESS RELEASE PRECEDING AND ACCOMPANYING A SUBMARINE ERUPTION IN NORTHERN TONGA

J.H. LATTER

Department of Scientific and Industrial Research, Geophysics Division, P.O. Box 8005, Wellington, New Zealand

ABSTRACT

In 1973 from July 11d 02h to 13d 18h (U.T.) an earthquake swarm, accompanied in its later stages by a volcanic eruption producing dacitic pumice, took place near Curaçoa Reef, at the extreme northern end of the Tonga arc. The swarm area was of the order of 10,000 km². Eruptions, identified by the nature of their T phases at seismograph stations nearby and at a hydrophone array near Wake Island, took place at two localities, which are correlated with observed eruptions at points about 3 km south-southeast of Curaçoa Reef, and 35 to 40 km to the northwest. The first eruption, which was by far the largest in the series, is assigned to the position near Curaçoa Reef. The principal earthquake in the series, on the other hand, took place near the other eruption source to the northwest. Eruptions at this latter site terminated the series. Differences in T wave arrivals at pairs of stations allow the majority of earthquakes in the swarm of magnitude $M_L \geqslant 3.5$ to be assigned to one or other of these two areas. The rate of occurrence of the earthquakes, the variation of the coefficient b in the frequency/magnitude relationship, and the pattern of stress release, have been examined for both areas for earthquakes of magnitude $M_L' \geqslant 3.3$. The same analysis has been carried out for the swarm as a whole, using more than 3500 events of estimated magnitude $M_L \geqslant 2.8$, recorded at a hydrophone near Wake Island. The rate of occurrence, inversely correlated with b, showed a peak in the number of earthquakes at about the time of the main shock. Each eruption, with the exception of the pair that terminated the series, was also associated with an increase in the number of events. The principal eruption took place during a period of sharp increase in the value of b. The pattern suggests a process of widespread intrusion, leading to failure of an obstruction and release of most of the seismic energy in the principal earthquake, and to a climactic eruption about 9 hours later.

INTRODUCTION

In 1973, beginning July 11d 01h 56m* and ending two days later, a notable earthquake swarm took place near Curaçoa Reef (15.48°S, 173.62°W) at the extreme northern end of the Tonga arc (Fig. 1). The later stages of the swarm were associated with observed submarine eruptions, at a point about 3 km south-southeast of Curaçoa Reef, and probably also at another locality about 35 km northwest of the reef, producing not less than 0.06 km³ of dacitic pumice (Simkin & Onyeagocha, 1973). Gibowicz et al. (in press) described variations in the rate of occurrence of some 500 of the earthquakes, and in the average value of the coefficient b (defining the frequency/magnitude relationship), and determined source dimensions for many of the events. These were found to be unusually large, and indicative of a major zone of weakness. In their study, all events were treated as a single set, as body-wave data were inadequate for location of most of the earthquakes. In this paper, evidence is presented suggesting that events detected at Afiamalu originated in two distinct areas. Records from a hydrophone near Wake Island are used to study a further 3500 events of the swarm.

No previous volcanic activity has been recognised in the Curaçoa Reef area. However, on 29 June 1973, a short series of 11 earthquakes, lasting a little over half an hour, took place in this area (R.H. Johnson, pers. comm.). The form of these events is indistinguishable from that of the 11 to 13 July series.

* All dates and times are given in Universal Time, which is 13 hours behind local Tongan time.

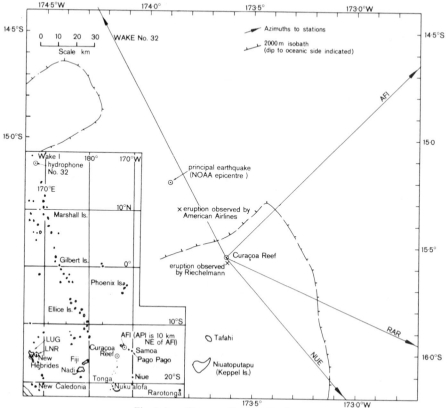

Fig. 1: Locality map, Curaçoa Reef area.

DATA

Records of the seismograph stations at Afiamalu (AFI), Niue (NUE), and Rarotonga (RAR), and of an array of four hydrophones near Wake Island, have been used to determine magnitudes, origin times, and relative locations for events of the swarm. For periods during which the AFI records were being changed, seismograms from Apia (API) were used. AFI, API, NUE, and RAR are seismograph stations controlled by the D.S.I.R. Seismological Observatory, Wellington, New Zealand, and AFI and RAR are part of the World-Wide Standard Seismograph Network. The Wake Island hydrophone network is part of the Pacific Missile Range Facility, operated by the U.S. Department of the Navy (Johnson, 1966). Figure 1 shows the locations of these stations.

Body waves were readable in only a minority of the events: 216 at AFI/API, 124 at NUE, and 26 at RAR. T waves (propagating through the ocean sound 'SOFAR' channel) were well recorded at one hydrophone (No. 32), situated about 180 km southeast of Wake Island, which detected more than 3500 events during the period of the swarm; other hydrophones of the array recorded T waves from only the larger events. T waves were also well recorded at the seismograph stations. In addition to events in which body waves were recorded, readable T wave arrivals were obtained in 354 events at AFI/API, 278 at NUE, and 266 at RAR.

ERUPTIONS

Activity was first observed from Tafahi (see Fig. 1), sometime between 17h and 23h on 12 July, when a rising cloud column was noticed to the north-northeast (T. Simkin, pers. comm.). The next observation of the eruption was made on 14 July at about 04h, when a white steam cloud was reported, rising to a height of about 5 km, at a position some 35 km northwest of Curaçoa Reef (Fig. 1). This was seen from an aircraft of American Airlines en route from Nadi to Pago Pago (Smithsonian Institution, 1973; Simkin & Onyeagocha, 1973), flying at an altitude of about 10 km. A large pumice raft, with dimensions estimated by flight personnel as about 12 ± 4 km by 8 ± 4 km, spread out from the eruption site, which was reported to be 'boiling', an observation interpreted by Simkin & Onyeagocha (1973) as indicating that pumice was still being actively supplied to the surface.

Between about 17h on 14 July and 05h on 15 July a vapour column was seen from Tafahi, rising from the sea, and on 15 July, between about 05h and 17h, a red glow was noticed in the clouds. Both these observations were in the same direction as that of 12 July (T. Simkin, pers. comm.). The final observation of the eruptions was made about 01h on 16 July by J.C. Riechelmann (pers. comm.), who flew over the area on a charter flight from Nuku'alofa and observed steam rising to a height of about 300 m from an area of hot pumice (estimated as 3 to 4 hectares). The point at which pumice was being supplied to the surface was about 3 km south-southeast of Curaçoa Reef. This position is in line with the direction of the sighting from Tafahi but does not agree with the American Airlines observation. Since the latter was estimated, at a distance of less than 10 km, from a position fixed by an inertial navigation system (Simkin & Onyeagocha, 1973), it is unlikely to be in error by more than 2 or 3 km. It seems probable, therefore, that two eruption sites were active during the swarm.

Certain events of the swarm are interpreted as submarine eruptions because they introduced anomalously large amounts of energy into the SOFAR channel to be propagated as T waves. Figure 2 shows the maximum trace amplitude at the hydrophone about 180 km southeast of Wake Island (No. 32) plotted against the local magnitude determined from body waves at AFI. Six events interpreted as eruptions, in which large trace amplitudes at Wake correspond to rather low magnitudes at AFI, are distinguished in the figure.

Not only are these events distinguishable on the basis of their relative T and body wave trace amplitudes, but their form is also strikingly different on the Wake hydrophone records. Furthermore, they are recorded at all stations of the array, whereas events interpreted as earthquakes are detectable at all stations only when their M_L magnitudes exceed 3.9. The majority of events of the swarm did not give rise to body waves large enough to be detected at the seismograph stations. For these, identification of eruptions is based on their characteristic form at all stations*, and particularly on the Wake hydrophone records. For the very large number of events detected only at hydrophone 32, those of appropriate form are tentatively identified as eruptions when their trace amplitude is such as to yield an equivalent magnitude of at least 3.5 at AFI (from the least-squares relationship in Fig. 2).

_____

* At AFI, NUE and RAR the onset of T phases in events interpreted as submarine eruptions is of markedly higher frequency than in other events (about 4 Hz as against 1.2 to 1.3 Hz at AFI).

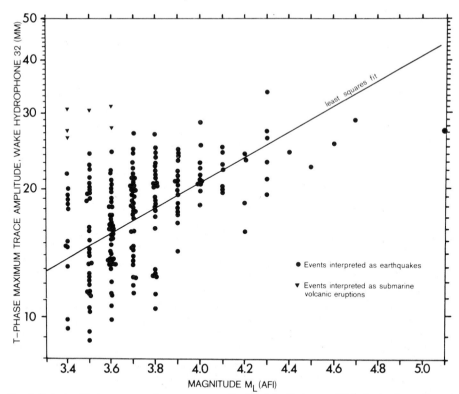

Fig. 2: Relationship between maximum trace amplitude of the T phase at Wake hydrophone 32, and magnitude M_L, determined from body waves at AFI. The least-squares fit to the data, and six events interpreted as submarine volcanic eruptions, are shown.

On the basis of this analysis, eight events, recorded by both seismograph stations and hydrophones, are confidently interpreted as submarine eruptions. Of these by far the largest was the first, with origin time 07h 55m on 12 July. The events are listed in sequence in Table 1. A further eleven events, listed in Table 2, are considered as possible eruptions.

The scarcity of body waves and the fact that the Wake hydrophone records were overloaded by the larger eruptions prevent reliable determination of radiated energy for the eruptions. Accordingly, only the relative size of these events has been noted. Nor have T wave strengths (Johnson, 1966) been computed. The fact that all events originated in a small area, relative to the distance to the Wake hydrophones, permits the use of 'trace amplitudes', read directly from the power level records, to be compared with local magnitudes at AFI, determined from the least-squares relationship (Gibowicz et al., in press)

$$\log (A/T) = 7.5 \, M_L - 6.14 \qquad (1)$$

where $(A/T)T$ represents the maximum trace amplitude, divided by the period, of the T phase at AFI. Magnitudes for events which only gave rise to T phases at hydrophone 32 are thus equivalent AFI magnitudes, determined from the extrapolated relation in Figure 2. The slope in this figure is equivalent to a change of about 15½ dB per unit on the M_L scale (about 25 dB on the unified or body wave

Table 1. Events in the Curaçoa Reef swarm interpreted as submarine volcanic eruptions.

No.	Date	Origin time (U.T.) h m	Zone	Rank in order of magnitude
1	July 12	07 55	A	1**
2		12 53	A	6=
3*		13 14	A	2
4		13 46	A	3
5		14 16	A	5
6		19 47	A	8
7	July 13	03 34	B	4
8*		03 38	B	6=

* Events which generated detectable body waves. In other cases only *T*-waves were recorded. The size of the eruptions is judged from the relative amplitudes of these *T*-waves.

** By far the largest eruption.

Table 2. Events in the Curaçoa Reef swarm interpreted as possible submarine volcanic eruptions.

No. (see Table 1)	Date	Origin time (U.T.) h m	Rank in order of magnitude **
a	July 12	07 38	2
1a		08 03	5*
1b		10 53	11*
1c		11 22	10
1d		11 28	6*
1e		11 51	1
1f		12 24	7*
3a		13 22	9*
5a		14 26	4
5b		16 59	3
5c		19 04	8*

* Recorded only at hydrophone 32.

** All these events rank much lower than those in Table 1.

scale). This is about 25 percent more than that given by Johnson & Northrop (1966).

LOCATION OF EVENTS

104 events of the swarm yielded sufficiently clear *P* and *S* arrivals at AFI/API and NUE for their epicentres to be approximately located, relative to the position given by NOAA for the principal earthquake (July 11d 22h 48m, M_L = 5.1*). These are shown in Figure 3. There is some evidence that the true position of the

* Local magnitudes are those derived from the AFI seismograms. For a fuller explanation of the methods used to determine magnitudes in this swarm see Gibowicz et al. (in press).

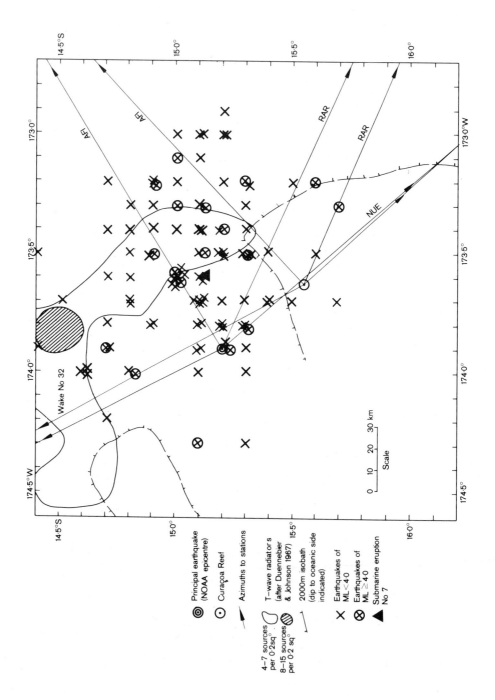

Principal earthquake
(NOAA epicentre)

Curaçoa Reef

Azimuths to stations

T–wave radiators
(after Duennebier
& Johnson 1967)

4–7 sources
per 0·2 sq°.

8–15 sources
per 0·2 sq°

2000m isobath
(dip to oceanic side
indicated)

Earthquakes of
ML < 4·0

Earthquakes of
ML ≥ 4·0

Submarine eruption
No 7

Wake No 32

Scale

0   10   20   30 km

main shock may have been east of the NOAA epicentre. In arriving at this epicentre, times at the three closest stations available to NOAA — AFI, with Lonorore (LNR) and Luganville (LUG) in the New Hebrides (Fig. 1) — were omitted from the calculations. The residual given for AFI is -6.8 s, and for LNR and LUG respectively +6.2 s and +5.1 s (U.S. Department of Commerce, 1973). These, had they been included in the computation, would have brought about a shift eastward of the epicentre amounting to about 0.5°. There is, however, insufficient evidence for further refinement of the epicentre, and the position given by NOAA has therefore been allowed to stand.

Available T-phase data are inadequate for determination of the T-phase radiators. Duennebier & Johnson (1967) have pointed out that a T phase must be detected by at least four well-spaced hydrophones for a reliable source location to be computed. It is likely that for events in this sequence T phases were generated at several different points in a given earthquake. Figure 3 shows radiators identified as active by Duennebier & Johnson (1967) in this area over the period December 1964 to January 1967. If, as Johnson's findings suggest (Johnson et al., 1963), T phases are generated at depths of the order of 2000 m on steep ocean-facing slopes, it is likely that energy in a given earthquake of the Curaçoa Reef swarm was propagated over varying distances as body waves before giving rise to T phases at widely different points on the paths to the three seismograph stations and to Wake Island.

The approximate locations obtained for those earthquakes in which body waves were observed, the considerable spread in S minus P intervals at AFI (24-37 s) and NUE (58-70 s), and the very low displacement values found by Gibowicz et al. (in press) for events of the sequence, alike imply that the swarm took place over a large region. While individual events in which only T phases were recorded cannot be located within this region, the intervals between T-phase arrivals at pairs of stations can be used to identify groups of like events which must have originated in closely spaced areas, activating the same T-phase radiators on each occasion.

Figure 4 shows the frequency distribution of intervals between T-phase arrivals at pairs of stations. The times are those of the arrival of the group containing the maximum T-phase energy, not, as described by Duennebier & Johnson (1967), the times of actual maximum amplitude in the T-phase group. The only station to show evidence of significant variation in the travel time of the T phase is AFI. Pairs involving other stations show little variation, and such as there is can well be explained by statistical scatter about a single value. In these cases a single T-phase radiator on the path to each station was probably activated by all events in the swarm, or by two radiators at comparable distances.

It would be possible to explain the observed variation in pairs of stations involving AFI by supposing much less accuracy in reading the arrival time of T at that station than at NUE, RAR, or Wake. This, to some extent, is expected (Johnson, pers. comm.) owing to the fact that AFI lies on the side of the large island of Upolu opposite to that facing the earthquakes. However, there is evidence

Fig. 3 (opposite): Approximate epicentres for events of the Curaçoa Reef swarm, relative to the position 15.2°S, 173.9°W, given by NOAA for the principal earthquake. T-wave radiators identified as active by Duennebier & Johnson (1967), during the period December 1964 to January 1967, are shown.

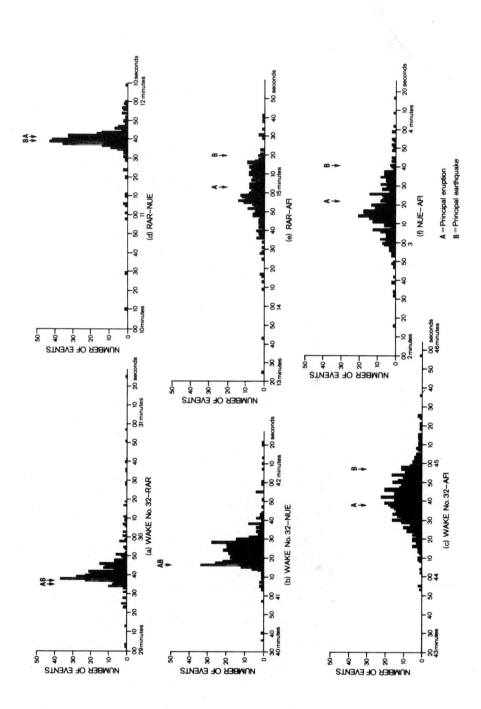

that variation due to this cause is only of the order of half what would be required to account for the observed differences. The interval between the onset* of the maximum T-phase group and the actual time of its maximum amplitude on the AFI records is 20 ± 8 s. Had the observed differences been due to reading error, a variation of at least twice this amount would have been expected. Furthermore, when observed differences are assigned to the positions of the epicentres in Figure 3, averaged, and contoured, the spatial variation for pairs of stations involving AFI is comparable to, or greater than that in pairs excluding AFI (see Fig. 5). This suggests that the observed spread of values at AFI is at least partly due to generation of T phases at different points for earthquakes in different parts of the swarm area.

On this interpretation, a significant difference exists between the travel time of the T phase to AFI for the principal eruption (marked A in Fig. 4) and for the principal earthquake (B in Fig. 4). The difference amounts to 19 s for the pairs Wake-AFI and NUE-AFI, and to 17 s for RAR-AFI. If this difference is attributed to that part of the path over which the energy was propagated as a T wave, so that the body-wave paths remain approximately equal, the T-wave radiator activated by the principal earthquake must lie some 25 to 30 km closer to AFI than that activated by the principal eruption (Johnson (1966) gives a velocity of about 1.48 km/s for the T phase in this area).

Location of events on the assumption that T-phase arrival differences can be interpreted geometrically, in terms of variation in the position of the earthquakes, leads to erroneous results. The considerable variation in arrival times at AFI (Fig. 4), and the distribution of recording stations, have the effect of elongating the source zone in the AFI direction after the manner described by Johnson & Norris (1968) and Johnson et al. (1971).

Differences between the arrival of the T phase at pairs of stations in events interpreted as volcanic eruptions in Zone A (Nos. 1-6, Table 1) are slight and may represent scatter about a single point. An indication as to where this point may lie is given by areas of maximum convergence of the contours representing differences between T-phase arrivals at pairs of stations, averaged and assigned to the positions of the epicentres in Figure 3 (see Fig. 5 for two such pairs). Figure 6 illustrates areas of convergence for the mean values determined from events Nos. 1-6 in Table 1. There are four possible areas for the location of these events. The closest fit of the contours is obtained for area 1, some 50 km west of the NOAA epicentre for the principal earthquake. A rather less close fit is obtained for area 2, immediately to the west of Curaçoa Reef, and much less precise fits for areas 3 and 4.

The only area of convergence close to one of the two points at which eruptions were observed is that to the west of Curaçoa Reef (area 2). It is therefore concluded that this was the most likely location of eruptions 1 to 6. If the point of maximum

* This onset is preceded by an earlier emergent T phase with a time difference, in more than 90 examples, of 26 ± 9 s at AFI.

Fig. 4 (opposite): Frequency distribution of intervals between T-phase arrivals at pairs of stations. a) Wake No. 32 — RAR; b) Wake No. 32 — NUE; c) Wake No. 32 — AFI; d) RAR — NUE; e) RAR — AFI: f) NUE — AFI.

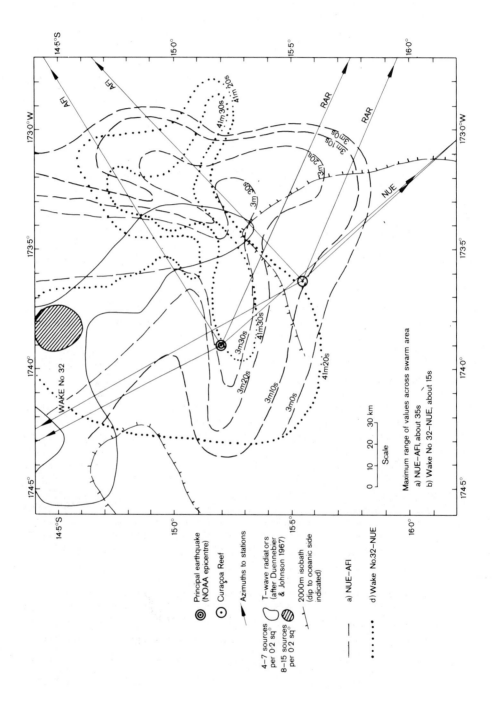

likelihood within this area is shifted some 10 km to the east in order to coincide with the point at which Riechelmann observed eruptions taking place, the location of the principal earthquake should also be shifted by the same amount to the point marked in Figure 6. T-phase radiators inferred from the bathymetry and located by Duennebier & Johnson (1967) fit these locations and the observed differences in T-phase arrivals well. The portions of these radiators considered to have been activated by the eruptions in Zone A and by the principal earthquake, on the paths to the four stations, are indicated in Figure 6.

Two of the inferred eruptions, Nos. 7 and 8, had intervals between T-phase arrivals at the stations similar to those of the principal earthquake. One of these was approximately located by P and S arrivals, relative to the main shock: its position is also marked in Figure 6. Following the interpretation of different source areas for the first six eruptions (Zone A) and for the principal earthquake and eruptions 7 and 8 (Zone B), an arbitrary division has been made for all events of the swarm. All those for which the interval between T-phase arrivals, Wake-AFI, is 44m 48s or greater (see Fig. 4) are assigned to Zone B, the estimated limits of which are shown in Figure 6. The rather close proximity of this area to the point at which eruptions were observed by American Airlines suggests that the latter was the source of eruptions 7 and 8 (Table 1).

Available evidence, though somewhat scanty, suggests that the June 29 1973 series of events also originated in Zone B. These are interpreted as earthquakes rather than eruptions. The M_L magnitude of the largest was about 4.3.

Variation in T-wave amplitudes relative to magnitudes determined from body waves is partly due to differences in the efficiency of T-phase generation, but may also be due to variation in depth of focus. Thus there seem to be a number of events which are intermediate in form between those interpreted as submarine eruptions and the bulk of the earthquakes. There are, however, insufficient data for actual depths to be assigned to the earthquakes. All are interpreted as shallow, within the upper crust.

RATE OF EARTHQUAKE OCCURRENCE

The rate of earthquake occurrence n is given by the number of shocks per unit time above a certain magnitude. Where ΔN is a fixed number of shocks, and δT the variable time interval during which they occur,

$$n = \Delta N / \delta T \tag{2}$$

This formula has been used to calculate n separately for Zones A and B, and for the swarm as a whole using the large number of shocks recorded at hydrophone 32. The calculations have been made for sets of 50 shocks, with a threshold level of M_L $\geqslant 3.3$ for the regionalised shocks and $M_L \geqslant 2.8*$ for the larger sample, spaced at intervals of 25 shocks, and assigned to the mean of each time interval. Results, in number of earthquakes per day, are shown in Figure 7, plotted against the time after the beginning of the swarm (taken as July 11d 01h 56m). 56 earthquakes of

* This figure represents an equivalent magnitude at AFI, extrapolated from the least-squares relationship in Figure 2.

Fig. 5 (opposite): Average values of intervals between T-phase arrivals at pairs of stations, a) NUE—AFI; b) Wake No. 32 — NUE, for epicentres determined from body waves.

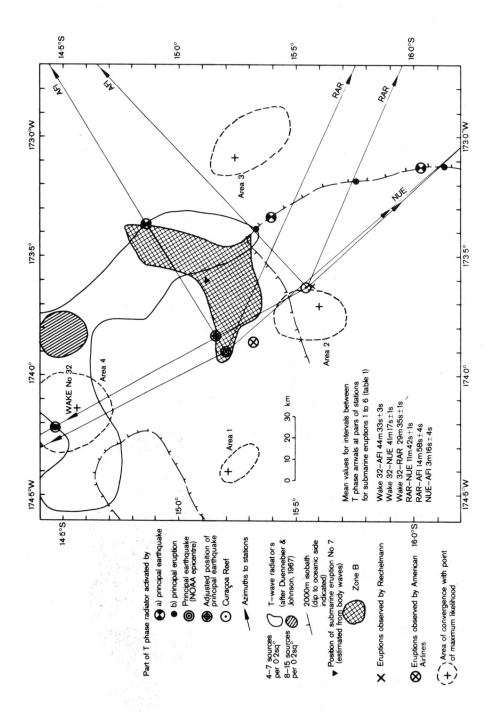

Part of T phase radiator activated by –

⊗   a) principal earthquake
●   b) principal eruption
◉   Principal earthquake
      (NOAA epicentre)
⊕   Adjusted position of
      principal earthquake
⊙   Curaçoa Reef

↗   Azimuths to stations

▢   T-wave radiators
      (after Duennebier &
      Johnson, 1967)

⬤   4–7 sources
      per 0·2sq°
▨   8–15 sources
      per 0·2sq°

⌇   2000m isobath
      (dip to oceanic side
      indicated)

▲   Position of submarine eruption No 7
      (estimated from body waves)

▨   Zone B

✕   Eruptions observed by Riechelmann

⊗   Eruptions observed by American
      Airlines

⤿+⤾ Area of convergence with point
      of maximum likelihood

Mean values for intervals between
T phase arrivals at pairs of stations
for submarine eruptions 1 to 6 (table 1)

Wake 32–AFI 44m33s±3s
Wake 32–NUE 41m17s±1s
Wake 32–RAR 29m35s±1s
RAR–NUE 11m42s±1s
RAR–AFI 14m58s±4s
NUE–AFI 3m16s±4s

magnitude $M_L \geqslant 3.3$ could not be assigned to either zone with available T- wave data: these were omitted from the calculations (although the argument is weakened, it is not invalidated by this omission, since the assumption that about an equal number of these shocks originated in each zone is probably not too far wrong).

There is no significant difference in the pattern of n between Zones A and B, apart from an increase in the number of shocks in Zone A associated with eruptions 1 to 5 (Table 1; zone B contributes to this peak owing to an aftershock series following the second largest earthquake in the swarm: M_L 4.7 at 07h 19m on 12 July). More earthquakes took place in Zone A than in Zone B (351 as compared with 165). Earthquakes began in Zone A about 10 hours before they began in Zone B, but persisted in Zone B for more than 6.5 hours after ending in Zone A. The dominant activity in both zones was clearly governed by the largest earthquake in the series rather than by the eruptions. The rate of earthquake occurrence increased until shortly before the largest earthquake and then began to decline, in a manner typical of a vigorous foreshock and aftershock sequence (Gibowicz, 1973b). The fact that the main shock occurred a little after the point of reversal, as shown in Figure 7, is a function of the grouping of the data rather than a significant feature of the distribution.

The complete sample, in which the bulk of events are those recorded only at hydrophone 32*, adds much detail to the picture. In particular, the peak associated with eruptions 2 to 5 (Table 1) is accentuated, and it is clear that each group of eruptions, with the exception of the final pair in the declining stage of activity, was associated with an increase in the number of earthquakes.

The overall timing of the swarm is of interest. Simkin & Onyeagocha (1973) suggested that it may have been tidally triggered. It began close to the time of maximum southern declination of the moon (July 12d 13h) and also close to the time of full moon (July 15d 12h). The time of maximum declination was therefore a mere five hours after the time of the principal eruption, and 35 hours after the beginning of the swarm.

FREQUENCY/MAGNITUDE RELATIONSHIP
The coefficient b in the frequency/magnitude relationship
$$\log N_M = a - bM** \quad \text{(Gutenberg \& Richter, 1954)} \quad (3)$$
has been calculated separately for Zones A and B, and, using the large number of events recorded only at hydrophone 32, for the swarm as a whole.

Gibowicz et al. (in press) determined values of b for the Curaçoa Reef swarm, using 564 earthquakes of magnitude $M_L \geqslant 3.4$. These show an inverse correlation with the rate of occurrence of the earthquakes, b decreasing until about the time of the main shock, when its value was about 1.1, and then increasing sharply to the

---

* There is of course no certainty that all these events originated in the swarm area. However, their similarity of form, and the fact that they were recorded only at this hydrophone, makes such an interpretation likely.

** N_M is the number of earthquakes within a given magnitude range $M + \Delta M/2$; M is any magnitude scale; a and b are constants. a depends on the range of magnitudes involved and is not an independent quantity.

Fig. 6 (opposite): Areas of convergence of the contours, representing mean intervals between T-phase arrivals at pairs of stations for eruptions 1 to 6 (see Table 1). The area defined as Zone B is shown.

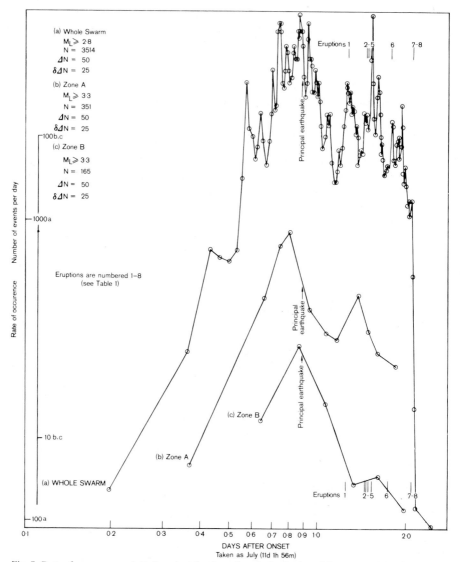

Fig. 7: Rate of occurrence, plotted against the time after the beginning of the swarm (July 11d 01h 56m), a) for the swarm as a whole; b) for Zone A; c) for Zone B.

very high level of 2.5 towards the end of the swarm. The average for the swarm as a whole was 1.77 ± 0.15 at 95 percent confidence level.

Values of b have been calculated in the same manner as described by Gibowicz et al. (in press), for sets of 100 shocks, spaced at intervals of 25 shocks, and assigned to the mean of each time interval involved. The magnitude threshold was $M_L \geqslant 3.3$ for events assigned to the two zones, and $M_L \geqslant 2.8$ for the larger sample, and, as in the determination of rate of occurrence, events (56 of them) which could not be assigned to zones were omitted from the regionalised

Fig. 8 (opposite): The coefficient b, plotted against the time after the beginning of the swarm (July 11d 01h 56m), a) for the swarm as a whole; b) for Zone A; c) for Zone B.

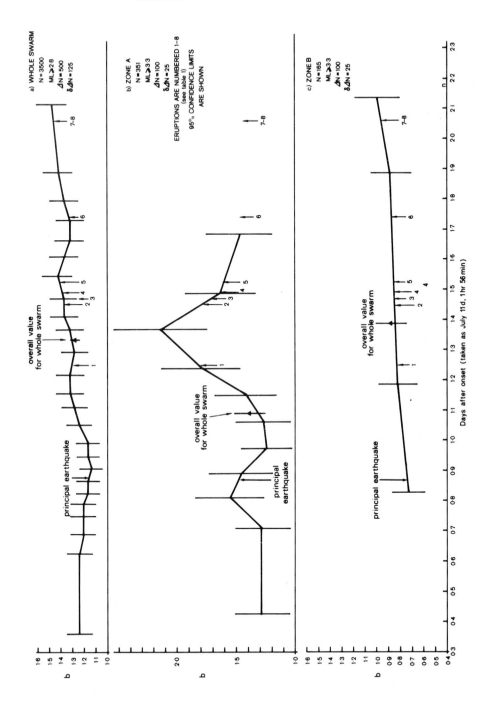

calculations (as noted above, this weakens but does not invalidate the argument). Results are shown in Figure 8.

The extended sample for the swarm shows a slight decrease in the value of *b* from the beginning of the swarm to about the time of the principal earthquake. Thereafter the value increased. The small variations shown are not statistically significant. Events interpreted as eruptions took place during the period of increase in the value of *b*. For the swarm as a whole, *b* is calculated as 1.28 ± 0.04.

Somewhat greater variation is observed for Zone A, although much of that shown is not statistically significant. About the time of the first and largest eruption, the *b* value increased sharply. Later eruptions appear to have taken place during a decrease in the *b* value. Very little variation in *b* took place in Zone B, and none is statistically significant. There is, however, a significant difference between values for Zones A and B over much of the period. The value of *b* for Zone A, for the swarm as a whole, is 1.29 ± 0.13, and that for Zone B, 0.88 ± 0.13.

The pattern is similar to that found by Gibowicz et al. (in press) from a smaller sample, although their values are higher. This is due to the fact that a large number of events of about magnitude $M_L = 3.4$ were detected at hydrophone 32, but not at the seismograph stations. Although the number of smaller shocks detected was large (almost 3000), it is likely that a disproportionate number passed undetected or were obscured by larger events. Furthermore, in assigning events to Zones A and B, many of the smaller shocks gave inconclusive results and were rejected.

The explanation for the increase in the *b* value during the course of the swarm, and particularly during the period of the eruptions, may lie in a growing heterogeneity or decreased strength of the source material, or in the development of a non-uniform stress field (Mogi, 1963). It may also be due to decreased stress, as suggested by Gibowicz (1973a) and by Wyss (1973), or to the presence of a stress field of thermal origin (Warren & Latham, 1970).

RATE OF ENERGY RELEASE

In contrast to the rate of occurrence of the earthquakes, which was higher in Zone A than in Zone B throughout the sequence, nearly twice as much energy in radiated seismic waves was liberated in Zone B than in Zone A.

Energies have been estimated by means of Richter's (1958) equation
$$\log E = 9.9 \, M_L - 0.024 \, M_L^2 \tag{4}$$
where E is in ergs and the logarithm to the base 10, and M is the value on the local magnitude scale.

During the period from the beginning of the swarm to immediately before the principal earthquake, energies liberated in the two zones were comparable (about 5 x 10^{11} J in Zone A, and about 4 x 10^{11} in Zone B). Most of the difference in energy release between the two zones took place during the period between the principal earthquake and the largest eruption, during which about 1.5 x 10^{12} J was liberated in Zone B and only about 4 x 10^{11} J in Zone A. After the eruptions commenced,

Fig. 9 (opposite): Energy release for Zones A and B, plotted against time.

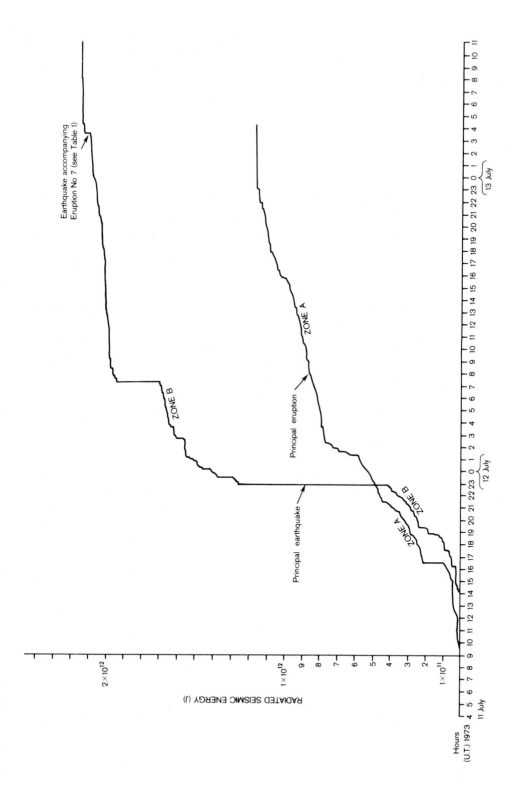

energies released in the two zones were comparable (about 3×10^{11} J in Zone A, and about 2×10^{11} J in Zone B). These are minimum figures, owing to the omission of the earthquakes in the swarm which could not be assigned to either of the two zones.

Variations in energy release in the two zones are shown in Figure 9, plotted against time.

CONCLUSIONS

The 1973 July Curaçoa Reef swarm took place over a large area, of the order of 10,000 km^2, which, from the evidence of seismic source size, has been identified by Gibowicz et al. (in press) as a major zone of weakness. All events within this region generated T waves at single radiators (or radiators at comparable distances) on the paths to the Wake hydrophone array, NUE, and RAR, but two radiators on the path to AFI were energised by events in different parts of the swarm area.

Eight events late in the series were identified as submarine eruptions from their anomalously strong T-wave signals. The largest, and the five that followed, are assigned to the position of observed eruptions near Curaçoa Reef, on the basis of intervals between T-wave arrivals at pairs of stations for earthquakes yielding body waves. The last of these eruptions took place about the time at which activity was first noticed at Tafahi, the nearest inhabited island. The last two eruptions in the series took place close to the principal earthquake. Their positions are tentatively identified with an observed eruption point about 35 km northwest of Curaçoa Reef.

Variations in the rate of occurrence of the earthquakes showed that activity in all areas depended principally on the largest earthquake, which took place nine hours before the first eruption. An increase in the number of earthquakes took place about the time of each of the eruptions in the Curaçoa Reef area, but the two final eruptions in the series, about 35 km northwest of the reef and within about four minutes of each other, were associated with a declining number of earthquakes and themselves marked the virtual cessation of the swarm. Gas emission continued in both areas after the end of seismic activity. All sightings, other than the first from Tafahi, belong to this period.

The coefficient b was inversely correlated with rate of occurrence, decreasing up to about the time of the principal earthquake and then increasing steadily to the end of the swarm, although considerable fluctuations took place in the zone of the principal eruption. A sharp increase in the value of b was associated with the first eruption in the series.

Swarm activity began in the zone in which the largest eruptions later developed, and only spread to the zone of the largest earthquakes after a lapse of ten hours. This suggests a process of widespread intrusion which first reached the surface in the weaker of the two zones (the Curaçoa Reef area). Activity did not end until an eruption had taken place in the stronger zone, which was the area in which the principal stress release had already occurred.

This stress release, and the reversal in *b* that accompanied it, may be seen as the failure of the last major obstruction to the upward movement of magma, which found its release in a strongly gas-charged eruption in a zone of weakness some 35 km away. Residual eruptive energy, in the form of weaker eruptions which terminated the series, reached the surface, in the area in which the main obstruction had been overcome, some 19.5 hours after this climactic eruption had taken place.

ACKNOWLEDGEMENTS

The author is indebted to R.H. Johnson, of the Hawaii Institute of Geophysics, for his generous provision of records from the hydrophones near Wake Island and for his comments on the first draft of this paper, and to T. Simkin of the Smithsonian Institution and J.C. Riechelmann of Tonga for information about sighting of the eruptions. Thanks are also due to S.J. Gibowicz and G.K. Sutton, formerly of Geophysics Division, D.S.I.R., who carried out some of the analysis of records on which this paper is based, to A.I. Tomlinson, N.Z. Meteorological Service, and A.C. Gilmore, Carter Observatory,Wellington, for information on weather and tides in the Curaçoa Reef area, and to J.C. Koot for typing the manuscript.

REFERENCES

Duennebier, F.K. & Johnson, R.H., 1967: T-phase sources and earthquake epicenters in the Pacific Basin. *Hawaii Inst. Geophys.,* HIG-67-24, 17 pp., appendix.

Gibowicz, S.J., 1973a: Stress drop and aftershocks. *Bull. seism. Soc. Am., 63,* pp. 1433-1446.

Gibowicz, S.J., 1973b: Patterns of earthquake swarm activity. *Ann. Geofis., 26,* pp. 637-658.

Gibowicz, S.J., Latter, J.H. & Sutton, G.K., in press: Earthquake swarm associated with volcanic eruption, Curaçoa Reef area, northern Tonga, July 1973. *Ann. Geofis.*

Gutenberg, B. & Richter, C.F., 1954: *Seismicity of the Earth and Associated Phenomena.* Princeton, Univ. Press.

Johnson, R.H., 1966: Routine location of T-phase sources in the Pacific. *Bull. seism. Soc. Am., 56,* pp. 109-118.

Johnson, R.H., & Norris, R.A., 1968: T-phase radiators in the western Aleutians. *Bull. seism. Soc. Am., 58,* pp. 1-10.

Johnson, R.H., Norris, R.A., Duennebier, F.K. & Northrop, J., 1971: T-phase data on Kamchatka/Kurils earthquakes: a reply. *Bull. seism. Soc. Am., 61,* pp. 791-794.

Johnson, R.H., Northrop, J. & Eppley, R., 1963: Sources of Pacific T phases. *J. geophys. Res., 68,* pp. 4251-4260.

Johnson, R.H. & Northrop, J., 1966: A comparison of earthquake magnitude with T-phase strength. *Bull. seism. Soc. Am., 56,* pp. 119-124.

Mogi, K., 1963: Some discussions on aftershocks, foreshocks, and earthquake swarms — the fracture of a semi-infinite body caused by an inner stress origin and its relation to the earthquake phenomena, 3. *Bull. Earthq. Res. Inst., Tokyo Univ. 41,* pp. 615-658.

Richter, C.F., 1958: *Elementary seismology.* San Francisco, W.H. Freeman and Co.

Simkin, T. & Onyeagocha, A.C. 1973, in press: Pumice drift and 1973 submarine eruption, north Tonga Islands. *Smithsonian Contrib. Earth Sci.*

Smithsonian Institution, 1973: Curacoa Reef submarine volcanic activity. *Event Notification Report, Center for Short-lived Phenomena, 1679,* July 19, 1973.

U.S. Department of Commerce, 1973: *National Oceanic & Atmospheric Administration, Earthquake Data Report No. 50-73,* Sept. 7, 1973.

Warren, N.W. & Latham, G.V., 1970: An experimental study of thermally induced microfracturing and its relation to volcanic seismicity. *J. geophys. Res., 75,* pp. 4455-4464.

Wyss, M., 1973: Towards a physical understanding of the earthquake frequency distribution. *Geophys, J. Roy. Astr. Soc., 31,* pp. 341-359.

PREDICTION STUDIES ON WHITE ISLAND VOLCANO, BAY OF PLENTY, NEW ZEALAND

R.H. CLARK and J.W. COLE

Department of Geology, Victoria University of Wellington, Private Bag, Wellington, New Zealand

ABSTRACT

White Island is an active volcano in the Bay of Plenty, North Island, New Zealand. The crater floor is easily accessible and the comparatively frequent tephra eruptions make it a convenient location to investigate methods of prediction. A level net was established in July 1967, magnetic studies were commenced in May 1968, and temperature probes were installed in September 1969. Measurements have been carried out by each method at regular intervals since these dates. The findings of these methods have been correlated, and changes occur which appear to be associated with eruptive activity. Prior to the eruption of 19 July 1971, bulging of the crater floor occurred; the ground temperature at the centre of the bulge decreased and a decrease was observed in the total magnetic field (slightly offset because of the Earth's magnetic field). After the eruption the bulge decreased, ground temperature at this location increased, and the magnetic field also increased. The tentative conclusion is reached that minor eruptions are caused by explosive clearing of blocked conduits below the crater floor.

INTRODUCTION

White Island is the emergent summit of a large submarine volcanic massif (16 x 18 km) in the Bay of Plenty, 50 km north of Whakatane, North Island, New Zealand (Fig. 1).

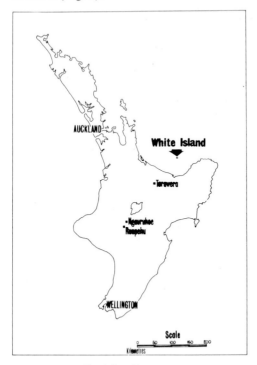

Fig. 1: Locality map.

The island consists of a complex of two overlapping stratocones (Black, 1970; Duncan, 1970). Remnants of the older cone form the western section of the island, centred on Mount Ngatoro (Fig. 2), and this is surmounted by the younger cone forming the large eastern and central part of the island. Both cones consist of lava flows, agglomerate breccia, and unconsolidated beds of tephra and tuff, all of andesitic or dacitic composition (Duncan, 1970; Cole & Nairn, in press). The top of the younger cone is occupied by a large crater approximately 1.2 km long and 0.4 km wide (Fig. 2), most of which is less than 30 m above sea level (Fig. 3).

HISTORICAL ACTIVITY

The volcano was named 'White Island' by Captain James Cook in 1769 because 'as such it always appeared to us'. He did not record an eruption, but earlier volcanic activity is evident from Maori legends (Luke, 1959). The first European landing was made in 1826, and since this time approximately 30 steam and tephra eruptions have been recorded. For most of the 19th century the crater floor was occupied by a lake, but this was drained in 1913 in order to work sulphur from the crater floor. Activity observed between 1826 and 1913 has been summarised by Luke (1959).

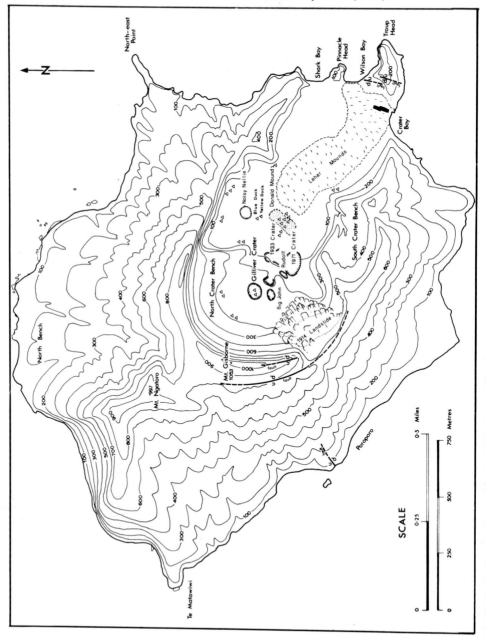

Fig. 3: White Island from the east.

On or about 10 September 1914, an extensive collapse of the crater wall occurred (Fig. 2). The debris became saturated with hot water, presumably from the blocked vents and then swept down the crater floor as a lahar overwhelming all 11 sulphur workers and burying their building and plant. It is the debris from this lahar that forms most of the present surface of the crater.

Since 1914 several vents have formed accompanied by tephra eruptions. Small eruptions occurred in 1922, 1924, and 1928 and a larger eruption occurred in 1933 forming '1933 Crater' (Fig. 4). This eruption deposited up to 1 m of tephra on the northern and southern slopes of the island. The next vent to form was 'Noisy Nellie' which had appeared by January 1947 and continued to erupt intermittently until 1959. 'Big John' was formed in 1962 and 'Gilliver' in 1966 (Fig. 4).

In February 1967 a party from Victoria University spent three days on the island. Gilliver continuously emitted tephra and occasionally ejected rocks. The northwest wall of 1933 Crater was seen to be clear of yellow sulphur and gave the appearance of having had the sulphur burnt off by heat from the numerous fumaroles. At night the mouths of these fumaroles glowed red-hot, indicating that the steam was emerging at a very high temperature. The steam from one fumarole, which later became 'Rudolf' crater, was established by optical pyrometer to be between 750° and 800°C, (Clark, 1970).

CHANGES IN ELEVATION OF CRATER FLOOR

On 12 July 1967 a precise level net was installed on the crater floor. Nine survey stations (III-XI inclusive in Fig. 4) were installed to form a closed loop

Fig. 2 (opposite): White Island (small triangles indicated sites of fumaroles).

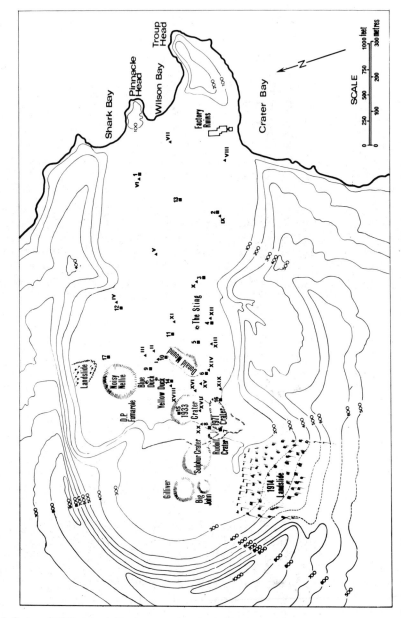

Fig. 4: Crater of White Island showing vents, locations of level stations (I-XX), and temperature probes (1-17).

about 2 km in circumference. Two stations (I and II) formed a hanging loop to the north of Donald Mound and three stations (XII, XIII, and XIV) another hanging loop to the south of Donald Mound. Measurements with a Zeiss self-adjusting level have given an error of less than 3 mm around the closed loop.

The first relevel survey was carried out on 1 November 1967. In plotting level changes, the lowest station, and that most distant from the active vents (VIII in

Fig. 4), was assumed to be fixed in height. It was found that between 12 July and 1 November an asymmetric bulge had formed, with its highest point centred on station XI to the east of Donald Mound.

On 9 February 1968 the crater floor was relevelled for a second time and the southeastern part of the bulge was found to be flatter than in November. Rudolf crater was emitting ash and steam, a loud roar accompanied the gas emission, and the orifice was glowing.

By the end of February the tephra eruptions were much larger. Clark (1970) recorded it as 'one of the largest [eruptions] from White Island in the last 100 years'. In view of this it was decided that aerial inspections should be carried out at intervals between landings.

During the next few months the vent of Rudolf was enlarged from about 7 m in March to 10 m by the end of April 1968, and to about 35 m by February 1969. During this period the temperature fell until in May 1968 a temperature of 450°C was recorded by a thermocouple (Clark, 1970).

From June 1968 onwards, vapour (mainly steam) was emitted continuously from Rudolf but tephra eruptions ceased. Relevel surveys on 29 April 1968, 10 September 1968, and 26 February 1969 indicate that following the development of Rudolf as a tephra vent there was considerable deflation of the crater floor northwest of Donald Mound, not far from Rudolf crater, and this deflation continued until about September 1969, well after the end of tephra emission.

On 5 December 1969 the level net was extended to include stations XV, XVI, and XVII, extending the hanging line south of Donald Mound across the floor of 1933 Crater to Rudolf crater (Fig. 4). It was found that a distinct and new bulge had developed, centred on station XI (Fig. 5). This had increased by June 1970, but by November* had diminished and by February 1971 the centre of tumescence had shifted to station XIV. From surveys in April and May 1971, the tumescence around station XIV was found to have increased. The area around 1933 Crater was depressed and by April was down 80 mm.

At approximately 1330 hrs on 19 July 1971, White Island was seen from the mainland to erupt again. Tephra was observed in the initial cloud but emission lasted only for a few hours. An aerial inspection on 20 July revealed a large new elliptical crater (100 m x 50 m), since named '1971 Crater', in the southwestern part of 1933 Crater. This new crater developed on the flank of the uplifted area around Donald Mound. It was clearly a collapse feature, following explosion, and had dropped about 4 m along ring fractures. Up to 1 m of tephra was deposited around the crater rim but the thickness rapidly diminished away from the crater.

Level stations XIII, XVI, and XVII were destroyed or buried by the eruption. However, sufficient of the net survived for a further survey to be made on 29 July,

* The party on 5 November 1970 included G.A.M. Taylor who made several useful suggestions for observational work on White Island. These included the analyses of gases and condensates from fumaroles, which are now an integral part of the programme.

when it was found that the closest surviving stations to 1971 Crater (stations XVIII and XIV) had dropped 25 mm and 4 mm respectively. A new bulge had started to develop to the east of Donald Mound, and continued to grow throughout 1971. Temperatures of the fumaroles around Donald Mound were high, and at night their orifices were glowing (Clark, 1973).

Little change had occurred by February 1972 although the bulge was slightly reduced, but by 9 July 1972 it had become depressed almost to the level of July 1971. A new bulge was developing in the northern part of 1933 Crater, and continued to rise in the surveys of 20 December 1972 and 13 June 1973. The area to the west of Donald Mound was subsiding during this period.

The next survey was not until 27 February 1974, when it was found that the new bulge was still rising, but perhaps more significantly station XI to the east of Donald Mound was also rising, recovering much of the depression it underwent after the 1971 eruption (Fig. 5). Both areas were still rising during a survey on 29 August 1974, and fumaroles were more active. Three large new fumaroles had developed in the western part of 1971 Crater, one of which was observed to eject occasional blocks and considerable quantities of mud during the survey.

Between this survey and one undertaken on 19 November 1974 a large fumarole had developed to the southeast of Donald Mound and had ejected some mud and tephra. During the survey it was discharging large volumes of wet steam which contained droplets of hydrochloric acid. It was thus named 'The Sting'. Despite this activity, station XI continued to show uplift (Fig. 5) at an increasing rate, perhaps indicating further activity in the future.

GROUND TEMPERATURE VARIATION

Measurement of ground temperatures began with the installation of 12 temperature probes on 1 September 1969 (1-12 in Fig. 4). Three additional probes (13-15) were installed on 7 December 1970 and two (16 and 17) on 29 July 1971. The probes consist of a glass tube, 16 mm in diameter, 1.4 m long, and protected by fibreglass. Each tube contains a chrome-alumel thermometer terminating below a rubber cork which is covered by a sealed plastic bag. The probes were installed to a depth of about 0.4 m at localities around the crater floor (Fig. 4). The voltage was measured during each survey with a high-impedance voltmeter. Field measurements were converted to nominal ground temperatures and air temperature added (Browne & Cole, 1973).

Several of the probes have remained at or near boiling point since the commencement of the survey. Others, however, have remained below boiling point since installation (Fig. 6), and over the 5-year period of observation it is apparent that there is a seasonal influence on these variations, temperature rising during the summer months and falling during winter months. Probes 3 and 13 at the eastern end of the crater and most distant from the active vents show this pattern most consistently.

Probes close to the active vents show an additional factor which may be related to eruptive activity. Between the surveys of 9 February and 6 April 1971,

Fig. 5 (opposite): Vertical movements of level stations XI and XIV relative to station XVIII from 17 July 1967 to 19 November 1974, plotted against time (surveys undertaken by R.H. Clark).

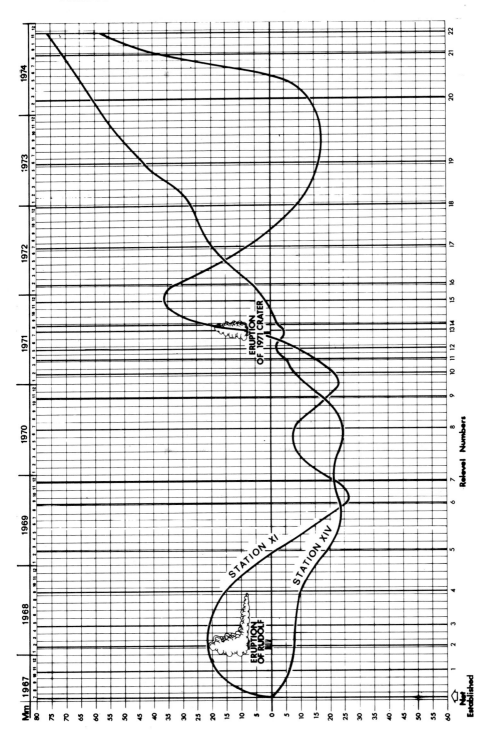

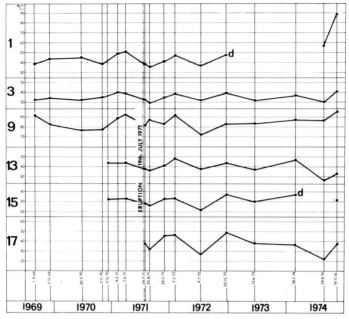

Fig. 6: Variation in ground temperature recorded by probes registering temperatures below 100°C between 1 September 1969 and 19 November 1974. (Temperatures recorded by P.R.L. Browne, J.W. Cole, and A.R. Duncan.)

prior to the 1971 eruption, most probes showed a slight increase in temperature. Probe 5, however, showed a decrease of 17°C. The siting of this probe corresponds to the centre of the bulge in the crater floor determined from the level survey of the same day. Between the surveys of 29 July and 26 August, following the 1971 eruption, probe 6 (probe 5 was destroyed by the eruption) showed an increase of 19.5°C whilst most of the other probes showed a slight decrease.

MAGNETIC VARIATION

Total-field magnetic surveys were begun on White Island in May 1968. A portable proton magnetometer was used, so the values obtained are independent both of the orientation of a sensing head and of the instrument used (Hurst & Christoffel, 1973). Measurements were taken at all 20 survey stations shown in Figure 4 together with some additional stations installed prior to the Victoria University programme. The sensing head was mounted on a special tripod 2.2 m long, and measurements were made with the bottom end of the tripod resting on the survey peg. Corrections made for variations of the Earth's magnetic field from external effects both during and between surveys are given in Hurst & Christoffel (1973).

Considerable changes in the total magnetic field were recorded between surveys, with a good inverse correlation between the magnetic and level changes (Hurst & Christoffel, 1973). The correspondence is not exact, however, but is offset slightly as a result of the angle of dip of the Earth's magnetic field.

DISCUSSION

From the data collected over the period July 1967 to December 1974, changes in level of the crater floor and changes of total magnetic force may be correlated with eruptive activity. Ground temperature changes appear to correlate both with eruptive activity and seasonal climatic change. Prior to the eruptions that formed Rudolf and 1971 Craters, bulges developed in the crater floor. After the eruptions these bulges subsided and new bulges developed elsewhere.

The magnetic field also changed before and after eruptive activity. After the eruption from Rudolf crater the total magnetic field increased, corresponding to a cooling of the rock in the vicinity of the crater. Before the 1971 eruption total magnetic field decreased indicating increase in temperature at depth (Hurst & Christoffel, 1973), while after the eruption the total magnetic field increased. By combining the magnetic measurements with the level surveys, Hurst & Christoffel (1973) estimated that in the 1971 eruption the centre of the area of 'heated rock' was at 200 m, the mass having a radius of 125 m.

Temperatures also show an inverse relationship with ground level. When a bulge occurs, as before 1971 eruption, temperatures generally decrease; conversely when a trough develops there is a relative increase.

These relationships are considered to be due to periodic blockage of channels and fissures permeating the solid material underlying the crater (Clark, 1973). In a zone below which blockage occurs, gas pressure is presumed to build up and cause local surface bulging. Temperatures at depth would increase, and if this increase is sufficient to raise the temperature of the rocks above their Curie Point, changes to the magnetic field would also occur. At the same time reduction of thermal gradient would cause surface temperatures to decrease slightly.

The opening of new channels during an eruption would clear the blockage, thus lowering the ground surface, increasing the surface temperature, and lowering the temperature at depth, with consequent increase in magnetic field. Channels would open through the weakest zones, which might be somewhat displaced from the centre of the bulge as found in the 1971 eruption.

COMPARISON WITH OTHER GEODETIC SURVEILLANCE PROGRAMMES

Correlation between volcanic activity and ground surface deformation has been demonstrated in many areas (Decker & Kinoshita, 1971; Decker, 1973). Most studies have, however, used permanently installed tiltmeters to measure changes in inclination of the ground surface on and around volcanoes. This is not at present practicable at White Island because of difficult access to the outer slopes of the cone, problems of corrosion within the crater, and difficulty of telemetering data to the mainland.

Optical levelling similar to that described in this paper has been done at Kilauea (Fiske & Kinoshita, 1969) and on Asama (Minakami, 1960) volcanoes, but in each example the area of elevation change was much larger than at White Island (of the order of 10 km diameter). These changes almost certainly reflect magma movements beneath the volcano. The study of White Island is on a much smaller scale, and the bulges and depressions recorded are generally less than 500 metres in diameter. These changes probably reflect hydrothermal activity relatively near the surface, probably related to magma activity at greater depth.

ACKNOWLEDGEMENTS

The work was made possible firstly through the owner of White Island, J.R. Buttle of Auckland, providing access, and secondly through the co-operation of the RNZAF and RNZN, which provided transport to White Island and whose personnel assisted with the work. Assistance with the level surveys was also provided by colleagues and students of Victoria University, and by DSIR. Temperature probes were installed by A.R. Duncan and P.R.L. Browne, and measurements were made by P.R.L. Browne on 10 occasions. Aerial observation of White Island was possible through the co-operation of the Civil Aviation Division of the Ministry of Transport. Financial support has been forthcoming from the Research Commitee of Victoria University of Wellington.

REFERENCES

Black, P.M., 1970: Observations on White Island Volcano, New Zealand. *Bull. Volcanol. 34(1)*, pp. 185-67.

Browne, P.R.L., & Cole, J.W., 1973: Surveillance of White Island volcano 1968-1972. Part 2 — Ground temperature variation between 1 September, 1969 and 1 December, 1971. *N.Z.J. Geol. Geophys., 16(4)*, pp. 959-63.

Clark, R.H., 1970: Volcanic activity on White Island, Bay of Plenty 1966-69. Part 1 — Chronology and crater floor level changes. *N.Z.J. Geol. Geophys., 13(4)*, pp 565-74.

Clark, R.H., 1973: Surveillance of White Island volcano 1968-1972. Part 1 — Volcanic events and deformation of the crater floor. *N.Z.J. Geol. Gephys., 16(4)*, pp. 949-57.

Cole, J.W., & Nairn, I.A., in press: Catalogue of the active volcanoes of the world — Part 22; New Zealand. *Int. Assoc. Volcanology.*

Decker, R.W., & Kinoshita, W.T., 1972: Geodetic Measurements. *in* Surveillance and prediction of volcanic activity. *UNESCO Earth Science Monograph 8, Paris*, pp. 47-74.

Decker, R.W., 1973: State-of-the-art in volcano forecasting. *Bull. Volcanol. 37(3)*, pp. 372-394.

Duncan, A.R., 1970: The petrology and petrochemisty of andesite and dacitic volcanoes in eastern Bay of Plenty, New Zealand. *Unpublished Ph. D. thesis, Victoria University of Wellington.*

Duncan, A.R. & Vucetich, C.G., 1970: Volcanic activity on White Island, Bay of Plenty 1966-69. Part 2 — Tephra eruptions — Stratigraphy and petrography. *N.Z.J. Geol. Geophys., 13(4)*, pp. 969-79.

Fiske, R.S., & Kinoshita, W.T., 1969: Inflation of Kilauea volcano prior to its 1967-1968 eruption. *Science, 165*, pp. 341-9.

Hurst, A.W. & Christoffel, D.A., 1973: Surveillance of White Island volcano 1968-1972. Part 3 — Thermo-magnetic effects due to volcanic activity. *N.Z.J. Geol. Gephys., 16(4)*, pp. 965-72.

Luke, J., 1959: History. *in* White Island (Eds. Hamilton, W.M. & Baumgart, I.L.). *Dept. Sci. Ind. Res. Bull. 127*, pp. 14-24.

Minakami, T., 1960: Fundamental research for predicting volcanic eruptions. Part 1: Earthquakes and crustal deformations originating from volcanic activites. *Bull. Earthquake Res. Inst., 38*, pp. 497-544.

PYROCLASTIC ERUPTIONS OF NGAURUHOE VOLCANO, CENTRAL NORTH ISLAND, NEW ZEALAND, 1974 JANUARY AND MARCH

I.A. NAIRN[1], C.A.Y. HEWSON[2], J.H. LATTER[3], and C.P. WOOD[4]

[1] N.Z. Geological Survey, Department of Scientific and Industrial Research, P.O. Box 499, Rotorua, New Zealand.

[2] Geophysics Division, Department of Scientific and Industrial Research, c/o Ministry of Works and Development, Private Bag, Taupo, New Zealand.

[3] Geophysics Division, Department of Scientific and Industrial Research, P.O. Box 8005, Wellington, New Zealand.

[4] N.Z. Geological Survey, Department of Scientific and Industrial Research, P.O. Box 30368, Lower Hutt, New Zealand

ABSTRACT

Pyroclastic eruptions of Ngauruhoe volcano in January and March 1974, were the largest and most violently explosive since the major eruptive period of 1954-55. Explosive eruptions were accompanied by strong volcanic earthquakes, visible atmospheric shock waves, and loud detonations, and ejected large blocks which were incandescent at night. The most voluminous eruptions were less explosive, and formed towering eruption columns accompanied by pyroclastic avalanches. Lack of significant volcanic earthquakes indicated that these voluminous eruptions occurred under relatively 'open-vent' conditions. The March eruptive sequence was shorter than that of January, but individual eruptions were more powerful, and a solid lava plug apparently rose in the vent. Seismic data and trajectory observations suggest a shallower focal depth for the March explosions. The ejecta from both January and March eruptions possess textural features indicating reheating and fusion of old lava, and are petrographically and chemically identical to the 1954 Ngauruhoe lava; this suggests that the 1974 eruptions were ultravulcanian in type.

INTRODUCTION

Ngauruhoe is an andesite stratovolcano of 2291 m elevation, rising above Tongariro volcanic massif (Fig. 1), within the Tongariro Volcanic Centre. Growth of the 900 m high Ngauruhoe cone apparently commenced about 2500 years B.P. (Topping, 1974). Many eruptions have been observed since 1900 A.D. (Cole & Nairn, in press). Two major eruptive episodes took place in 1948-49 and 1954-55, when lava flows were extruded and a subsidiary inner cone was built up on the western side of the summit (Gregg, 1960). Since 1955 there have been only small intermittent pyroclastic eruptions from a deep summit vent, until the major explosive activity of January and March 1974.

Ngauruhoe crater lies 1130 m above and 9.3 km northeast of the Chateau Volcano Observatory (Fig. 1), where a permanent station (CNZ) of the New Zealand Seismological Observatory network is located (see Appendix for details).

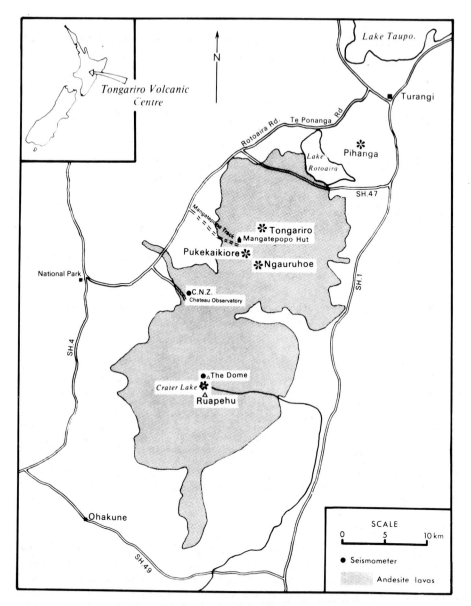

Fig. 1: Tongariro Volcanic Centre.

PRECEDING EVENTS

Observations

During 1973, Ngauruhoe erupted ash on a number of occasions, most vigorously when an eruptive period extending from 29 December 1972 climaxed in a spectacular eruption of ash and incandescent material on the evening of 2 January 1973*. Small eruptions also occurred in September and October 1973.

*All dates and times are N.Z. Standard Time, 12 hours ahead of Universal Time.

Petrographic examination showed that ash samples were derived from old tephra and comminuted lava which had been partly altered, milled, and mixed with secondary minerals in the vent, without any fresh magmatic component.

Ngauruhoe crater was examined on 27 November 1973, during a period of low emission. The crater was then at least 180 m deep, with nearly vertical walls and a secondary vent at the bottom (R.B. Glover, pers. comm.).

The next eruptive period occurred between 7 and 15 December 1973. Ash samples differed from all others collected in recent years, in containing up to 30 percent of fresh glassy material: this suggested that new magma or fresh andesite had contributed to the tephra output.

Seismic activity

Volcanic tremor and earthquakes, attributed to Ngauruhoe, were recorded at CNZ during several periods in 1972 and 1973. These are listed together with 1974 activity in Table 1, which gives maximum power output and overall and daily average energy output of the volcanic tremor, estimated using the relationship (Dibble, 1972)

$$\text{seismic power in kW} = 0.19 \ A^2,$$

where A is the zero-to-peak trace amplitude in mm of the CNZ seismogram: this relationship was established for a source at Ruapehu, but as Ngauruhoe is at a similar distance from CNZ, and as the uncertainties of the method are so considerable, the relationship is here used for Ngauruhoe tremor also.

By far the most seismically active period was in July to October 1973, although only minor eruptive activity took place. This period alone contributed 52 percent of the total energy output of Ngauruhoe tremor for the years 1972-1974 inclusive, and is considered a precursor of the powerful 1974 eruptions. C. Blot, who has postulated a causal connection between certain deep and intermediate earthquakes and volcanic eruptions (Blot, 1964) predicted (pers. comm., 10 April 1973) an eruption in July or August 1973 from the occurrence of a magnitude M_L = 6.7 intermediate-focus earthquake on 6 January 1973, located 40 km west of Ngauruhoe at a depth of about 170 km.

Between 1 January 1972 and the January 1974 eruption period, 105 volcanic earthquakes originated at Ngauruhoe, with magnitudes (M_L) ranging from 2.1 to 3.1. These include both apparently shallow and deeper events, provisionally identified as B- and A-type volcanic earthquakes (Minakami, 1960). The largest occurred on 2 January 1973, accompanying the major observed eruption. Treating this set of data as a single series, the coefficient b has been calculated (following the method described by Gibowicz, 1973) from the magnitude/frequency relationship,

$$\log N_M = a - bM^* \text{ (Gutenberg \& Richter, 1954),}$$

as 3.20 ± 0.62. This is an unusually high value, even for volcanic sequences.

---

*N_M is the number of earthquakes within a given magnitude range $M + \Delta M/2$: M is a particular scale of magnitude (in this case M_L, the local magnitude scale) and a and b are constants.

Table 1. Summary of estimated energy output of volcanic tremor at Ngauruhoe Volcano, 1972-1974.

	Maximum power output of volcanic tremor (kW averaged over 1 hour)	Overall energy output of volcanic tremor (Jx10⁶)	Average rate of energy output of volcanic tremor/day (W)	Date/Time (N.Z.S.T.) of maximum activity during the period	Principal types of activity
1972 March 13-July 12	0.60	301.8	28.9	April 9 0735-1300	Tremor and eruptions
1972 November 21-22	0.06	0.3	3.5	November 21-22 2315-0045	Tremor and eruptions
1972 December 27-1973 mid-January	0.16	88.0	(48.6)	January 7 1845-2300	Tremor and eruptions
1973 March 27-May 30	0.18	20.0	3.5	April 8-9 1100-2000	Tremor
1973 June 19-24	0.03	0.3	1.2	June 20 1200-1800	Tremor
1973 July 22-October 19	1.85	574.0	74.1	September 28 0400-1100	Tremor and minor eruptions
1973 December 4-27	0.27	14.1	7.0	December 6 2100-2230	Tremor
1974 January 20-29	0.21	26.5	31.2	January 22 1830-2130	Tremor and eruptions
1974 February 13-March 14	0.04	2.6	10.5	February 21 0800-1800	Tremor
1974 March 20-April 2	4.42	79.3	66.0	March 28 0845-1040	Tremor and eruptions
1974 April 7-May 8	0.06	11.0	3.5	April 19 0020-0700	Tremor and B-type shocks
1974 May 20-June 2	0.01	—	—	June 1 0000-1200	B-type shocks
1974 August 14-15	0.01	—	—	August 15 1200-1800	B-type shocks

JANUARY 1974 ERUPTIONS

Observations*

Eruptive activity at Ngauruhoe, estimated power output of Ngauruhoe tremor, and principal volcanic earthquakes recorded during the January eruption sequence are shown schematically in Figure 2. The first eruption was reported at about 13h 55m on 22 January, when a convoluting black ash column rose 600 m

*Detailed observations on the January 1974 eruptions are contained in an unpublished N.Z. Geological Survey report (Nairn, 1974a).

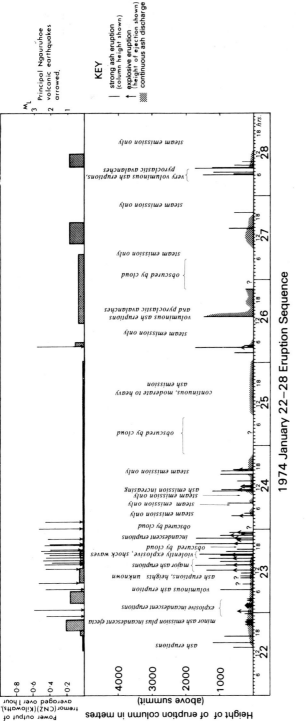

Fig. 2: Schematic diagram summarising Ngauruhoe eruptive activity, power output of Ngauruhoe tremor, and principle volcanic earthquakes during January eruption sequence. 'Explosive eruptions' (see key) are defined by high initial ejecta velocities, usually associated with visible atmospheric shock waves, or loud detonations, or both.

above the crater. Eruptions continued until darkness fell, when small amounts of incandescent ejecta could be seen thrown above the crater. A particularly explosive event occurred at 00h 29m on 23 January, when a dark column of ash erupted violently, its base lit by a red glare; many large orange-red glowing blocks rose rapidly through the cloud to several hundred metres above the summit and fell back onto the cone, where they broke up to outline the whole mountain top with glowing rock. A loud report and substantial earthquake ($M_L = 3.0$) accompanied the explosion. Similar incandescent eruptions followed during the night, while ash was almost continuously emitted through into the morning of 23 January. An aerial inspection was made during which it was possible to see about 100 m into the crater. The floor could not be seen, and no sign of lava was present.

A large explosive eruption occurred at 12h 12m on 23 January, about 7 minutes after a regional earthquake (see later discussion). This, and other explosive eruptions during the afternoon, followed a cycle in which a period of quiescence and low seismic activity prior to each eruption was broken by a violent explosion throwing out many large blocks which rolled and bounced to the foot of the cone. A dark, very dense, eruption 'slug' of gas-fluidised pyroclastics would be violently projected above the crater (see Fig. 5 for example) before expanding into a convoluting ash column, commonly reaching 1300 m above the summit. Vigorous ash emission followed for some minutes after each explosion.

Particularly violent explosions were accompanied by visible atmospheric shock waves forming rapidly expanding arcs of brightness which passed through steam and thin cloud a few hundred metres above the summit. Strong volcanic earthquakes accompanied all the explosions, and loud reports and rumblings were often heard at the Observatory.

Similar but less violent cyclic activity continued into 24 January, when ejecta were sampled from the inner cone which was covered with small, impact-shattered, angular lava blocks. The vent interior was completely filled by the strongly convoluting ash column and a collar of pale brown steam, but no heat was felt or sounds heard at the crater rim.

Moderate to heavy ash emission continued during 25 January, declining to practically zero during the morning of 26 January. Later, a small B-type earthquake at 13h 10m 45s was followed at 13h 12m 5s by commencement of the largest eruption yet seen, when densely ash-laden convoluting clouds slowly rose to 1500 m above the summit, and large blocks rolled down the west slopes of the volcano. The ash column above the summit gradually thickened and then started to collapse, moving down the slopes of the inner cone and the outer crater rim. Sheet-like ash flows covered the western slopes of the inner cone (Fig. 3), and thin rope-like pyroclastic avalanches flowed through avalanche chutes formed by southwestern and northern low points on the vent rim. The avalanches continued 800 m down the cone in gullies, as widening cascades of turbulent ash and blocks moving at about 16 m/sec, while the ash column thickened above the summit (Fig. 4). The whole eruption sequence lasted for about 10 minutes, and was accompanied by strong tremor but no major volcanic earthquakes, indicating that fairly open-vent conditions were allowing voluminous but non-explosive degassing. A temperature of 91°C was measured within an avalanche deposit about 90 minutes after its eruption.

Fig. 3: Sheet-like pyroclastic flows covering upper slopes of Ngauruhoe, while more confined flows extend down gullies; 13h 15m on 26 January 1974. Photo from Chateau Observatory by D.L. Homer.

Similar eruptions at 06h 14m and 06h 20m on 28 January again followed a period of no emission, but commenced rather more violently with the sudden projection of a very dense eruption column to 500-600 m above the summit (Fig. 5). Further ash emission then widened the eruption column which remained stationary above the summit, before large convoluting clouds rolled down the western slopes of the cone, preceded by avalanches in the gullies. Intense volcanic tremor, but no major earthquakes accompanied these eruptions, which were the last major events of the January sequence.

Fig. 4: Pyroclastic avalanches flowing down gullies; 13h 20m 26 January 1974. Photo from Chateau
Observatory by D.L. Homer.

Further ejecta samples were collected during a ground inspection on 12
February, and an aerial inspection was made on 22 March, but gas and steam
emission prevented observations to any depth in the crater on both occasions.

Seismic accompaniment to the January eruptions

Volcanic tremor. Volcanic tremor originating at Ngauruhoe commenced
gradually at about 12h 30m on 22 January (see Fig. 2), reaching a peak between
18h 30m and 21h 30m; it then decreased sharply. Practically no tremor took place
during the period of the large eruptions on 22 January. Tremor power output never
reached as high a level during the remainder of the January active period, except
for short periods of about 10 minutes immediately following the voluminous
eruptions at 13h 12m on 26 January and 06h 14m on 28 January.

Fig. 5: Violently projected eruption 'slug' to 600 m above summit; 06h 14m 28 January 1974. Photo from Chateau Observatory by D.L. Homer.

Volcanic earthquakes. Sixty-eight volcanic earthquakes of magnitude M_L = 2.1 or greater took place at Ngauruhoe during the January sequence, giving a b value of 1.75 ± 0.42. All showed a detectable air-wave phase about 24 s after the P onset (the theoretical travel-time of sound from Ngauruhoe to CNZ in still air at 5⁰C is about 25.9 s).

All the larger volcanic earthquakes (M_L greater than, or equal to 2.1) took place at the same time as observed eruptions, except for the initial M_L = 2.3 earthquake at 22h 46m on 22 January and four events on 23-24 January during a period when Ngauruhoe was obscured by cloud. However, many observed eruptions took place with little or no associated seismic activity, including the first eruptions at 13h 55m and 15h 24m on 22 January and most of the sharp increases in ash emission. A number of observed large eruptions were preceded or accompanied by very small volcanic earthquakes, of magnitude less, and sometimes much less, than 2.1. These included the largest eruptions of the sequence on 26 and 28 January.

Most of the large volcanic earthquakes in January accompanied the early explosive eruptions, whereas only weak seismic activity accompanied the most voluminous eruptions late in the series. This indicates that early eruptions were vent-clearing, while the later ones took place in more open-vent conditions.

Two major eruptions accompanied by volcanic earthquakes, at 12h 12m on 23 January and 06h 14m on 28 January, followed 7 to 8 minutes after regional tectonic earthquakes. The first of these earthquakes (M_L = 4.6) took place in the upper crust about 110 km to the northwest, and the second (M_L = 4.0) about 20 km to the north-northeast of Ngauruhoe. However, a third very similar shallow earthquake (M_L = 4.0), located close to the second event at about 09h 57m on 28 January, was not followed by any significant eruption. Source radii of these three tectonic earthquakes have been estimated as 2.5, 2, and 0.7 km respectively, using a method established by Gibowicz (1975). None of the earthquakes was close enough or large enough to have included the Ngauruhoe sub-structure in its source volume, so there is no evidence that the apparent triggering was other than coincidental.

1974 MARCH ERUPTIONS

Observations*

Eruptive activity at Ngauruhoe, power output of Ngauruhoe tremor, and principal volcanic earthquakes during the main March eruptions are shown schematically in Figure 6. The eruption commenced at 14h 32m on 27 March, following about 10 minutes of strong steam emission. A convoluting ash cloud rose 1700 m above the summit. Practically continuous ash eruption continued during the afternoon with dense fallout onto Tongariro. Incandescence of ejecta became apparent as darkness fell. Activity followed a consistent cyclic pattern in which the power of tremor fell sharply, accompanied by dwindling emission, before volcanic earthquakes marked the onset of each eruption.

An eruption began at about 07h 30m on 28 March, ash emission and power of volcanic tremor then steadily increasing into the most rapid and voluminous period of degassing observed from Ngauruhoe since 1954. Pyroclastic avalanches and convoluting ash clouds flowed down the western slopes, very widespread avalanches reaching 750 m below the summit between 09h 15m and 09h 25m, as a dense eruption column spilled over the western crater rim (Fig. 7). The eruption column rose 4700 m above the summit as a very strongly convoluting, rapidly expanding but coherent pillar. No major volcanic earthquakes occurred during this period, indicating little hindrance to gas emission.

Violently explosive eruptions followed during the morning, accompanied by air-waves, which shook the Chateau Observatory, and the ejection of very large blocks. One explosive eruption observed from the air at 10h 59m, threw out blocks up to 5 m across, some first impacting half-way down the cone. Incandescent blocks were thrown up to about the crater rim before falling back into the vent, while a dull red glow was reflected from the base of the ash column. Part of the northwestern crater wall had collapsed into the vent, lowering the rim in a V-shaped notch. The crater floor was briefly observed sloping down from west to east at an estimated 50-100 m depth, and covered by dark brown lava blocks with several incandescent patches and glowing fissures. Configuration of the crater interior had evidently changed considerably from the near-vertical walled hole more than 180 m deep, reported in November 1973.

*Detailed observations on the 1974 March eruptions are contained in an unpublished N.Z. Geological Survey report (Nairn, 1974b).

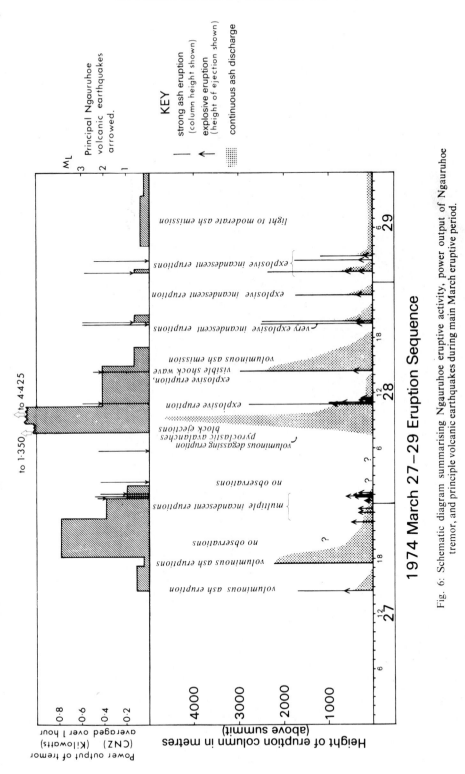

Fig. 6: Schematic diagram summarising Ngauruhoe eruptive activity, power output of Ngauruhoe tremor, and principle volcanic earthquakes during main March eruptive period.

A highly explosive eruption at 14h 25m again shook the Observatory and was heard at National Park and Turangi. A visible atmospheric shock wave emerged from the vent only just ahead of the violently projected eruption slug.

The largest explosive eruptions occurred during the night of 28-29 March, following periods of low volcanic and seismic activity and accompanied by sharp volcanic earthquakes (up to M_L = 3.0). They were considerably more powerful than the explosive eruptions observed in January. Time-exposure photographs (e.g. Fig. 8) show that many incandescent blocks were ejected in the initial explosive outbursts, some vertically to 450 m above the summit and others at low trajectories which just cleared the crater rim. Loud rumbles were heard at the Observatory. Ejecta broke up on impact, and the western and southern slopes of the cone appeared to be covered with incandescent blocks. Some very large blocks were still visibly glowing 30 minutes after coming to rest on the slopes of the inner cone. The initial explosive bursts were followed by dark ash columns which rose 2500 m above the summit, their interiors lit by bright lightning flashes. Seismic activity quickly fell off after each explosion, while ash emission also diminished quickly, the tall ash columns becoming detached from the summit by wind, and drifting northwest as vertical pillars. These eruptions marked the end of the main events in the March sequence.

Ejecta samples were collected from the summit on 29 and 31 March. Lava blocks, some still hot, covered the southern slopes of the inner cone while the northern and eastern slopes were thickly coated with ash. Large craters on the southwest slopes of both the inner cone and outer volcano slopes had been formed by impact of blocks up to several metres across (Fig. 9). Gas at high pressure was jetting through a large fumarole in the southern crater wall, producing a loud roaring noise.

Seismic accompaniment to the March eruptions

Volcanic tremor. Very weak Ngauruhoe tremor recommenced at 09h 15m on 26 March, and continued at a low level until about 19h 00m. Significant tremor commenced at 14h 36m on 27 March, just after the first reported eruption. The strongest tremor during the period accompanied the major degassing eruption on the morning of 28 March.

Volcanic earthquakes. Twenty-three volcanic earthquakes (M_L 2.1 to 3.0) took place during the March eruptions, beginning with an event at 03h 10m on 27 March, and ending with one at 02h 26m on 29 March. Almost all accompanied observed eruptions. Air-wave phases were again conspicuous on the CNZ records. b in the magnitude-frequency equation was determined at 1.24 $\pm$ 0.51, a figure substantially lower than any previously determined for Ngauruhoe volcanic earthquakes.

As in January, a number of significant eruptions took place without any marked seismic expression, including the initial voluminous ash eruption at 14h

Fig: 7 (opposite): Major degassing eruption, 09h 20m on 28 March 1974, showing eruption column rising 4700 m above summit, and pyroclastic avalanches and convoluting ash clouds flowing down western slopes of cone. Photo from Chateau Observatory by J. Scobie.

Fig. 8: Time exposure (about 20 seconds) of explosive incandescent eruption 01h 90m on 29 March 1974. Note divergent trajectories of incandescent blocks, and lightning flashes which occurred in the ash column. Photo from Chateau Observatory by D.L. Homer.

32m on 27 March. Unlike the situation in January, all the principal volcanic earthquakes occurred during the middle to end of the observed eruption series: only earthquakes of magnitude 2.4 or less took place before 10h 58m on 28 March. There were no instances, during the March episode, of near or regional earthquakes apparently triggering eruptions.

NATURE OF EJECTA

Lithology and petrography

Some small lava spatter bombs (up to 25 cm) were thrown out in January, but most of the ballistic ejecta of both January and March eruptions consisted of old, but unaltered, lava blocks and secondary scoria. Altered lava and breccia blocks were also present. The larger blocks up to 4 m across (Fig. 9) were of dense, non-vesicular andesite ejected in a solid state. Scoria formation could be traced by textural changes from solid lava, through expanded blocks showing plastic

Fig. 9: Impact crater and remnants of a solid incandescent block that fell and shattered within a few metres of the crater lip in 01h 09m eruption on 29 March 1974. Photo by D.L. Homer.

deformation and vesiculation (Fig. 10), to highly vesicular, amorphous scoria. The range of textures showed that most of the eruption products were either old solid lava or could have been derived therefrom by heating, softening, vesiculation, and refusion. Even the spatter bombs could have been formed by localised complete refusion.

All the unaltered andesite ejected in 1974 was olivine-bearing, low-Si andesite with identical petrography to the 1954 lavas described by Steiner (1958).

Petrochemistry

Though the 1949 and 1954 lavas were both of olivine-bearing andesite, chemical analyses (Steiner, 1958, analyses 6 and 9) showed the 1954 lava to be slightly more basic than the 1949, with higher MgO, CaO, and total-iron oxides, but lower SiO_2 and alkalis. Except for K_2O the differences are small, but they are petrologically consistent and could be used to distinguish between the 1949 and 1954 lavas. With this in mind, new analyses were made of four 1954 lavas and compared with five bombs and blocks ejected in 1974 (Table 2). The 1974 material ranged from a solid, unexpanded block (analysis 1), through expanded scoria, to a lava spatter bomb (analysis 2), and gave analyses identical within the limits of error. Further, the differences between the average 1954 lava and average 1974 andesite are statistically insignificant.

DISCUSSION

Explosive eruptions

The major explosive eruptions commonly followed a pattern in which:

a) low and/or declining volcanic tremor immediately preceded the eruption, while emission of gas and ash was also declining, or sometimes nil;

b) a sharp, instantaneous explosion occurred, accompanied by a volcanic earthquake and on several occasions by a visible atmospheric shock-wave passing above the crater;

c) the explosions threw out numerous large incandescent blocks, followed by short periods of violent ash emission producing a tall eruption column;

d) finally, emission and tremor would gradually decline to low or zero levels before onset of the next eruption.

This cyclic activity may be explained by gas pressure repeatedly building up beneath obstructions within the volcano until explosively released. Rapid degassing could then follow through the rupture until loss of pressure allowed resealing of the obstruction and recommencement of the cycle.

Crater conditions and depth of explosions

Eruption observations. The well-collimated, near-vertical trajectories of ejecta slugs and individual blocks erupted in January (e.g. Fig. 5) indicated that the crater

Fig. 10: Expansion cracks in bulging surface of an expanded block. The block was brittle, suggesting that expansion occurred after impact. Photo, D.L. Homer.

Table 2. Chemical analyses of Ngauruhoe andesites [a].

	1	2	3	4
SiO_2	55.48	55.80	55.73	55.83
TiO_2	0.76	0.76	0.77	0.76
Al_2O_3	17.27	17.31	17.23	17.03
Fe_2O_3	1.67	2.47	2.13	2.14
FeO	6.76	6.24	6.47	6.36
MnO	0.15	0.16	0.16	0.16
MgO	4.83	4.87	4.89	4.79
CaO	8.55	8.57	8.52	8.37
Na_2O	2.90	3.00	2.86	2.93
K_2O	1.06	1.08	1.08	1.20
P_2O_5	0.15	0.17	0.15	0.15
H_2O	0.06	0.18	0.10	0.14
Total	99.64	100.61	100.09	99.86
Rb ppm	29	30	30	33
Sr ppm	230	230	230	231

1. P39864[b]. Solid andesite from block ejected in 27-29 March 1974 eruptions.
2. P39851. Lava spatter bomb from January 1974 eruptions.
3. Average analysis of 5 bombs and blocks ejected during January and March 1974.
4. Average analysis of four 1954 lavas.

[a] All analyses determined at NZGS Lower Hutt by X-ray fluorescence except for FeO (dichromate titrimetry), alkalis (flame photometry) and total H_2O (Penfield method). Analysed by C.P. Wood, D. Clyma, I.N. Smolnicki.

[b] P numbers refer to samples in N.Z. Geological Survey Petrology Collection, Lower Hutt.

was still relatively deep. If explosions had been occurring high in the crater then ejecta would have been sprayed over a much wider angle — as clearly shown by the divergent trajectories of many incandescent bombs erupted on the night of 16 September 1954, when lava was overflowing the crater rim (see photograph in Gregg, 1956, p.678).

When the crater interior was first observed from the air on 28 March it was considerably shallower than the depth reported in November 1973 and inferred during the January eruptions, a largely solid lava plug having apparently risen in the volcano. The explosive eruptions which followed originated higher in the vent, as ejecta trajectories were more divergent, some blocks being thrown almost horizontally, and the visible atmospheric shock wave observed at 14h 25m on 28 March was a much shorter distance ahead of the solid ejecta than were the shock waves seen in January. The very much larger size of the blocks ejected in March also suggests that explosions were occurring higher in the vent than in January, and were thus able to throw larger material over the crater rim.

Depth of volcanic earthquakes. Records at CNZ of volcanic earthquakes in both the January and March eruptions (Table 3) invariably showed the presence of an air-to-ground coupled wave. The amplitude of this phase varied widely, being as much as five to six times that of the maximum ground waves in some events, and as little as a fifth in others. Although this provides a qualitative picture of the relative

Table 3. Summary of volcanic earthquakes at Ngauruhoe used for depth analysis.

Event No.	N.Z.S.T. Date	Time	Magnitude (M_L)	Minakami type	Accompanying eruption (if any)
1	1973 Jan 2	2115	3.1	A	Incandescent eruption
2	1974 Jan 22	2246	2.3	?	—
3	Jan 23	0023	2.3	A	—
4	Jan 23	0029	3.0	?	Major eruption, incandescent
5	Jan 23	0627	2.3	(A)	May have preceded observed eruption by 2 minutes
6	Jan 23	0810	2.7	?	?Ash eruption
7	Jan 23	1212	2.3	B	Large eruption
8	Jan 23	1339	2.4	B	Observed eruption
9	Jan 23	1344	2.0	(B)	Small ditto
10	Jan 23	1448	2.5	?	Eruption, ashfall to north
11	Jan 23	1537	2.3	(B)	Eruption, visible shock wave, ashfall for 10 minutes
12	Jan 23	1628	2.1	B	Eruption, visible shock wave
13	Jan 23	1648	2.4	(B)	Ash eruption, with blocks
14	Jan 23	1738	2.6	?	Large ash eruption
15	Jan 23	1742	2.0	(B)	Nil observation
16	Jan 23	1818	2.2	?	Break in observations (cloud)
17	Jan 23	1907	2.4	(B)	Break in observations (cloud)
18	Jan 23	1913	2.1	B	Break in observations (cloud)
19	Jan 23	1945	2.0	(B)	Break in observations (cloud)
20	Jan 23	2052	2.2	?	Break (NB event 2217 hours not used in
21	Jan 23	2308	2.5	?	Ash column seen analysis, $M_L = 2.8$)
22	Jan 23	2333	2.1	?	Nil observation
23	Jan 23	2344	2.8	?	Nil observation
24	Jan 24	0146	2.4	(B)	In cloud
25	Jan 24	0347	2.0	?	—
26	Jan 24	0349	2.0	?	Incandescent eruption
27	Jan 26	0420	2.9	?	Observed eruption, accompanied by strong tremor
28	Jan 28	0620	2.5	?	Hot avalanches. Strong eruption
29	Mar 28	0040	2.3	B	—
30	Mar 28	0044	2.4	(B)	Audible explosion, incandescent eruption
31	Mar 28	0057	2.1	B	Incandescent eruption
32	Mar 28	0106	2.2	B	Incandescent eruptions report-
33	Mar 28	0112	2.1	B	ed at about these times
34	Mar 28	0200	2.1	B	No observation
35	Mar 28	0226	2.2	(B)	No observation
36	Mar 28	0426	2.0	(B)	No observation
37	Mar 28	0555	2.3	(A)	No observation
38	Mar 28	0726	2.0	(B)	Moderately large ash eruption
39	Mar 28	1049	2.4	(B)	Large explosion
40	Mar 28	1058	3.0	(B)	Violently explosive eruption, large blocks
41	Mar 28	1425	2.5	B	Violently explosive eruption and visible shock wave
42	Mar 28	1931	3.0	(B)	Violently explosive eruption
43	Mar 28	1944	3.0	(B)	Violently explosive eruption
*44	Mar 29	0109	2.9	(B)	Strong eruption, incandescent
45	Mar 29	0226	2.5	(B)	Moderate explosive eruption

*NB a large earthquake $M_L = 2.9$ at 2239 hours on 28 March has not been used in the analysis as the record was obscured.

amount of energy being propagated as ground and air waves, and therefore some idea of the coupling of an apparently explosive source within the volcano, it is not directly related to the depth of focus of the earthquakes. Depth of focus is only one parameter in relationship which depends also on the degree of coupling in the vent, and on the energy and directionality of the explosions. However, for events in which air amplitude was more than twice ground amplitude, air amplitude in the March series exceeded ground amplitude by a factor of 4.46 $\pm$ 0.93, whereas in January the comparable figure was 2.57 $\pm$ 0.66. This variation suggests shallower explosion earthquakes in the March series, propagating a greater proportion of their energy into the air above the vent.

Slight differences also exist in the time interval (Air-P) at CNZ between events in January, and those in March which were identified by their low-angled ejecta trajectories as of very shallow origin. (Air-P), corrected for wind and temperature effects, averaged 24.2 $\pm$ 0.75 s in the January events, against 23.9 $\pm$ 0.55 s in the case of the March shallow events. The difference in these averaged values is not statistically significant although it lies in the expected direction, that of lengthening of the (Air-P) interval with increasing focal depth. Assuming propagation in the vent wall at a velocity of the order of 1 km/s, and conversion to air waves at the summit, this would suggest, if taken at its face value, depths of origin of the March events of the order of 0.3-0.4 km less than those of January.

Pyroclastic avalanches

The most voluminous 1974 eruptions did not commence explosively; instead emission steadily increased until the dense eruption column widened to spill over the crater rim, forming pyroclastic avalanches which flowed down gullies on the cone. Avalanches may also have originated from the simple gravity fallout of ejecta on the summit of the cone, to accumulate and avalanche downwards.

Clouds of fine ash 'boiled off' the descending avalanches so that they resembled nuées ardentes. However, the avalanches were not at very high temperatures, and the term 'pyroclastic avalanche' is preferred. The avalanches did not flow far (not reaching beyond the foot of the cone), or very fast, even though moving down slopes of 30-35°. The degree of fluidisation, either from expanding gas or heating of entrapped air, cannot have been great.

Ejecta

To observers, the early 1974 eruptions appeared typical of the vulcanian type (Macdonald, 1972), with violent ejection of solid, or very viscous, hot fragments of new lava. However, liquid lava neither flowed from, nor ponded in, the crater, so there was no proof that new magma had actually reached the surface from deep chambers. Incandescence of ejecta was observed only at night and during a close daytime aerial inspection on 28 March, whereas true lava fountains were clearly incandescent in daylight in the 1954 eruptions. Such andesitic lava would have been at about 1000-1100°C (Scarfe & Eldefield, 1975), but the glowing material of the 1974 eruptions need only have been at 700-900°C. Support for this view comes from Clark (1970), who reported that old lava and tephra forming the orifices of gas-heated fumaroles at 850-900°C were incandescent in daylight on White Island in 1968.

While a column of magma is undoubtedly present beneath Ngauruhoe, providing the gas generating the eruptions, the nature of the ejected blocks and bombs suggests that they are derived from an overlying plug of solid andesite lava which has been reheated to incandescence and locally fused. Over 6×10^6 m^3 of lava flowed from Ngauruhoe in 1954 and presumably much also congealed in the vent and conduits. As the 1954 lava and 1974 ejecta are petrochemically identical, such a plug is inferred to consist of 1954 andesite, reheated and explosively disrupted in the 1974 eruptions. On this interpretation the 1974 eruptions would be classed as ultravulcanian.

Reheating of the plug would occur both by conduction from an underlying magma column and by the passage of hot gases through fractures and fumaroles. Under such conditions local temperatures and viscosities would vary greatly, but the eruption products suggest that much of the plug was solid or in a highly viscous plastic state. Melting to magma fluid enough to make lava bombs probably took place along channels for gases coming directly from the deep primary magma.

Keller (1969, and pers. comm.) has suggested that the 1888-90 Vulcano eruptions — the holotype vulcanian events — were also in fact ultravulcanian. Breadcrust bombs which were previously regarded as the new magmatic products of the Vulcano eruptions, now appear to have originated as reheated blocks of older volcanic rocks. However, the typical Vulcano breadcrust blocks that expanded by vesiculation after landing (Walker, 1969) were absent at Ngauruhoe, though some blocks showed features that are interpreted as post-impact distension and flow (e.g. Fig. 10). It is believed that most of the vesiculation and expansion phenomena occurred in the vent before eruption.

CONCLUSIONS

The 1974 pyroclastic eruption sequence was the most vigorous activity at Ngauruhoe since 1954, when lava flowed from the volcano. Textural features of the 1974 ballistic ejecta suggest their derivation by reheating of a solid lava plug, an interpretation supported by petrographical and chemical identity of the 1954 lava and 1974 ejecta. Unless new magma within Ngauruhoe is identical to that of 1954, the 1974 ballistic ejecta appear likely to be 1954 lava, reheated to incandescence and locally fused. When lava again flows from Ngauruhoe, and can be analysed, it will be possible to check this interpretation.

ACKNOWLEDGEMENTS

Observations of the eruptive activity at Ngauruhoe were assisted by the following people: G.T. Hancox, B.R. Paterson, B.D. Hegan, N.Z. Geological Survey, Turangi; T.R. Grammer, Ministry of Works and Development, Turangi; R.B. Glover, L. Klyen, P.M. Otway, A. Slatter, D.J. Dickinson, G. McDowell, D.S.I.R., Wairakei; O. Espanola, J. Healy, S. Nathan, B.J. Scott, N.Z. Geological Survey, Rotorua; P.R.L. Browne, D.L. Homer, N.Z. Geological Survey, Lower Hutt; P.C. Whiteford, Geophysics Division, Wellington; M. Lankow, Information Service, D.S.I.R., Wellington; R.R. Dibble, M. Michael, Victoria University of Wellington; J. Scobie, Chateau Service Station.

The authors' responsibilities mainly comprised: eruption observations and co-ordination by Nairn and Hewson; seismic analysis by Latter; and petrographical and petrochemical studies by Wood. The paper was compiled by Nairn. We are indebted to J. Healy, S. Nathan, R.D. Adams and W.I. Reilly for their criticisms of early drafts of this paper.

REFERENCES

Blot, C., 1964: Origine profonde des séismes superficiels et des eruptions volcaniques. *Publs. Bur. cent. seism. int. U.G.G.I., Ser A, Trav. Sci., Fasc. 23*, pp. 103-121.

Clark, R.H., 1970: Volcanic activity on White Island, Bay of Plenty, 1966-69. *N.Z. J. Geol. Geophys, 13*, pp. 565-74.

Cole, J.W., & Nairn, I.A., in press: Catalogue of the active volcanoes of the world, including solfatara fields. Part 22; New Zealand. *Naples, Int. volc. Assoc.*

Dibble, R.R., 1972: Seismic and related phenomena at active volcanoes in New Zealand, Hawaii, and Italy. *Unpublished Ph.D. Thesis, Victoria Univ. Wellington.*

Gibowicz, S.J., 1973: Variation of the frequency-magnitude relation during earthquake sequences in New Zealand. *Bull. seism. Soc. Am., 63(2)*, pp. 517-528.

Gibowicz, S.J., 1975, in press: Determination of seismic moment and related source parameters of shallow local earthquakes from Wellington records. *D.S.I.R., New Zealand, Rep. No. 99*, Geophys. Div.

Gregg, D.R., 1956: Eruption of Ngauruhoe 1954-55. *N.Z. J. Sci. Techn., B37*. pp. 675-688.

Gregg, D.R., 1960: The geology of Tongariro Subdivision. *N.Z. Geol. Surv. Bull. n.s. 40.*

Macdonald, G.A., 1972: *Volcanoes.* Englewood Cliffs, Prentice-Hall Inc.,

Minakami, T., 1960: Fundamental research for predicting volcanic eruptions. Part 1. Earthquakes and crustal deformations originating from volcanic activities. *Bull. Earthq. Res. Inst., Tokyo Univ., 38(31)*, pp. 497-544.

Gutenberg, B., & Richter, C.F., 1954: *Seismicity of the Earth and associated phenomena.* Princeton, Univ. Press, reprinted 1965 by Hafner Publishing Co. Inc., New York.

Keller, J., 1969: Fossa di Vulcano 1888-1890. Eruption without magma emission. *in* Symposium on volcanoes and their roots (abs. vol.). *Oxford Internat. Assoc. Volc. and Chem. Earth. Int.*

Nairn, I.A., 1974a: Observations of Ngauruhoe eruptions, January 22-January 31 1974: Preliminary Report. *Unpubl. Rep. N.Z. Geol. Surv.*

Nairn, I.A., 1974b: Observations of Ngauruhoe eruptions, March 27-April 2 1974: Preliminary Report. *Unpubl. Rep. N.Z. Geol. Surv.*

Scarfe. C.M., Eldefield, H., 1975: Structure of two silicate rock melts charted by infrared absorption spectroscopy. *Chem. Geol., 15*, pp. 77-80.

Steiner, A., 1958: Petrogenetic implications of the 1954 Ngauruhoe lava and its xenoliths. *N.Z. J. Geol. Geophys., 1*, pp. 325-63.

Topping, W.W. 1974: Some aspects of Quaternary history of Tongariro Volcanic Centre. *Unpubl. Ph.D. Thesis, Victoria Univ., Wellington.*

Walker, G.P.L., 1969: The breaking of magma. *Geol. Mag., 106*, pp. 166-73.

Appendix. *Geophysical equipment operating during the 1974 eruptions**

1) Single-component vertical seismometer, Willmore Mk I (magnification 44,980 at 0.3 sec period). Permanent Chateau station (CNZ) of the New Zealand Seismological Observatory network: photographic drum record at speed of 1 mm/sec.

2) Single-component vertical seismometer, Willmore Mk II (magnification about 300 at 0.3 sec period) at The Dome, 0.7 km from Ruapehu Crater Lake, linked by cable to Chateau Observatory. Photographic drum record at speed of 1 mm/sec.

3) Seismic power recorder, amplifying the CNZ signal logarithmically and recording on two channels — one responding to signals of duration longer than about 3 seconds, the other to signals of duration longer than about 60 seconds (P.C. Whiteford, pers. comm.). Inked chart record.

The seismic power recorder provides an almost immediate record of seismic activity for correlation with eruption observations. However, it cannot indicate the source of seismic acticity, which is estimated from the CNZ and Dome seismograph records.

*Magnitudes of the larger events were estimated from records of seismograph stations elsewhere in New Zealand.